The Rise of
Industrial Society in
England
1815–1885

S. G. CHECKLAND

PROFESSOR OF ECONOMIC HISTORY
THE UNIVERSITY OF GLASGOW

LONGMANS

LONGMANS, GREEN AND CO LTD
London and Harlow

*Associated companies, branches and representatives
throughout the world*

© *S. G. Checkland 1964*
First published 1964
Fourth impression 1969
Printed in Great Britain
in the City of Oxford at the Alden Press

Social and Economic History of England

EDITED BY ASA BRIGGS

Contents

Maps

Acknowledgements

We are grateful to The Clarendon Press, Oxford, for permission to use material from 'The Mind of the City' by S. G. Checkland published in the *Oxford Economic Papers* of 1957, to the Economic History Society for permission to use material from 'Growth and Progress: The Nineteenth Century View in Britain' by S. G. Checkland published in the *Economic History Review*, 2nd Series, Vol. XII, No. 1, to Methuen & Co. Ltd. for permission to redraw three maps from Wilfrid Smith, *An Economic History of Britain*, and to Oliver and Boyd Ltd. and the editors of *Urban History* for permission to reproduce the maps of Glasgow from the author's article 'The British Industrial City as History: the Glasgow Case'.

Introductory Note

Interest in economic history has grown enormously in recent years. In part, the interest is a by-product of twentieth-century preoccupation with economic issues and problems. In part, it is a facet of the revolution in the study of history. The scope of the subject has been immensely enlarged, and with the enlargement has come increasing specialization. Economic history is one of the most thriving of the specialisms. Few universities are without an economic historian. New research is being completed each year both in history and economics departments. There are enough varieties of approach to make for frequent controversy, enough excitement in the controversy to stimulate new writing.

This series, of which Professor Checkland's volume is the second, is designed to set out the main conclusions of economic historians about England's past. It rests on the substantial foundations of recent historical scholarship. At the same time, it seeks to avoid narrow specialization. Economic history is not lifted out of its social context, nor are the contentious borderlands of economics and politics neglected. The series is described as 'a social and economic history of England'.

The bracketing together of the two adjectives is deliberate. Social history has received far less scholarly attention than economic history. A child of the same revolt against the limited outlook of the political historian, it has grown less sturdily. Its future depends on the application of greater discipline and more persistent probing. Developments in recent years are encouraging, and many of them will be reflected in these volumes. So too will developments in historical geography and, where they are illuminating, in demography and sociology. There is hope that just as the economist has provided useful tools for the study of economic history, so the sociologist may be able to provide useful tools for the study of social history and the demographer valuable quantitative data. There is no need, however, for economic and social historians to work in separate workshops. Most of the problems with which they are concerned demand cooperative effort.

However refined the analysis of the problems may be or may become, however precise the statistics, something more than accuracy and discipline are needed in the study of social and economic history. Many of the most lively economic historians of this century have been singularly undisciplined, and their hunches and insights have often proved invaluable. Behind the abstractions of economist or sociologist is the experience of real people, who demand sympathetic understanding as well as searching analysis. One of the dangers of economic history is that it can be written far too easily in impersonal terms: real people seem to play little part in it. One of the dangers of social history is that it concentrates on categories rather than on flesh and blood human beings. This series is designed to avoid both dangers, at least as far as they can be avoided in the light of available evidence. Quantitative evidence is used where it is available, but it is not the only kind of evidence which is taken into the reckoning.

Within this framework each author has complete freedom to describe the period covered by his volume along lines of his own choice. No attempt has been made to secure general uniformity of style or treatment. The volumes will necessarily overlap. Social and economic history seldom moves within generally accepted periods, and each author has had the freedom to decide where the limits of his chosen period are set. It has been for him to decide of what the 'unity' of his period consists.

It has also been his task to decide how far it is necessary in his volume to take into account the experience of other countries as well as England in order to understand English economic and social history. The term 'England' itself has been employed generally in relation to the series as a whole not because Scotland, Wales or Ireland are thought to be less important or less interesting than England, but because their historical experience at various times was separate from or diverged from that of England: where problems and endeavours were common or where issues arose when the different societies confronted each other, these problems, endeavours and issues find a place in this series. In certain periods Europe, America, Asia, Africa and Australia must find a place also. One of the last volumes in the series will be called 'Britain in the World Economy'.

The variety of approaches to the different periods will be determined, of course, not only by the values, background or special interests of the authors but by the nature of the surviving sources and the extent to which economic and social factors can be separated out from other factors in the past. For many of the periods described in this series it is extremely difficult to disentangle law or religion from economic and social structure and change. Facts about 'economic and social aspects' of life must be supplemented by accounts of how successive generations thought about 'economy and society'. The very terms themselves must be dated. For the more recent history with which Professor Checkland is concerned one of the many problems is the huge volume of evidence. Another is the tendency of specialists to fit the evidence into rather narrow categories.

Where the facts are missing or the thoughts impossible to recover, it is the duty of the historian to say so. Many of the crucial problems in English social and economic history remain mysterious or only partially explored. This series must point, therefore, to what is not known as well as what is known, to what is a matter of argument as well as what is agreed upon. At the same time, it is one of the particular excitements of the economic and social historian to be able, as G. M. Trevelyan has written, 'to know more in some respects than the dweller in the past himself knew about the conditions that enveloped and controlled his life'.

ASA BRIGGS

Preface

Certain acknowledgements of a general personal kind must be made: to W. H. B. Court who first roused my interest in economic history, to M. M. Postan who was responsible for my conversion to the subject and whose provocativeness has been so valuable, and to Asa Briggs for his encouragement and advice as general editor. My wife has both pushed the work along and attempted to restrain a tendency to make 'dangerous statements'.

A great deal is owed to my colleagues at the University of Glasgow, both in terms of continuous contact and for their assistance on special aspects of the work. In particular I would thank, among the political economists, Donald J. Robertson, Ronald L. Meek, Maxwell Gaskin, and George Houston, among the geographers, Cyril Halstead and Derek Diamond, and among the economic historians, Peter L. Payne, John R. Kellett, and Roy H. Campbell, now Professor of Economic History in the University of East Anglia. To the latter three I owe particular thanks for their part in forming the Department of Economic History here, a process concurrent with the writing of this book.

Another colleague, Derek H. Aldcroft, has undertaken the heavy task of compiling the index.

On aspects of science and technology I have gained much from discussions with Andrew Kent and John Lamb; Andrew Maclaren Young has performed the same service over art and design. Upon the University Librarian R. Ogilvy MacKenna, and his staff, I have leaned heavily, being one of their more troublesome clients, and especially upon Miss Elizabeth G. Jack, reference librarian. The tedium of typing and the simultaneous complications of departmental administration have been borne by Margaret E. Davies, with the assistance of Jean P. Clark.

Other debts are owed to Thomas Ferguson, and to John Butt.

To the Institute for Advanced Study at Princeton I am grateful for a term in 1960 during which much of the present work took shape; generous leave of absence from the University of Glasgow made this possible.

The editors of the *Economic History Review* and *Oxford Economic Papers* have kindly given permission to incorporate portions of articles they have published; thanks are also due to the editors of *Urban Studies* for permission to reproduce the plans of Glasgow.

The staff of Longmans, Green & Co. have been unfailing in their patience and help.

The University of Glasgow   S. G. CHECKLAND
8 April 1964

List of Abbreviations

Generally accepted abbreviations such as J. for Journal, Proc. for Proceedings Trans. for Transactions, R.C. for Royal Commission are not listed below.

Ag.H.R. : Agricultural History Review.
A.H.R. : American Historical Review.
B.H. : Business History.
B.I.H.R. : Bulletin of the Institute of Historical Research.
B.J.S. : British Journal of Sociology.
B.M. : British Museum.
C.R. : Contemporary Review.
D.A. : Dissertation Abstracts.
D.N.B. : Dictionary of National Biography.
Econa. : Economica.
E.H. : Economic History (Supplement to Economic Journal).
Econ. H.R. : Economic History Review.
Econ. : The Economist Newspaper.
Econ. C.H. & R. : Economist Newspaper Commercial History & Review.
E.R. : Edinburgh Review.
E.H.R. : English Historical Review.
F.R. : Fortnightly Review.
H.L.Q. : Huntington Library Quarterly.
H.J. : Historical Journal.
I.R.S.H. : International Review of Social History.
I.T.A. : Index to Theses Accepted for Higher Degrees in the Universities of Great Britain and Ireland, London, 1950 in progress.
J.A.H. : Journal of Agricultural History.
J. Econ. H. : Journal of Economic History.
J.H.I. : Journal of the History of Ideas.
J.M.H. : Journal of Modern History.
J.R.A.E. : Journal of the Royal Agricultural Society of England.
J.R.S.S. : Journal of the Royal Statistical Society (1887 and after).
J.S.S. : Journal of the Statistical Society (until 1886).
J.Tpt.H. : Journal of Transport History.
M.S. : The Manchester School.
N.C. : Nineteenth Century.
O.E.P. : Oxford Economic Papers.
P.P. : Parliamentary Papers.
P. and P. : Past and Present.

List of Abbreviations

P.S. : Population Studies.
Q.J.E. : Quarterly Journal of Economics.
R.B.A. : Reports of the British Association for the Advancement of Science.
S.J.P.E. : Scottish Journal of Political Economy.
T.I.B.G. : Transactions of the Institute of British Geographers.
T.N.A.P.S.S. : Transactions of the National Association for the Promotion of Social Science.
T.N.S. : Transactions of the Newcomen Society.
V.S. : Victorian Studies.
Y.B. : The Yorkshire Bulletin of Economic and Social Research.

PART ONE

The Evolving Economy

CHAPTER ONE

The New Potential

SOME time in the early eighteenth century a scattering of men, driven by the urge for self-assertion, no less than by the prospect of money gain, became active over the face of Britain, making new combinations of ideas, of things, of forces, and of other men. This had happened before, and was happening in other places. But in Britain economic initiative was on a new scale. Moreover it proved capable in the nineteenth century of generating a cumulative wave of new initiatives. The inventor and organizer, each individually finding his own area of freedom in an old and sophisticated society, lifted British productive capacity clear ahead of the rest of the world.

The idiom of the men responsible for this new manipulative power was disruptive; each sought to alter society at a particular point. The process was by nature microcosmic — a remaking of society from within, by industrialists, traders, bankers and the rest. By the last decade of the eighteenth century they were ready to carry Britain, within eighty years, to a state of advanced industrialism. So were set in train many complex phenomena.

The larger world perspective must always be borne in mind.[1] Since the sixteenth century the centre of initiative in Europe had been moving to the societies of the north and west. Italy and Germany had lost their renaissance leadership; further east, the rise of serfdom on the great plains of Poland and the Ukraine had been part of the response to the threat from the east.[2] This, together with the intrusion of Turkey into south-eastern Europe, had created a situation in striking contrast to that of western Europe. At the same time as the area of humanism and

---

[1] For an introductory attempt at a systematic approach see F. Mauro, 'Towards an "Intercontinental Model": European Overseas Expansion between 1500 and 1800', *Econ. H.R.*, 1961.

[2] G. R. Elton, ed., *The New Cambridge Modern History*, vol. II: *The Reformation 1520–1559*, Cambridge, 1958, pp. 15, 21.

volition was contracting in eastern European lands, the western states began their preparation for world ascendancy by colonization and example.[1]

This western European and Atlantic initiative culminated in the nineteenth century. It is shown in most striking terms by the changes in the pattern of world population. In 1800 Europe contained somewhat less than 200,000,000 people, but the forces at work in her caused the number to increase until a century later there were about twice as many. In addition, Europeans overseas, to a considerable extent through the continuation of the 'swarming' process, had increased from about 30 million to above 150 million.[2] If the peoples governed by Europeans are added, the world by the end of the nineteenth century becomes predominantly a European affair.

The new initiatives and knowledge released in Britain were clearly not an exclusively British product. They were the outcome of forces at work in the wider world. Although there survived within the various European societies many barriers to quick renovation, they yet contained many elements of initiative that were to gather momentum—especially under the stimulus of British example. The economy of Britain was not isolated; it was closely linked with the external world, helping to renovate the countries with which it dealt as new ideas and products spread, and itself receiving return impulses from rival sovereign competitive states abroad. Moreover the ability of Britain to take such an initiative in the nineteenth century was dependent not only on the societies to which she had long stood in intimate relationship but, increasingly, on the new and almost empty world to the west, and on the ancient and heavily peopled societies to the east.

The *Pax Britannica* was unique. Britain, in the three generations or so after the mid-eighteenth century, was able quietly to generate within itself the conditions making it possible for the country to dazzle the world with its industry. This almost secret gestation could occur because of the inadequacy of international communications, and the great lags in the responses of other countries. So large was Britain's lead, by the eighteen-sixties, and so divided by national rivalries were the other great states, that she could play upon the Concert of Europe and maintain an equipoise. Yet, impressive though British power might

---

[1] For the European economy in 1800 see Rondo E. Cameron, *France and the Economic Development of Europe, 1800–1914*, Princeton, 1961, part I.

[2] United Nations Organization, *The Determinants and Consequences of Population Trends*, New York, 1953, pp. 11, 12.

be, it was not of such an order that Britain could aim at world dominance in her own right. Other powers were neither greatly affronted nor frightened as they accepted the verdicts produced by Britain's adroitness, and might even have been grateful. It was within this situation that British statesmanship enjoyed that rarest of opportunities – it could be fully effective in the promotion both of world peace and the national interest.

As we turn from the external context and look more closely within Britain we discover the variety of her regional elements.[1] We find that even within so tiny an area there were great differences in natural endowment, found in association with long established regional differences in the character and outlook of the people. So it is possible to trace many distinct domestic patterns of change, components of a larger whole.

---

[1] See the *Victoria History of the Counties of England.*

CHAPTER TWO

The Growth and Stability of the System

I. THEORY AND HISTORY

THE first and most difficult task of the economic historian of the nineteenth century is to think in terms of Britain as a complex of elements that so interacted as to produce both expansion and diversification of output. A few crude physical comparisons between 1815 and 1885 indicate the scale of the change to be explained.[1]

	1815	*1885*
Raw Cotton Imports (U.K.)	81 million lb.	1,298 million lb.
Raw Wool Imports (U.K.)	7·5 million lb. (1816)	505 million lb.
Pig Iron Output (G.B.)	·243 million tons (est. 1806)	7·4 million tons
Coal Output (U.K.)	13 million tons (est.)	159·4 million tons
Railways (G.B.)	—	16,594 miles
Steamships (U.K.)	—	3·9 million registered tons
Population (England and Wales)	10,164,000 (1811)	25,974,000 (1881)

This growth of output and population did not occur in a simple untroubled way. The components of the economic system — enterprises in industry, agriculture, transport, banking — were in a state of continuous adjustment through the estimates and actions of controllers

---

[1] See the invaluable volume, B. R. Mitchell with the collaboration of Phyllis Deane, *Abstract of British Historical Statistics*, Cambridge 1962, pp. 6, 115, 131–2, 179, 192–3, 218, 226.

of resources. These adjustments took place most often in successive pulsations of optimism and pessimism, initiative and purge, of a more or less decennial kind. In this way the parts of the expanding whole were caused to grow at rates which made it possible for them to assimilate to one another, maintaining a kind of moving ecological equilibrium between enterprises. Moreover the case has been argued for a longer cycle of an average length of some fifty years.[1] The question is thus posed whether the dynamics of a capitalist economy involved wave motions, long and short, of a kind well known in physical science. Of the reality of the shorter fluctuations there can be no doubt, though their periodicity was not precise and their nature was far from uniform. There is evidence, too, for longer cycles; the exceptional and persistent difficulties of the later thirties and early forties were in some ways comparable with those of the seventies and eighties, though there were notable differences, both of a short-term and of a structural kind.

On an even more extended scale the question has been raised whether it is possible to assimilate the economic experience of Britain in the nineteenth century to a universal theory of economic growth.[2] To such problems no satisfactory answer yet seems possible, so that the historian, though gaining illumination from such attempts at system, must still proceed as in the past, phasing his account into such periods as seem to have historical unity, labelling each in terms of the experience dominant within it. Yet contemporaries tried to explain to each other what was happening; their efforts are an important part of the history of the times.[3]

2. PEACE AND BREAKDOWN: 1815–21

Prosperity in England had mounted steadily from 1811 as Napoleon's continental blockade broke up. Collapse came with the return of peace

---

[1] Nikolai D. Kondratieff, 'The Long Waves in Economic Life', *Review of Economic Statistics*, 1935, reprinted in *Readings in Business Cycle Theory*, American Economic Association Series, London, 1950; George Garvy, 'Kondratieff's Theory of Long Cycles', *Review of Economic Statistics*, 1943, reprinted in A. H. Hansen and R. V. Clemence, eds., *Readings in Business Cycles and National Income*, London, 1953; Jeffrey G. Williams, 'The Long Swing: Comparisons and Interactions between British and American Balance of Payments, 1820–1913', *J. Econ. H.*, 1962.

[2] Such an attempt is made in W. W. Rostow, *The Stages of Economic Growth*, Cambridge, 1960, and is developed in W. W. Rostow, ed., *The Economics of Take-Off into Sustained Growth*, London, 1963. For a unique study of growth and stability on the basis of an exhaustive collation of data see A. D. Gayer, W. W. Rostow, and A. J. Schwartz, *The Growth and Fluctuations of the British Economy 1790–1850*, Oxford, 1953.

[3] See below, Chapter 10, section 10.

in 1815. It was succeeded by a hectic but short post-war boom as exports of cotton goods, hardware, and cutlery, bar and pig iron were rushed to starved markets, coming to its climax in 1818, and plunging to the depths in the following year. The extravagant hopes raised on the prospect of free access to Europe and the Americas were dashed.[1]

The sudden dampening of the flood of government spending, along with the rigorous reduction of the volume of money and credit in order to resume cash payments in gold at the pre-war parity, reversed the upward trend of prices and brought to a sudden halt the expansion of new capacity that had been virtually continuous since 1782.[2] The commercial discounts of the Bank of England were £20 million in 1810; by 1815 they were £15 million; by 1817 they had sunk to £4 million.[3]

For a third of a century the economy of Britain had been subject to a powerful stimulus. Industrial output had greatly increased.[4] Agriculture had extended; so too had trade.[5] Now there was a new atmosphere: arrestation, bewilderment, and fear. It had become apparent that at least some of the new investments since 1793 had been in directions inappropriate to peacetime needs. More frightening still, it appeared as though the lift had gone out of the system, and that expansion was capricious and incapable of being sustained.

The great question of the day became: could the principle of un-controlled private initiative, on the basis of which the expansion had begun, and which had sustained it in its response to the government's needs throughout the war, respond to the challenge of faltering growth?[6] There were plenty of people who called for official action to support credit and employment.

---

[1] W. W. Rostow, 'Adjustments and Maladjustments after the Napoleonic Wars', *American Econ. Rev.*, *Supplement*, vol. XXXII, no. 1, March 1942.

[2] See N. J. Silberling, 'British Prices and Business Cycles, 1779–1850', *Review of Economic Statistics*, 1923; P. Rousseaux, *Les Mouvements de fond de l'economie anglaise, 1800–1913*, Louvain, 1938.

[3] Leland Hamilton Jenks, *The Migration of British Capital to 1875*, New York, 1927, p. 26.

[4] W. G. Hoffmann, *British Industry 1700–1950*, trans. by W. O. Henderson and W. H. Chaloner, Oxford, 1955, p. 29 *et. seq.* This work contains a discussion of the sources available for the historical measurement of industrial output and their difficulties. See also W. A. Cole, 'The Measurement of Industrial Growth', *Econ. H.R.*, 1958; J. F. Wright, 'An Index of the Output of British Industry since 1700', *J. Econ. H.*, 1956.

[5] Werner Schlote, *British Overseas Trade from 1700 to the 1930s*, trans. W. O. Henderson and W. H. Chaloner, Oxford, 1952, p. 14 *et. seq.*; also Albert M. Imlah, *Economic Elements in the Pax Britannica*, Cambridge, Mass., 1958.

[6] For contemporary discussion and legislation see William Smart, *Economic Annals of the Nineteenth Century, 1801–1830*, London, 1910–17.

Such pleas were of two kinds: some were made by those who saw the problem merely as a transitional one to better days, and some came from those who really sought to arrest the pace of industrialization, and thus preserve some kind of controlled balance between industry and agriculture, with the roots of political power still remaining in the land. But though the agriculturalists pressed successfully for a measure to protect their incomes after the catastrophic fall in farm prices, and gained the famous Eighty Shilling Corn Law of 1815, the predominant tenor of opinion was against any attempt to rehabilitate general state control.[1]

Falling prices meant a general discouragement to new initiative; few men were prepared to chance their arm in such a situation. For who could guess what form recovery would take, even if its coming could be anticipated? Food prices came tumbling down with the rest. Bad weather conditions had prevailed from 1811 to 1813, thus partially masking the new productive potential of enclosures, new investment, and new methods. But in 1813 nature perversely smiled on agriculture. A series of bumper crops flooded onto the market, just as war scarcity ended and general prices fell. Natural plenty and wartime improvements in agricultural method combined to shrink farm incomes. Formerly optimistic farmers, now encumbered with heavy debts and long leases at inflated rents, found themselves wedged between fixed costs and falling prices. The annual average price of wheat was 96s. 11d. in 1817; during 1821 it was only 56s. 1d., a severe blow to farmers' earnings.[2] The local Poor Rates added to the difficulties, shooting upward as over a third of a million men from the forces returned to their homes, and often failed in their search for employment.[3] Select Committees on Agriculture sat in three successive years from 1820.

The tax burden was now very serious in absolute terms; it was made worse than ever by the price collapse. The fund-holders (the owners of government debt), and indeed all creditors, were now to collect their interest and capital in an appreciated currency; the landowners, on whom taxation lay particularly heavily, were to be obliged to accept

---

[1] See Chapter 10, section 1, below; D. G. Barnes, *A History of the English Corn Laws from 1660–1846*, London, 1930, chapter VIII; C. R. Fay, *The Corn Laws and Social England*, Cambridge, 1932.

[2] Smart, op. cit., vol. I, p. 566; vol. II, p. 19.

[3] Norman Gash, 'Rural Unemployment, 1815–24', *Econ. H.R.*, 1935; G. E. Fussell and M. Crompton, 'Agricultural Adjustments after the Napoleonic Wars', *E.H.*, 1939.

the new burden.[1] The ending of the Property Tax in 1816 was a dubious aid to the landowners for it relieved others even more.

Profitable outlets for investment almost ceased to exist. Even the boom of 1817–18 did not do much for the capitalist. Long-term interest rates fell sharply, but cheapness of borrowing was an insufficient incentive to persuade business men to embark on new ventures. Canals and the enclosure and development of agricultural land were no longer tempting, nor were mining or engineering, or the metal trades. Textile capacity, too, seemed more than ample.[2] Indeed, investors were far more eager to take up the loans of foreign governments than to face the risks of sinking their funds at home.

The Bank of England was blamed by many for the difficulties of the day.[3] But though it had exerted deflationary pressure to return to gold in 1821, after that date the Bank was probably more willing to lend than entrepreneurs were to borrow. Nor were the country bankers much more successful in extending their loans. Some, like Thomas Attwood, agitated with great vigour that the government should take measures to replace its own contracted demand by policies of spending and lending. His opponents replied that he had both under-estimated the distortion brought by war, and overestimated the power of the government to correct it through continued inflation.

The clamours raised by those in trouble were loud and persistent, especially in those trades in which there was a high degree of organiza-tion. France, though defeated in war, was still Britain's greatest industrial and trading rival; she sought competitive rehabilitation after 1815 by adopting high protection. Birmingham, having enjoyed a wartime boom, was heavily hit. The cotton trade in Lancashire and the west of Scotland, it was now clear, had developed to a point at which it was highly sensitive to fluctuations in demand. Shipbuilding was depressed for twenty years after Waterloo.

All this was bound to produce labour unrest on a frightening scale. Those who were able to stay in employment often found their real wages rising, for prices, to some extent at least, led wages downward. But even this was an erratic gain, and the feeling of insecurity among

---

[1] For the size and distribution of the national income in 1812, and the problems of calculation, see P. K. O'Brien, 'British Income and Property in the early Nineteenth Century', *Econ. H.R.*, 1959.

[2] G. W. Daniels, 'The Cotton Trade at the Close of the Napoleonic War', *Trans. of the Manchester Statistical Society*, 1917–18.

[3] E. V. Morgan, 'Some Aspects of the Bank Restriction Period, 1797–1821, *E.H.*, 1939; E. Cannan, *The Paper Pound of 1797–1821*, London, 1919.

those in employment was added to the blank frustration of the many who, in all industries, were without jobs. The trade unions, never strong, were quite unable to help the workers in this baffling post-war situation.[1]

The government doggedly relied upon a combination of containment in the political sphere, embodied in the notorious 'Six Acts' of 1819, with the 'system of natural liberty' in the economic. Eventually justification seemed to come from events. By 1821 there were real signs of general recovery. It seemed as though the private initiatives that had operated so strongly before and during the war were now able to resume their expansive role.

3. RECOVERY AND EXPANSION: 1821–36

Real growth in wealth took place between 1821 and 1836. Indeed it has been suggested that the industrial revolution in Britain, far from tapering off in the twenties and thirties, was then at its height, in the sense of invoking and applying new capital. Men of business showed vigour and daring in the promotion of trade and industry. Real wages in many trades began to rise again.[2] The unskilled labourer whose lot had deteriorated from the 1790s to about 1821 probably made some gains, though they were by no means dramatic. Industrial output accelerated to produce the most substantial relative increase of the entire century.[3] It seems likely that average real incomes, after a period of stagnation, 1800–12, showed a marked rise over the next ten years, and a significantly high rate of increase between 1822 and 1831.[4]

Investment abroad went on continuously to some degree and reached something of a peak in the thirties in financing the often dubious projects of American state governments. To some extent also, British exports in the thirties were financed by short-term lending abroad.

---

[1] See below, Chapter 9, section 1, p. 326.

[2] T. S. Ashton, 'The Standard of Life of the Workers in England, 1790–1830', *J. Econ. H.*, *Supplement*, 1949; 'Some Statistics of the Industrial Revolution in Britain', *M.S.*, 1948; A. J. Taylor, 'Progress and Poverty in Britain 1780–1850: A Reappraisal', *History*, 1960; R. M. Hartwell, 'The Rising Standard of Living in England, 1800–1850', *Econ. H.R.*, 1961. For a different point of view see E. J. Hobsbawm, 'The British Standard of Living, 1790–1850', *Econ. H.R.* For further discussion see below Chapter 7, section 4.

[3] Hoffmann, op. cit., pp. 30, 31.

[4] See Phyllis Deane, 'Contemporary Estimates of the National Income in the First Half of the Nineteenth Century', *Econ. H.R.*, 1956, p. 353; also 'The Implications of Early National Income Estimates for the Measurement of Long-Term Growth in the United Kingdom', *Economic Development and Cultural Change*, vol. IV, no. 1, 1955.

But in general, British long-term capital between 1825 and the fifties was placed at home, in textiles, mining, chemicals, metal working, and engineering, and in various public utilities now of high priority – especially the provision of gas and water in the towns, and the improvement of harbours. This is not to say that there was a failure of British initiative abroad; quite the contrary, for new opportunities for investment were continually being sought.

Yet activity abroad and new investment at home, with the consequent bidding for resources, impressive though it was, was not enough to reverse the downward price trend. Only a very strong and sustained demand could do this, a condition that was met only briefly, in the later phase of a boom; prices continued their long-run decline until 1853. Much of the new investment was of a kind that fructified quickly, adding to the flow of consumable goods within a year or so of taking place. There was a great deal of technological progress in the period, of a dispersed kind, taking place on many fronts: machine tools, Cort and Neilson's contributions to metallurgy, constructional and civil engineering, the Leblanc process in chemicals, and so on. Most of such inventions took the form of increasing efficiency in established lines of activity. Cheap food, made possible by the adoption by British agriculture of new crops and new methods on the lately enclosed fields, together with somewhat better access to foreign supplies in times of scarcity, following the modification of the Corn Law in 1828, made a significant contribution to falling prices.

But though falling prices in a particular country may be the result of increasing mastery over nature, or may be the result of favourable changes in other countries with which trade is conducted, such improvements may themselves be the cause of serious difficulties for many.

New methods meant the obsolescence of formerly prosperous producers and the reduction or even the destruction of their incomes. If the cost of production of a particular article falls drastically, those who produce it, though they have more to sell, may find their total incomes declining. The farmers in post-Waterloo Britain with their newly enclosed and improved fields were the leading example of this, for in spite of the protective Corn Law, food prices were generally low, dragging down agricultural incomes and strengthening the landed interest in clinging to its tariff protection. In addition, many farmers were bound by obligations contracted during the period of high prices – long rent contracts, and heavy mortgages. The West Indian sugar interest found that in spite of rising population and rising consumption per head, there was a net surplus of world productive capacity, bringing prices down. To the end of the Napoleonic Wars the sugar interest had enjoyed a virtually pre-emptive right to one of the City of London seats in Parliament, but this was soon to cease. The kelp industry, gathering seaweed used in the production of alkali for soap and glass, upon which the Scottish Highlands so heavily depended, had enjoyed prices at over £20 per ton. By 1828 kelp sold for less than £5. This latter was the critical price: below it profits wholly departed and the industry rapidly declined, dragging down the kelp magnates and ruining much of the population of the Scottish Highlands.[1] In East Anglia and in the west of England woollen textile manufacture largely succumbed to the greater vigour and natural advantages of Yorkshire. The new gaslight companies roused the fear and hostility of the brass workers, the oil lamp producers and the whale fishery men. Bristol, so long the great outport of the west, yielded place to Liverpool. Owners of canal shares soon felt the effects of the new railways. Rates for carriage were reduced by at least one-third to one-half; no new inland canals were constructed in England after 1834. Most tragic of all, the handloom weavers, who had held their own against powered looms for so long, were suffering the slow and painful death of their craft in the thirties and forties. But the purge of those whose functions were being eclipsed went relentlessly on.

Those overtaken by obsolescence in one form or another were not the only casualties of the rapid growth in productivity. There were also the victims of unpredictability and error. To share in the fruits of expansion it was necessary to take risks. The crisis of 1825 demonstrated how great the risks now were.

The much increased role played by the capital market in the financing of trade and industry has caused some scholars to regard the excitement of 1825 as the first cyclical boom of the modern sort. There was a proliferation of new joint-stock companies after the repeal in 1825 of the Bubble Act of 1720. With the removal of this ambiguous, arbitrary, and restrictive statute, promoters of assurance and mining companies vigorously pressed upon their clients all kinds of new projects with transferable shares. Investment on an altogether new scale was funnelled into docks, railways (including the famous Stockton and Darlington),

[1] Malcolm Gray, *The Highland Economy, 1750–1850*, Edinburgh, 1957, pp. 155–8. For Highland clearances see Ian Grimble, *The Trial of Patrick Sellar: The Tragedy of Highland Evictions*, London, 1962.

gas and water companies, shipbuilding, and especially in lending abroad by the purchase of mining shares and government securities. Loans to South American republics stimulated exports. Some were for the financing of governments in revolt against Spain, some were for mining speculations, for it was hoped that the steam engine operated by Cornish miners would put the production of gold and silver onto an altogether new basis. A boom in the United States contributed to the general buoyancy. The iron industry reached capacity and raised prices still further, often by monopoly agreement. Heavy speculation in stocks of commodities took place, especially in cotton.

But by the summer and autumn of 1825 optimism had gone too far. A bout of liquidation and failures occurred, beginning in the United States, and spreading to Liverpool and London. The South American mining ventures collapsed. Thus ended the first great rush to employ the mounting capital of industrializing Britain. Though much of it was wasted, much was successfully embodied in new equipment, laying the basis for a further increase in national output.

For some seven years after 1825 the economy, though it made progress in the sense of a continued increase in output due to greater efficiency, and perhaps to greater effort, was not able to reach full employment. Signs of recovery appeared twice, but both in 1829 and 1831 they perished in loss of confidence. Partly this was due to poor harvests in 1828 and 1830. New investment, both at home and abroad, was at a very low ebb.

Yet there was another full employment boom, beginning in 1832 and culminating in 1836. Again there was heavy investment at home, in railways after the astonishing success of the Liverpool and Manchester line, and in joint-stock projects of great variety and extent, both in industry and in banking. Exports were stimulated by loans to America (the chief element in the now reduced lending abroad). Between 1830 and 1836 Anglo-American trade had doubled.[1] By 1836 the American trade was in a state of great excitement, with much speculation based upon accommodation bills, with the London merchant bankers and the Bank of England taking much too casual a view of the speculative danger. There was a considerable rush to buy American securities. To the stimulus of new investment was added that of demand, caused by good harvests from 1833 to 1835, raising the incomes of food producers and strengthening the exchanges, helping the recovery to mount to boom. But nemesis overtook the unwary in 1836.[2]

[1] Jenks, op. cit., p. 84.   [2] R. C. McGrane, *The Panic of 1837*, Chicago, 1924.

Since 1815 it had become increasingly clear that for British industry to continue the expansion that it had begun, and upon which the support of the mounting population depended, markets abroad were of the greatest importance. But though the development of woollen cloth exports since early Tudor times had held, in embryo, the lessons of trade based upon industrial specialization, they had not been relevant to the whole society. Between the twenties and forties of the nineteenth century, however, the situation was so changing that the economic well-being of all became increasingly involved in the trading position.[1]

Here the pattern was rapidly altering. Between Waterloo and the forties the proportion of the total value of British exports going to the traditional markets of north and south Europe fell slightly from 46 to about 44 per cent.[2] But north Europe retained its place as the largest single buyer. The other great areas of trade lay in North and South America, and in the East. In Africa, Asia and Central and South America, cheap cotton exports were especially appropriate to climate and general standards of living. In the case of India her own cotton crop helped her to provide the means of purchasing finished goods and yarn. But volatility was the great characteristic of the new non-European markets. This meant that, as trade grew, so too did the liability to fluctuations. In a general sense it was speculation in Latin-American markets that dominated the expansion and contraction of British exports in the twenties; the American trade, especially on the import side, was the more significant in the thirties; by the forties Asia was a very important determinant of market behaviour.

Thus it was that the economy of Britain became highly sensitive to the conditions under which pioneer societies were being formed and sought to exploit their magnificent natural endowments, and also to the renovation of ancient societies. When these conditions were propitious, and a sustained demand for British goods, especially textiles, occurred, new investment took place in mills and machinery, with consequent incentives to other kinds of capital formation, eventually stimulating all elements of the economy.

This sensitivity to conditions abroad was heightened by the rapid rate of capital accumulation at home, together with the limited experience and lack of caution of the bankers. Thomas Tooke, J. S. Mill and others argued that the low earnings available to investment

[1] A. Redford, *Manchester Merchants and Foreign Trade, 1794–1858*, Manchester, vol. I, 1934, *passim*.
[2] Imlah, op. cit., pp. 129, 130.

on sound security combined with bankers' optimism in granting credit, together with the restrictive effect upon expansion of the Corn Laws, had caused capitalists to seek more dubious ventures, and by so doing had generated a speculative boom which, for a time, could feed upon itself and upset the judgement even of men of skill and caution.[1]

There was a third kind of sufferer from the overall trend toward expansion. It was beginning to become apparent that, among all the productive marvels of the age, the construction of houses for the labourer was less profitable than other tasks, for those able to do it well. The skilled and semi-skilled workers made some gains in real wages, often in the later stages of the boom, and began to exert an effective demand for houses. But the mass of the workers gained little in wages, and suffered harshly from the general failure to provide reasonable conditions for urban life. The market mechanism, for all the splendid equipment with which it was providing society, could not at this stage, as the great trek from country to town gathered speed, greatly improve the supply of houses for the many.[2] Yet it should be noted that by the rather crude standard of the number of persons per domicile (which takes no account of the size or quality of the accommodation), overcrowding was somewhat reduced in the later twenties and the thirties, and again in the fifties.[3]

Much pain was suffered by the casualties of obsolescence and error, and by the humble victims of the inadequacies of the market. But all these forms of discomfort seemed, in the twenties and thirties, to be necessary conditions for the very rapid expansion of new equipment embodying the latest skill and perception, as the economy of Britain jolted forward to ever higher levels of output.

The period from the twenties to the forties is difficult to summarize. In terms of capital formation, the development of new skills, and the increase of total output, it was a time of great progress. But in terms of improvement of real wages, though many workers were gaining ground, it is highly doubtful whether the mass of men enjoyed any great material advance. Certain groups suffered heavy direct blows, the

---

[1] G. S. L. Tucker, *Progress and Profits in British Economic Thought, 1650–1850*, Cambridge, 1960, p. 188.

[2] See below, Chapter 4, section 9, p. 166; Chapter 7, section 6. For the provision of houses see A. K. Cairncross and B. Weber, 'Fluctuations in Building in Great Britain, 1785–1849', *Econ. H.R.*, 1956; H. A. Shannon, 'Bricks – a Trade Index, 1785–1849', *Econ.*, 1934; J. Parry Lewis, 'Indices of House-Building in the Manchester Conurbation, South Wales, and Great Britain, 1851–1913', *S.J.P.E.*, 1961.

[3] Hoffmann, op. cit., p. 77.

prelude to their diminution or eclipse. Prices as a whole fell continuously, except for hectic boom intervals, suggesting in a *prima facie* way that the system was not reaching its full potential output. Finally, the economy as a whole produced two major breakdowns, in the aftermath of which the great waves of social protest of the early nineteenth century arose: Owenite socialism and mass trade unionism, and Chartism.[1]

4. CRITICAL DEPRESSION: 1836–42

Things were very bad down to 1842; especially as the depression in industry deepened in the last two years after the severe financial crisis of 1839.[2] All the way from 1836 to 1842 nature was niggardly with her harvests. Heavy grain imports were necessary, especially in 1838 and 1839, with consequent loss of gold abroad and monetary stringency at home. But such imports did little to reduce the very high price of food. Political difficulties at home and abroad further helped to keep the cost of borrowing high. Disillusionment with foreign borrowers in the United States and elsewhere discouraged foreign investment abroad, which in turn reduced British exports.

The taking up of available investment opportunities had gone very far by the later thirties and early forties. This was true in cotton, in coal, and in iron.

In cotton, as early as 1833, it began to look as though capacity was excessive, for profits fell away sharply, with only the most efficient firms making any gains. The millowners, having involved themselves in costly plant and equipment, could not afford, in a competitive situation, to contract output in order to keep prices up. Indeed, capacity actually increased, for spinners, in the attempt to improve their individual positions, built weaving sheds and installed powered looms. This precarious situation was saved by a recovery of demand in 1834, bringing better profits and a new wave of investment, this time of great size. New capital poured into all sections of the industry not only for the improvement of old mills, but for the construction of new ones. By 1837 there was, once more, depression in the cotton industry. It was not serious at first, but conditions deteriorated until, in 1841–42, near disaster came. The owners of new cotton factories in Belgium,

---

[1] See below, Chapter 9, section 6.
[2] See R. C. O. Matthews, *A Study in Trade Cycle History: Economic Fluctuations in Britain 1833–42*, Cambridge, 1954; Jenks, op. cit., p. 128.

Prussia, and Saxony, heavy buyers of British yarn, had reached a crisis and so reduced their consumption. For the first time since Arkwright's day, there were two successive seasons in which the consumption of raw cotton actually fell. As selling prices abroad tumbled, British cotton manufacturers pumped more and more goods into foreign markets in the attempt to restore earnings, thus heightening the calamity by further depressing prices.

New mills were also shooting up in the woollen industry.[1] The worsted sector grew more rapidly than the woollen branch for it was more suitable for mechanization. Within three years down to 1838, the West Riding was provided with an increase in the number of its woollen mills by 30 per cent, and its worsted mills by 65 per cent. Like cotton, and perhaps to an even greater degree, wool was in deep depression by 1842, burdened with heavy excess capacity.

The enormous untouched resources of south and south-east Durham were opened up, with the aid of new railways, new harbour facilities, and new sinkings.[2] So, too in Scotland, and South Wales, where output grew greatly. New capital poured into the coalfields in response to rising coal prices, to the grave embarrassment of all producers by the later thirties.[3] In the period 1836–43 coal output rose inexorably (by some 60 to 70 per cent) while the sale of coal increased by only 30 per cent. Attempts by the colliery owners to control output by quotas failed miserably as each enterprise sought to attract a larger share of the inadequate demand.

The heaviest user of coal, the iron industry, followed the same pattern. South Wales and south Staffordshire expanded output. But the most dramatic growth was in Scotland. Neilson's hot-blast, invented in 1828, made it possible to work the blackband ironstones of central Scotland. Between 1830 and the mid-forties, Scotland's share of total British iron output rose from 5 to 25 per cent — an amazing feat. But it meant that Scotland embarrassed English and Welsh producers at home, and especially in foreign markets, which they soon virtually monopolized. This involved a general re-orientation of the Scottish economy around iron and coal and heavy industry generally. Iron-masters elsewhere made attempts to maintain prices and profits in the traditional way, by agreed limitation of output, but without Scottish

[1] J. H. Clapham, *An Economic History of Modern Britain*, vol. I. *The Early Railway age 1820–1850*, Cambridge, 1950, p. 192.

[2] See below, Chapter 4, section 8, pp. 159, 160.

[3] Matthews, op. cit., p. 156.

cooperation this was impossible. Thus by 1840 the industry found profitability very low, though Scotland suffered less than other areas.

The railway and shipbuilding boom, in the sense of financial speculation, broke in 1836. But though many proposed railway schemes were abandoned, many were carried through, so that construction continued, reaching a peak for completion of lines in 1840. Shipbuilding had suffered a long period of low profits, but like railway building, it enjoyed a boom in the formation of new concerns in 1836. Again, as with the railways, new ships were being built long after the crisis had passed. By 1840 both railway construction and shipbuilding had lost their power to generate employment.

It seems that a situation had been reached by the later thirties in which the home demand for the products of industry, together with available exports markets, was insufficient to consume the whole of the new potential. The system, having adjusted itself to the new rate of growth of its industrial sector, was now heavily dependent upon its continuance. Coal depended upon iron, iron depended upon industrial expansion, industry depended upon increasing consumers' demand at home and abroad, demand depended upon willingness to invest, and upon earnings. So far as foreign countries were concerned their willingness or ability to consume textiles was not increasing fast enough to keep the economy of Britain growing at the necessary rate.

To this, Britain's own trading policy contributed substantially. Tariffs of many kinds had been raised during the war against France, in order to produce revenue. The enthusiastic abandonment of the income tax in 1816 made the pressure upon tariff revenues even greater, so that by 1840 taxes on imports comprised some 46 per cent of total revenue. Moreover, the prices of British exports fell far faster than those of imports, so that the relative cost of foreign raw materials and food was increasing. But there were all manner of vested interests, among both agriculturalists and industrialists, to resist tariff reductions, so that in spite of Huskisson's reforms in 1823–25, British tariffs were still high in the early forties, at an average rate of about one-third the value of net imports, with some articles much higher. Such protection had many effects. The import of goods with a low duty grew much faster than was the case with those bearing high duties. Fortunately raw cotton and wool were lightly taxed, so that these fundamental industrial materials did not have their prices raised. But foodstuffs were heavily encumbered. Cereals were taxed under the Corn Laws; tea, tobacco, sugar, and wines bore heavy burdens.

Much of the ineptitude of tariff policy down to the forties lay in the effect upon the workers' capacity to consume, and in consequence, their incentive to work, which in turn was reflected in the cost of production of exports. With food and drink costs maintained at artificial levels, the ability to spend on the products of home industry was limited. So far as foreign suppliers were concerned, British tariffs limited their ability to sell, and consequently to buy.

It was natural enough that foreign countries in Europe and the Americas should respond to British tariffs by retaliation. A stiffening in the American tariff in 1842 sharply reduced British exports to that country. This action was followed by an enormous American repudiation of British debt by state governments. All this caused severe discouragement, since much was hoped for in trade with the new world. There is evidence also that British shipowners, looking to government support in the form of the Navigation Laws, rather than to their own efficiency, were losing ground to foreigners. Growth was especially slow on the protected empire routes.

In 1836 Britain entered upon a deficit phase on trade and services account. In the period 1816–20, there had been an annual average surplus of £5·48 million; between 1836 and 1840 imports predominated to the extent of £5·36 million per year.[1] Though interest and dividends on earlier investment abroad more than compensated for this, it was a serious warning that British sales abroad were less in value than purchases. Even more worrying, the best available estimates suggest that real income per head, the basic test of economic progress, was, for the first time in the modern period, actually falling between 1831 and 1841.[2]

Manpower, by the forties, was moving in considerable numbers from the countryside to industry, especially in the north and in Scotland. But arrested prosperity slowed this movement, damming up this excess manpower in the rural areas. The very rapid increase in population in the second decade of the century, now come of working age, flooded into the labour market in the later thirties and early forties, just as employment opportunities were lessening. Marriage and birth rates, especially in the industrial areas, already in decline, fell faster than ever, increasing social frustration.

The impression of breakdown thus gained is strengthened when we look to the political sphere. These were the years of greatest difficulty and unrest. The workers were disillusioned with the outcome of the

[1] Imlah, op. cit., p. 70, n. 1.
[2] Deane, op. cit., *Econ. H.R.*, 1956, p. 353.

Reform agitation of 1832. Crop failures and high food prices strengthened free trade sentiment; the first Anti-Corn Law Association was founded in London late in 1836. The agriculturalists founded counterassociations in each county of England. The sting of the new Poor Law of 1834, hidden by two prosperous years, was now fully felt, especially in the industrial north. It was notable that though the Commissioners could make drastic reductions in poor relief in the farming counties, this was impossible in the seats of industry in Lancashire and Yorkshire. The Chartists began their four years of maximum agitation in 1838. These were the years of the scandalous disclosures in the Blue Books; the years of Engels's somewhat unscrupulous account of English social conditions.[1]

The contrast between economic potential and the condition of the people was at its sharpest. For these derangements struck a society much of the social equipment of which was deplorable in terms of schools, hospitals, homes, and sanitation. But a real attack on these problems, difficult in times of prosperity, became impossible in times of depression. The new and now excessive capacity in cotton, railways, coal, iron, and engineering, could not be switched readily to the provision of houses and consumer goods for the multitude of denizens.

5. THE CONDITIONS OF EXPANSION AFTER 1842

The future of a free enterprise economy with high capital accumulation in real terms is equivocal. Such capital is a necessary (though not a sufficient) condition for cumulative growth. If the other required elements are not present, the economy must undergo such alterations as make the excess capital accumulation impossible. The fundamental question becomes: can business men find new investment outlets over time such that the surplus return on former investment can be continuously embodied in new projects capable of earning, in their turn, satisfactory profits?

It may be that the familiar chop and lop of cyclical depression is not enough; something more far-reaching may be required. This could involve an explosion: a general breakdown in society, which drastically lowers the total product out of which capital is formed. There were some signs and much fear in the thirties and forties that this might occur. Or it

[1] Friedrich Engels, *The Conditions of the Working Classes in England*, 1844, new ed., trans. and ed. by W. O. Henderson and W. H. Chaloner, Oxford, 1958; also below, Chapter 9, section 6.

may take the form of an increase in the share of national income going to wage earners, so that profits and capital accumulation are diminished, with the workers consuming more of the product as it is created. The workers, in fact, struggled with vigour and bitterness to bring this about, but without much success.

On the other hand, it may prove possible to continue the growth process in a free enterprise economy without substantial immediate change in the relationship existing between capitalists and employees. Two general kinds of possibility may be present. It may be that an institutional adjustment, capable of being carried out by governments, is available. Or non-governmental factors, inherent in the evolution of the national economy, or in that of the world generally, may produce a new situation making continued growth possible.

As to the first possibility, governments may either erect or dismantle techniques of control. Britain had developed an elaborate system of tariffs and prohibitions, and, in spite of some modifications by Huskisson, the system was seriously inhibitive by the forties. In 1839 1,146 articles were dutiable; only 741 actually mounted the barrier and entered Britain. Of these, 531 yielded some ·004 per cent of the tariff revenue, and a mere 17 brought in no less than 94·5 per cent of total yield.[1]

Peel with his budget of 1842 took a great step forward in the removal of these archaic restrictions.[2] He sought an alternative to the old Tory stronghand methods of rule, for it was clear by the forties that the pressures engendered in society could not be contained, but must be relieved. Earlier, any attempt at reducing protection had provoked storms of protest from vested interests in both industry and agriculture. But by the early forties opinion had altered a good deal. The only policy available to the government was to try the effect upon trade of an extension of the ideas that Pitt had received from Adam Smith before the French Revolution. Certainly no policy resting merely on improving the system of control could be entertained.

To make up for the deficiency in the revenue the income tax was brought back at sevenpence in the pound. This had more merit than mere expediency, for it meant that the damaging and distorting effects of taxing trade could be lessened; it meant further that the tax system as a whole could be made to bear more heavily on larger incomes and more lightly upon the smaller.

Thus began the acceleration towards free trade. In 1845 further great

[1] Imlah, op. cit., p. 148.   [2] See below, Chapter 9, section 7, p. 353.

reductions and abolitions of duties took place; in 1846 the import of corn was freed, except for a registration duty of a shilling a quarter. The Navigation Laws were repealed in 1849 for foreign trade; in 1854 the coasting trade was also thrown open. The budget of 1860 virtually completed the transition to freedom. The Cobden-Chevalier Treaty of the same year between France and Britain was the furthest point reached in Europe in the direction of free trade.[1] America joined in the liberating trend with the Walker Act of 1846, introducing a phase of tariff moderation lasting down to the Civil War.[2]

Protection to colonial interests was also being reduced. Both West Indian and Canadian producers were to lose their preferred positions in British markets, upon which they had greatly depended.[3] All elements of the Empire, so far as Britain was concerned, were to follow the free trade principle; only when they were sufficiently powerful, as with Canada, to gain control of their own trade policy, was it to be otherwise.

There followed a remarkable and sustained growth of trade. In the forties and fifties the rate of expansion of foreign commerce proportionate to the rate of growth of population was at a maximum. Down to 1850 overseas trade had increased more or less at the same rate as industrial production; but from the fifties onward the rate of growth of trade outran that of industry.[4]

One school of thought, both among contemporaries and among later scholars, has attributed the great days of this post-forties period almost entirely to these free trade principles. James Wilson, founder and first editor of *The Economist*, was convinced of this; the liberals of the age of expansion which followed took it as the great justification of their creed.[5] In one sense this view was undoubtedly true. In so far as political wisdom and policy could contribute to expansion at this juncture, the policy adopted was the right one.

But the policies of statesmen were not the only changing element in the situation. Three principal kinds of new opportunity were appearing, as Britain and the world experienced the implications of the new forces that industrialization had released.

[1] A. L. Dunham, *The Anglo-French Treaty of Commerce of 1860*, Ann Arbor, 1930.

[2] F. W. Taussig, *The Tariff History of the United States*, New York, new ed., 1931, chapter VII.

[3] See below, Chapter 9, section 5, p. 343.

[4] Schlote, op. cit., p. 50; see also H. J. Habakkuk in *Cambridge History of the British Empire*, Cambridge, 1940, vol. II, chapter XXI, 'Free Trade and Commercial Expansion'.

[5] Scott Gordon, 'The London Economist and the High Tide of Laissez-Faire', *J.P.E.*, 1955; Anon., *The Economist, 1843–1943, A Centenary Volume*, Oxford, 1943.

Population growth both in Britain and abroad was creating new demand; the appearance of new inventions provided profitable new applications of capital; access to new territories and natural resources served to lower the costs of food and raw materials, and called for their exploitation. The action of any one of these factors might have been capable of bringing the investing process, for a time at least, to a level capable of consuming the available capital. On the other hand, though some combination of such elements may preserve continuity in one sense, in another they themselves must bring change, for mounting population, new methods, and new resources, inevitably alter society and may eventually do so in directions inconsistent with continued rapid growth.

For the dominant clues to expansion we must look to inventions and new territories. The railway and the iron-hulled steamship were the keys to both.[1] The former produced renewed activity on an extraordinary scale at home, in the rush after 1845 to lay down new lines. The railway has been the classic capital user of all time; it appeared in its greatest strength in Britain at the very moment when the need was greatest. Thereafter it was exported, and the same process of laying down capital equipment on a heroic scale took place in Europe, in the Americas, and in Asia. Coal mining and iron manufacture responded with mounting output. Societies formerly inaccessible were drawn into the trading orbit as the iron steamship added to the marvels of the age. Emigration could assume new proportions and a new significance in the nation's life.[2] The supplying of foreigners with consumer goods, and with the means for their own industrialization, so eagerly sought after, accompanied the transport revolution. Invention, too, responded: in chemicals, in metallurgy, in mining, and in many other lines, processes were improved.[3] Indeed, the countries that were seeking to industrialize, like America, Germany, and France, all contributed to the inventive process, and often took the initiative. Such new techniques both caused the writing off of much old capital embodied in obsolete equipment, and provided outlets for new investment.

Perhaps it is not too much to say that the frontier within Britain was closing by the forties, but the railway and the steamship produced a break-out into new territory, and so saved the economy from attrition

---

[1] E. A. Pratt, *A History of Inland Transport and Communication in England*, London, 1912; Christopher I. Savage, *An Economic History of Transport*, London 1959.

[2] W. S. Shepperson, *British Emigration to North America*, Oxford, 1957.

[3] See below, Chapter 4.

or collapse and made possible the resumption of expansion. As Emerson put it in 1844: 'Railroad iron is a magician's rod, in its power to evoke the sleeping energies of land and water.'[1] Britain, as industrial and trading leader, had reached a point at which, if further growth was to be achieved, there must be cumulative change abroad: this was forthcoming.

Even internally the impact of the transport revolution of the third quarter of the nineteenth century upon the economy of Britain was not a simple one. Throughout the century, in spite of the rise of other industries, textiles continued to dominate British exports. This meant that whatever changes were occurring, Britain's trading position continued to depend largely upon the foreign demand for articles of clothing. But the great difference in the new situation was that railways and steamships built in Britain greatly reduced production and delivery costs, both of manufactures and of raw materials. Britain exported metals and machinery, and these had effects abroad greatly in excess of their quantitative scale, in helping forward foreign trade and industry, so that other countries could take an increasing share in expansion. But the great sustaining influence for British trade in the two or three decades after the forties was lower costs in an almost domestic sense: railways and steamships meant that British manufactured consumer goods could be delivered much more cheaply in foreign markets.

The transport revolution also contributed very greatly to the power of Britain to earn abroad through invisible exports. The earnings of British shipping, with the ancillary services of insurance and finance, were very great from the mid-century onward. For this provision of steamship facilities Britain was peculiarly suited in terms of natural, industrial, and human resources. Such earnings made a major contribution to the net surplus of Britain's trade, out of which her mounting foreign investments were created.

But the contribution of transport investment must not cause us to overlook other kinds of invention going forward at the same time. The sewing-machine and its impact afford an instructive contrast to the railway and the steamship.[2] It was in the later forties that the manufacture and use of the sewing-machine began to affect Britain. Its adoption called for no great indivisible investment in factory or plant. Individual machines appeared in the miserable garrets where needlewomen, the

---

[1] *The Works of Ralph Waldo Emerson*, London, 1890, vol. I, p. 296.
[2] 'Efficient Production and Effectual Distribution — The Results of the Invention of the Sewing Machine', *Econ.*, 26 May 1877, p. 601.

wives of seamen and dockers, stitched stays and corsets for three or four shillings per week. Soon it spread to the clothing industry generally and the degradation of the needle, so bitterly attacked by Thomas Hood in *The Song of the Shirt*, was vastly relieved. Workers acquired their own machines, and a new kind of home industry resulted. Wages rose, recruitment to the garment trades increased, causing a real rise in the wages of workers. In 1857 the sewing-machine was adapted to the boot and shoe trade, and after a contest between masters and workers, was adopted. New factories sprang up as the demand for shoes increased as production was cheapened.[1] In this way the clothing and footgear trades were revolutionized by a new principle of assembly, and as incomes rose, a more generous standard of clothing became universal. Though not of the same order of importance as the railway and the steamship, the sewing-machine stands as a symbol of the myriad other renovating forces at work.

But British industry and commerce could hardly have attained such splendid results had there not been a vigorous response by British agriculture to its new, and after 1846, unprotected situation. Critics of free trade had prophesied a deluge of foreign food that would ruin home producers and work adversely upon the balance of payments. Instead, British farming responded to the challenge, using new techniques, including the application of guano from Peru, and attained new levels of productivity. Though the farm labourer did not gain greatly, agricultural output, stimulated by the new situation, contributed significantly to general expansion.

Finally there was the question of the means of payment as the transactions of the world mounted. Where Britain was obliged to make payments in markets in which she was in deficit she could do so either in securities or other paper claims, or in gold. As the world's business grew the availability of the former increased as part of the growth process. The volume of the precious metals also responded to the needs of commerce, for by the fortuitous circumstances of discovery of gold in California (1848) and Australia (1851) the problems of payment were eased.[2]

Thus the elements fell together to produce an astonishing surge forward in the capacity of Britain to create and consume wealth. Policy was appropriate to potential, and in a manner and degree far beyond

[1] See below, Chapter 9, section 8, p. 370.
[2] See below, also Chapter 6; A. Del Mar, *A History of the Precious Metals*, London, 1880, chapters XX and XXI.

what its authors could foresee. It was between the fifties and the seventies that the absolute increase in production in Britain became so staggering, with all the accompanying effects upon social life and outlook. Moreover the new wealth came forward in an atmosphere of exhilaration among men of business, in rosy contrast to the long struggle against falling prices. By 1853 the general trend of prices, downward since 1819, was reversed; prices mounted all the way to 1873.[1] On the workers' side, too, things were brighter, with real wages rising convincingly from the sixties. Attention could switch from the attempt to offer an alternative formula for society, to organizing to secure a larger share of expansion.

6. POPULATION, URBANIZATION, MIGRATION

Though we can hardly award the palm for the resumed advance of the British economy in the forties to demographic factors, we can never ignore them for long.

The most striking characteristic of the population of Britain in the nineteenth century is that of increase.[2] In England and Wales it more than doubled between the censuses of 1811 and 1871 (from 10 to 22·7 million), and had multiplied 2½ times by 1881, to reach nearly 26 million. From 1811 to 1831 the decennial increase in numbers was just under two million; in the thirties, forties and fifties, it was just over two million. Then there was a jump: in the sixties no less than 2⅔ million were added to the population, and in the seventies the enormous total of 3·2 million, almost the exact equivalent of another Greater London of 1861. Thus, the absolute increase, dramatically and continuously upward, became even greater in the sixties and seventies.

But in some ways it is the relative rate which is the more revealing, showing as it does the capacity of a given generation of people to alter subsequent numbers. The curve of the rate of increase had two peaks, 1811–21, and 1871–81. The former decade, in spite of war and post-war troubles, had the highest recorded decennial rate of growth in English history — no less than 18·06 per cent.[3] Decline followed; then

---

[1] Mitchell and Deane, op. cit., pp. 470–2; Sir Walter Layton and Geoffrey Crowther, *An Introduction to the Study of Prices*, London, 1935, chapters VI and VII.

[2] For population statistics see the *Annual Reports of the Registrar General of Births, Deaths and Marriages*, continuous from 1839. The introductions to these reports often contain much of interest. For the history of the census, and some caveats, see A. J. Taylor, 'The Taking of the Census, 1801–1951', *British Medical Journal*, 1951.

[3] R. Price-Williams, 'On the Increase of Population in England and Wales', *J.S.S.*, 1880.

in the sixties, as the great commercial, industrial and agricultural expansion mounted, came recovery, rising to a second, though lesser, peak of 14·4 per cent in the seventies.

It is not easy to explain these variations in the rate of growth of the population. An intricate interplay of factors will always be at work, both on the death rate and on the birth rate.[1] Some of these will reflect the operation of individual choice, and some the working of changes in society as a whole. In addition, as conditions altered, society drew people from abroad, and extruded some of its own number, causing a divergence between the rate of actual increase and the rate of natural increase.

The death rate was, of course, a matter outside the control of the individual member of the population, except in so far as self-debauchery may be held to be a matter of individual responsibility, or in so far as it can be argued that longevity may be promoted by wise spending and continent living. The great variations in the death rate were the outcome of changes in the age distribution of the population, of public policies of sanitation and medicine, of changes in conditions of working and living, and of changes in nutritional standards.

It might seem, on the birth rate side, that men and women had much more ability to control the situation. Indeed we find that adults of child-bearing age were adjusting their conduct in this regard, as in all others, to the economic and social changes at work upon them. But Malthus and many contemporaries believed that they would increase their families as soon as opportunity permitted; Malthus regarded variations in births as merely the result of changes in prevailing circumstances, permitting or curtailing the operation of this fundamental urge. There is room for much argument here, and for much difference in conduct between, for example, skilled artisans with high wages, who might prefer a higher living standard to a maximum number of offspring, and the unskilled labourer, who could make no such calculation.[2] But even if we simplify the problem, as Malthus did, taking breeding behaviour to follow a fixed rule, with the outcome determined by prevailing conditions, it is still full of difficulties, for the changes in circumstance were complicated enough.

The Registrars General of the nineteenth century saw the marriage rate as the mechanism of adjustment of the birth rate.[3] Difficult times

[1] See below, Chapter 7, section 3.
[2] For the population debate see Chapter 10, section 2.
[3] G. U. Yule, 'On the Changes in the Marriage and Birth-Rates in England and Wales in the Past Half-Century', *J.R.S.S.*, 1906.

meant that in many cases marriage was delayed or did not take place at all. This is shown statistically in a fall in the number of marriages per thousand, and a rise in the average age of marriage. This meant that a decline in births followed in a year or so, and persisted, depending upon how long the discouragement to marriage had continued. During the earlier part of the century, in a society still mainly agricultural, the copiousness of the harvest, because of its importance for the incomes of the farming community and also on occasion because of losses of gold abroad, could affect the marriage rate considerably. But as the economy moved from the traditional agricultural base in the later nineteenth century, though the yield of the harvest was by no means without its effect, variations of the marriage rate were not related to it in an identifiable way, but rather followed indices of the general level of economic activity.

Perhaps the most baffling aspect of the nineteenth-century demographic story is the extraordinarily high rate of increase in the second decade of the century, in a period dominated by war followed by painful adjustment to peace.[1] The uniquely high rate in this decade seems inconsistent with the difficulties of the time, though there were indeed some favourable short-term factors at work upon the birth rate. Commercial conditions improved as Napoleon's Continental System, so damaging to British trade, broke up. There was the optimism of peace and boom and the return of great numbers of men from the armed forces. The increasing labour demands of the factory masters made child-bearing profitable, and the great ease with which Poor Law relief could be obtained had made couples feckless for more than a generation. But if we look at the longer course of birth rates, we find that the late eighteenth century had produced a very high one and that this high rate continued into the new century.[2] It fell only slightly in the period 1811–21.[3] In short, the favourable factors in this decade, in their conflict with difficulties of the time, were not quite able to support the high birth rate that had previously operated.

The behaviour of the death rate in the last decade of the eighteenth century and the first two of the nineteenth remains obscure. There were a number of circumstances ostensibly favourable to the extension

[1] T. H. Marshall, 'The Population of England and Wales from the Industrial Revolution to the World War', in E. M. Carus-Wilson, ed., *Essays in Economic History*, London, 1954.

[2] H. J. Habakkuk, 'English Population in the Eighteenth Century', *Econ. H.R.*, 1953, p. 128.

[3] T. H. Marshall, 'The Population Problem during the Industrial Revolution', in Carus-Wilson, ed., op. cit., p. 316.

of life. The violent intemperance centred on the eighteenth-century gin shops was declining. But much more important was the rise of preventive medicine. Smallpox had long been the great destroyer of infants, but with the coming of inoculation after Jenner's discovery of vaccination in 1798, its effect upon the death rate was reduced, though it is hard to say to what degree. Summer diarrhoea and winter fever, deaths in childbed, and scurvy, also appear to have yielded to improved medical knowledge. But new calculations, taking greater account of the possibility that the extent of failure to register deaths in these early decades has distorted the picture, suggest that no distinct fall in death rates occurred.[1] (Civil Registration came only in 1837; previously the Parish ministers of the Church of England were the recorders.) This has caused the medical revolution of the turn of the century to be sharply discounted. But the profound effect of new medical techniques after 1820 can hardly be denied, helping to bring the death rate from well over 25 per thousand to the lower twenties by 1840.

As the difficulties of the later thirties and early forties multiplied, maintenance of the very high rate of increase in numbers during the second decade proved no longer possible. Inherent in the survival of so many more infants in the early part of the century was a fall in the proportion of the total population capable of childbirth, during the infancy of the new generation. This in itself depressed marriage rates and birth rates. But by the later thirties and early forties these new-comers were ready to marry and bear children. Nevertheless the increase that might have been expected on this ground of changing age distribution did not occur. These years showed a decline in the proportion of women of child-bearing age marrying, and a decline in their fertility. It was noticeable that the industrial districts were more affected than the rural, not only in terms of a falling marriage rate, but in the decline of savings and the rise of pauperism.[2] The most probable explanation is that the economic difficulties of the times caused a delay in marriage, so strong as to overbear the countervailing trend. The Census of 1851 showed that of all women in Great Britain between the ages of 20 and 40, forty of every hundred were unmarried; of both men and women in this age group there were 1·4 million of each, making nearly 3 million adult persons, unwed.[3]

[1] D. V. Glass, 'A Note on the Under-Registration in Britain in the Nineteenth Century', *P.S.*, 1951; J. T. Krause, 'Changes in English Fertility and Mortality, 1781–1850', *Econ. H.R.*, 1958.

[2] John Towne Danson, *Economic and Statistical Studies, 1840–1890*, London, 1906, p. 92.

[3] T. E. Cliffe Leslie, *Essays in Political and Moral Philosophy*, Dublin, 1879, p. 4.

Certain obvious negative factors were at work on marriage and birth rates. The new Poor Law of 1834 made public relief much more rigorous; the Factory Act of 1833 lowered the value of children by reducing their short-term earning capacity. Moreover, and perhaps of greatest importance, these were difficult times. For five or six years wages were low and unemployment fairly general. It would seem that the stiffening of economic conditions, together with legislative action, were powerful enough more than to counterbalance the demographic factors.

To these elements on the side of birth rate must be added the adverse trend of the death rate: this rose again to a highish plateau in the later forties. Though, even after increase, the death rate of the forties was a good deal less than the late eighteenth-century level, the increase, together with the fall in births, explains why the natural increase in this half-decade was very low.

The rise in deaths took place in step with the great expansion of the towns and of the factory system. There seems little doubt that as society continued to move toward its new urban-industrial form the casualties of disease increased once more, overbearing the effects of the new medicine. In the early forties the urban death rate revealed what was happening, with vulnerability to infection increasing more rapidly than medical and other improvements. Liverpool and Manchester in the early forties had a rate twice as high as that of Anglesey or the Isle of Wight.[1]

But the pattern of increase again altered radically from the fifties to the seventies. Births recovered and the death rate was held. It is tempting to associate the increase in births with economic recovery. But the increase came before this factor could have had much time to operate. The hard times of the early forties could not depress the marriage rate indefinitely against increasing numbers of men and women attaining a marriageable age.[2] But once the increase in the birth rate had begun, it is reasonable to assume that improved incomes sustained it. The evidence for this becomes increasingly convincing as the great boom of the early seventies is approached. Very high marriage rates from 1872 to 1874 meant that the highest birth rate in the recorded experience of England and Wales (36·3) came in 1876.

The death rate remained more or less stable during this quarter-

---

[1] For some regional differences in the early forties see Table, 'Relative Mortality of Different Places', *J.S.S.*, 1843, p. 367.

[2] T. H. Marshall, 'Population of England and Wales', op. cit., p. 335.

century of expansion. It might seem remarkable that no real improvement seemed possible here. But maintaining a constant death rate in the

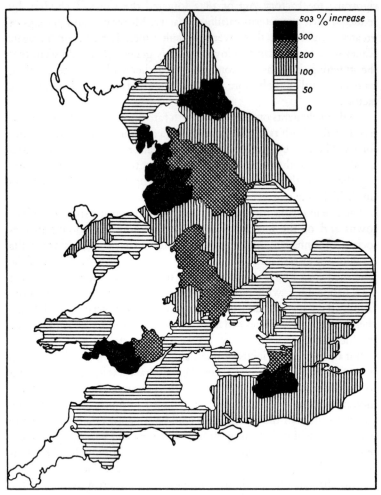

POPULATION: PERCENTAGE INCREASE BY COUNTIES,
1811–81

Source: General Report of Census of England and Wales, 1881, Vol. IV.

face of a population increasing so rapidly, especially in absolute terms, was no passive business.

Urbanization, the necessary condition for the increase of numbers on such a scale, was proceeding rapidly from the forties to the seventies. Though quantification is made difficult by problems of definition and demarcation, the outline is clear. Proportionate to population, the greatest movement to the urban centres occurred in the forties; large numbers of immigrant Irish joined the native English and Scots in the gravitation townward. By 1851 the urban population of Britain had just overtaken the rural; by 1861 the proportion of those living under urban conditions stood to their rural counterparts in the ratio of 5 to 4; by 1881 the urban figure was more than double the rural.[1]

The population of the countryside was not, of course, stationary.[2] Like the country as a whole its maximum rate of increase was in the second decade of the century; indeed it was the countryside which provided most of this expansion. But from the twenties to the fifties rural numbers grew only slowly — by some 28 per cent in thirty years. It is clear that the towns were drawing strongly upon the rural areas. There was some recovery in the countryside in the fifties and sixties, corresponding to the golden years of agriculture. Yet the degree of agricultural prosperity was not the primary element governing the distribution of population between town and country. Rather it was the growth of urban industry, aided by the railways. By the late sixties and seventies the rate of urbanization was at its height, even in the face of a prosperous agriculture (London received new citizens, to add to its own natural increase, at a rate of 300 per day). The rural population was once more approaching stability, having lost its power to grow absolutely in the later sixties.[3]

Migrants had moved in and out of Britain as far back as we can see, often in such groups as were qualitatively of high significance. But so far as noticeable effects on the numerical trend were concerned, they first became important in the second quarter of the nineteenth century, when transport improvements began to make large-scale movements possible. The relationship between men, resources, and techniques, could now express itself significantly in terms of those who opted to enter or to leave Britain.[4] But the data is not very satisfactory. Estimates of migration

[1] J. Clapham, in G. M. Young, *Early Victorian England, 1830–1865*, London, 1934, vol. I, p. 3; Arthur Redford, *Labour Migration in England, 1800–1850*, Manchester, 1926; A. K. Cairncross, *Home & Foreign Investment, 1870–1913*, Cambridge, 1953, p. 68 *et seq.;* John Saville, *Rural Depopulation in England and Wales*, London, 1957.

[2] R. Price-Williams, op. cit., p. 470.

[3] A. L. Bowley, 'Rural Population in England and Wales, *J.R.S.S.*, 1914, p. 607.

[4] For emigration, down to 1856, 'Seventeenth General Report of the Emigration Commissioners', 1857; thereafter the *Statistical Abstract*. Also Brinley Thomas, *Migration and Economic Growth, A Study of Great Britain and the Atlantic Economy*, Cambridge, 1954.

must be based upon the passenger lists of ships sailing from United Kingdom ports, whatever the origin of those leaving, and whether or not they were making a permanent change of domicile for, down to 1870, there is no quantitative information about those entering Britain. The Irish and Scottish elements, and foreigners also, are difficult to distinguish from the English and Welsh. Consistent with the liberal outlook, the state kept no adequate record of its human exports and imports.

Yet it is clear that the United Kingdom as a whole suffered great losses in numbers in the ten years from 1845 to 1855 with emigrants pretty well equalling the natural increase. Three-quarters of those leaving sought new homes and incomes in the United States. These enormous losses were largely due to the tragic evacuation of Ireland after the failure of the potato crop. The potato had caused Irish population to explode from less than six million in 1811 to more than eight million in 1841. When the potato was blighted in the forties, in a country suffering from minute subdivision of holdings, English repression of industry, and eviction by landlords, the stage was set for one of the greatest mass migrations relative to total population ever known.[1] The population of Ireland decreased between 1841 and 1851 by no less than 1,659,000 persons. Of these, about 700,000 were by death and the rest by emigration. Almost a million and a half people left Ireland in the seven years from 1849 to 1856, with the peak emigration coinciding with the Great Exhibition of 1851. Thus the potato, with its combination of ease of cultivation and vulnerability to disease, provided a vast addition to the unskilled labour force of America, helping expansion forward at a time when labour was the most necessary element for growth. Moreover the transplanted workers called for a much higher investment of capital per man in the new setting than in the old.

If England and Wales alone are taken, rather than the United Kingdom, there was, in the forties, net immigration of well over a quarter of a million, as Irishmen and Highland Scots abandoned their native countries and moved east and south. This increase in numbers had the affect of adding to the disturbance of the times, and further discouraging native marriage and child bearing. But in the fifties there was considerable net emigration from England and Wales, almost to the point of producing labour shortage. It was for this reason that, though

---

[1] K. H. Connell, *The Population of Ireland, 1750–1845*, Oxford, 1950; R. Dudley Edwards and T. Desmond Williams, eds. *The Great Famine: Studies in Irish History, 1845–52*, Dublin, 1956; Cecil Woodham-Smith, *The Great Hunger*, London, 1962.

the minimum *natural* increase in Britain occurred in the forties, actual increase was least in the fifties.

By the late fifties and early sixties there was very great difficulty in manning the Navy adequately. Its senior officers argued, in the traditional manner, that an intensification of the system of bribery or cajolery would solve the problem. But the difficulties were not to be removed by such means: there had been a basic change in the relationship between men and jobs. The average annual increase in population in the United Kingdom had been largely drained by emigration; in the ten years down to 1845 only 750,000 had left to go abroad, in the next ten years the figure was no less than 2,750,000. Between 1811 and 1846 natural increase had been at the annual rate of 300,000; between 1846 and 1860 it was only 116,000. On the other hand, new jobs had appeared at an accelerated rate, the railways alone employing some 100,000 in construction and well over that number in operation and maintenance, with industry and agriculture offering employment to some multiple of this. At best, the crimp and the recruiting sergeant were faced with a doubling in the rate of increase in the demand for labour, and a halving of the growth of supply.[1]

In this situation workers' earnings at home were bound to improve, with the result that emigration fell to a very low level. After 1865, especially, there was a substantial element of re-entry into Britain of many who had left. The reflux from America was due, in the main, to the disruption caused by the Civil War. From the Continent many returned seeking to avoid the unsettled conditions in France and elsewhere. Indeed, in the decade of the sixties, especially its latter half, Britain probably lost little on balance between those leaving and those entering. For, in addition to the adverse conditions abroad, high incomes at home offered a powerful attraction. So much so, that it was argued that the virtual reversal in the sixties of the declining rate of population growth which had been apparent from the second decade of the century, was partly to be explained in terms of re-immigration.[2]

7. GROWTH BY LEAPS AND BOUNDS: 1842–73

The later nineteenth century, with its great expansion of output, dramatized the pulsative character of economic growth. It had manifested itself in the first forty years of the century, culminating in

---

[1] 'The Real Difficulties of Manning the Navy', *Econ.*, 11 February 1860, p. 141.
[2] 'The Re-Immigration into England', *Econ.*, 11 April 1874, p. 442.

the very bad period ending 1842.[1] In the second half of the century fluctuations persisted. In a sense they became increasingly obvious as, in the great quarter-century of expansion down to 1873, the system at the same time became more sophisticated and quicker in its reactions. But with the general trend in the direction of expansion, the impact of crises was lessened.[2]

The boom which began in 1844 crowned the second great phase of railway construction in Britain, bringing the virtual completion of the main-line system. The Bank of England, after the discouragements of the years down to 1842, was prepared to provide loans at a low rate of interest. Investors, so long starved of profitable outlets, noted the success of lines built in the thirties and rushed to sponsor new undertakings. Inevitably there was a mania. Between 1844 and 1849 mileage in operation in Britain rose from two to six thousand, with an investment of some £150 million.[3] In addition British railway building initiative spread across the Channel to France, beginning with the Paris–Rouen railway, and continuing from 1844 to 1847. In Belgium, too, British initiative was active in branch-line development. Railway building meant rising money wages, with no immediate increase in consumer goods, resulting in rising prices. Brickworks, mines, machine making, iron production, all boomed; the wages of heavy labourers rose; a general stimulus followed. Traders felt the same exhilaration, both in the West and East.

Credit for other projects was cut to a minimum by the claims of the railways, so that the supply of other commodities could not respond in step. Speculation, as usual, appeared in the commodity markets, especially in indigo from India. With the system fully extended in this unbalanced fashion, catastrophic crop failure, both of wheat and potatoes, struck in 1846. Out flowed gold, threatening the liquidity of the banking system and enforcing credit contraction.[4] The American cotton crop, too, was well below expectation, sending prices up. The

[1] See Sir William Beveridge, 'The Trade Cycle in Britain before 1850', *O.E.P.*, 1940.

[2] In 1854 the Board of Trade presented to Parliament a *Statistical Abstract for the United Kingdom in Each Year from 1840 to 1853*, it continued to do so annually thereafter. This first attempt at systematic presentation of national information was quite unasked for, and revealed how incoherent was the data presented to Parliament, so that 'it is impossible, we believe', said an informed source, 'to find a greater mass of confusion than pervades our administrative accounts'. 'Annual Statistical Tables', *Econ.*, April 1854, p. 339.

[3] Jenks, op. cit., p. 129.

[4] C. N. Ward-Parkins, 'The Commercial Crisis of 1847', *O.E.P.*, 1950; D. M. Evans, *The Commercial Crisis, 1847–48*, London, 1849; *The City; or, the Physiology of London Business*, London, 1845.

blow to confidence was devastating. The soundness of the multitude of railway companies was called in question, and frightening revelations of unwise and sometimes dishonest promotion, culminating in the destruction of the colossus George Hudson, completed the rout.[1] Among the casualties were no less than six of the large agency houses in India, brought down by the failure of the Union Bank. In October 1847 the Bank Act was suspended, for so great was the rush to acquire liquid assets that the Bank of England was no longer able to redeem its notes in gold.[2]

This was perhaps the classic example of a domestic fixed investment boom: a sudden rush to create an enormous addition to the fixed capital of the country, greatly in excess of the available savings. For though the mania had spread to France and Belgium, it was activity in Britain that provoked the boom.[3] The financial panic ended quickly when it became known that the government had temporarily relieved the Bank of England of its obligation to pay gold for its notes. Moreover, as in the years after 1836, work continued to completion on many of the railway projects, though the value of railway stock declined continually throughout 1848; indeed as a consequence of the continuance of railway building there was almost unrelieved pressure in the money market as late as 1850. Thus was the shock cushioned for the iron and coal trades, but not for cotton. Depressed conditions persisted. In 1847 all Europe and America suffered from bad harvests; Britain suffered a further heavy drain of bullion to pay for imported food. The political disturbances on the Continent, associated with the year of revolutions in 1848, also brought derangement and insecurity.

But in the second half of 1849 failures greatly diminished, industrial activity increased and trade revived, all aided by an abundant crop and low corn prices. By 1850 prosperity had been resumed.

If 1847 was the classic culmination of a domestic investment boom, 1857 was a notable climax of an export boom.[4] Since the early forties British exports, both to foreign and colonial countries, had mounted briskly, increasing by over 130 per cent in the fifteen years down to

[1] R. S. Lambert, *The Railway King, 1800-1871, a Study of George Hudson and the Business Morals of his Time*, London, 1934.

[2] E. V. Morgan, 'Railway Investment, Bank of England Policy and Interest Rates, 1844-8', *E.H.*, 1940.

[3] H. G. Lewin, *The Railway Mania and its Aftermath*, London, 1936.

[4] J. R. T. Hughes, 'The Commercial Crisis of 1857', *O.E.P.*, 1956, p. 194; G. W. van Vleck, *The Panic of 1857; an Analytical Study*, Oxford, 1943. For a wider view see J. R. T. Hughes, *Fluctuations in Trade, Industry and Finance: A Study of British Economic Development, 1850-1860*, Oxford, 1960.

1857.[1] In 1856 alone they rose by over £20 million, an increase of over one-fifth. Cotton and wool exports, iron and steel, shipping figures – in fact almost all elements – responded.

Four main factors had contributed to this. The continued fall in the real cost of production in Britain, due to better methods, was perhaps the most important; mechanization and inventions brought down the cost of production, especially in textiles, in an amazing way; demand responded vigorously. In the early fifties there were magnificent additions to the number, size, and efficiency of cotton mills. Secondly, the general reduction by the British government of obstacles to trade made a significant contribution, for since the repeal of the Corn Laws in 1846, the task of removing trade obstacles and simplifying and reducing tariff and excise levies had gone on apace.[2] Exports of textile goods increased in the four years down to 1856 from £40 to £50 million. There was also much railway building abroad – in France and on the Continent generally, and in Canada and in the United States; British iron and steel exports rose from £6·6 million in 1852 to £12·9 in 1856. By 1857 British nationals held some £80 million-worth of American railway securities. The electric telegraph, by 1858, covered much of Europe and India. Thus the third element of growth, the external world, was feeling the drive of industrial acceleration. Finally the increasing relative world shortage of gold, that might well have made the smooth working of free trade impossible, was greatly relieved by the new discoveries in California and Australia. The bankers may well have been made over-sanguine by the inflow of new gold. Between the beginning of 1853 and the end of 1857 no less than £125 million of gold was shipped from California and Australia making it possible for European and American bankers to increase their holdings, or causing them to have more confidence that they could do so, should the need arise.[3] As is usual with gold rushes, it was the suppliers of transport, food, and tools, selling at prices that inflated with the fever, who gained most, so that much of the new gold quickly moved to the bankers in the great commercial centres, especially in Britain.

The crisis came in 1857; it reflected the general characteristics of the expansion. At home opportunities for large-scale railway investment

---

[1] 'The Result of Recent Commercial Legislation', *Econ. Supplement for 1859*, p. 25.

[2] See below, Chapter 9, section 7.

[3] 'The Production of Gold and the Bank Reserve of Bullion', *Econ.*, 23 October 1858, p. 1176; also Jenks, op. cit., p. 159.

had not appeared to top off the boom. This time the final wave of speculation took the form of buying and selling commodities, especially involving the products of the East and West Indies — sugar, wool, silk, copper and oils. In this, American traders took a full part. A new sophistication had appeared in putting short-term accommodation paper afloat.[1] In the five or six years preceding 1857 many new firms were formed, in response to expanding trade, carrying on an enormous traffic abroad on slender capital. The possibility of incorporation with limited liability, available since 1855, perhaps contributed to irresponsibility. As an additional strain much capital went into land drainage and farm improvement.

On the Continent bills drawn upon London had become an important part of the available circulating capital. Many had no real basis, and were used to finance speculation in the produce markets.[2] When the crisis came German demands upon London for support added to the general strain, and afforded an early warning of the price of being the world's money market.

There had been much building of new large high-class ships in order to participate in the Californian and Australian gold rushes. The owners of the excess capacity thus created were saved from the consequences of their error by the Crimean War in 1854 which restored high profits in shipping.[3] Great gains were made, but with peace in 1856 these dried up and there was much redundant capacity. In addition, many foreigners, encouraged by the repeal of the Navigation Acts in 1849, rushed ships into the British shipping market, hastening the collapse of freight rates.

Certain of the new banks were outrageously loose in their discounting. The joint-stock banks, undergoing enormous development in these years, took to offering interest for deposits, in order to extend their lending powers.

The whole fabric had shaken badly in 1855, both in America and in Britain. Such speculation could only go on so long as each dealer could pass on his holdings at rising prices, but eventually a point was reached at which the consumer refused to pay more, demand fell off, prices in the commodity markets collapsed, and depression came, flooding in with announcements of failure in New York, Boston, Philadelphia, and Baltimore. In terms of hectic liquidation, the crisis of

[1] See below, Chapter 6, p. 205.
[2] Jenks, op. cit., p. 191.
[3] *Report of Select Committee on Merchant Shipping*, 1860.

1857, both at home and abroad, was probably the most intensive ever experienced. The extremely rapid extension of credit and banking facilities in the previous decade made this possible.

But the speed of the recovery was astonishing. The economy had not, in fact, produced any very serious structural distortion as had been the case in 1847; there were few large-scale long-term ventures that had to be written off.[1] Soon America was buoyant again, European trade was restored, there was rapid growth in trade with India, with more hoped for from China and Japan under treaties lately concluded. The Anglo-French Treaty of Commerce of 1860 caused the export of British manufactures to France virtually to double;[2] other treaties were soon made with Turkey, Italy and the German Zollverein. On the other hand the Morrill Tariff Act of 1861, passed just before the firing upon Fort Sumter began the Civil War, brought the American return to policies of exclusion of the products of other countries. Production rose at a rate that astonished the world: between 1857 and 1859 Britain increased her coal output by 15 million tons – equal to the total production of Russia, Austria, Spain, Portugal, Australia, and India, or about her own total output in 1816.[3]

In fact it is not going too far to suggest that all the way from 1847 to 1866 the expansion of the British economy suffered no really serious check; for some twenty years conditions were propitious for a move forward involving no great disproportionalities in the system.

One response to this condition was apparent in the behaviour of owners of capital. Their willingness to lend both at home and abroad became much more general as their confidence and knowledge grew. It is sometimes thought that the effect of ignorance is to promote commercial rashness, and this is no doubt true in the case of those who have nothing to lose. But in those with substantial assets ignorance was a great impediment to investment, causing men of substance to prefer security – the sweet simplicity of 3 per cent. The general knowledge of the world outside Britain extended greatly from the fifties onward. With it men were prepared to move their money over a much wider area of opportunity, and to calculate returns and risks with much greater refinement. In the forties the country gentry held government debt (with an occasional share in some wildcat scheme), but by the

---

[1] 'When will Trade Revive? The Convulsions of 1847–1857', *Econ.*, 1 May 1858, p. 477.
[2] A. L. Dunham, op. cit., chapter XI.
[3] 'Four years (1873–6) of Coal Mining', *Econ. C.H. and R. of 1876*, 1877, p. 48.

sixties, with the aid of brokers and bankers, their range of lending was much extended.[1]

This halcyon phase did not last indefinitely. By the early sixties there were signs of the operation of factors that were, in 1866, to bring a new crisis, in which the element of structural difficulty was to be serious once more.[2] This phase has aptly been called the 'extension mania'. The time had come for another great burst of railway investment, both at home and abroad. In addition the cotton industry built and extended on a great scale.

The cotton shortage due to the American Civil War and its aftermath certainly caused grave problems. There was heavy and painful regional unemployment, especially in Lancashire. But the difficulties of the cotton industry did not generalize themselves to bring the economy down. Nor did the Civil War, though it caused American railway building to stagnate, seriously affect migration to the United States. Yet it did have important general effects of an indirect kind upon the British economy. In 1860 British manufacturers had produced enormous quantities of cloth, so much so that, had there been no interruption of the cotton supply, there would have been collapsed prices and serious losses. Instead prices soared, and men who, on a strictly commercial calculation, had made a grave error, in fact made fortunes. This cash, combined with the subsequent near-stoppage of work, created a plethora of capital in Lancashire and London. The Lancashire cotton men, undeterred by the war, undertook vigorous investment in all kinds of plant, in order to meet the resumed competition when peace came in America.[3]

The effects of the American stoppage did not end here. A switch to India in search of cotton resulted. This produced a prodigious prosperity among the growers and merchants of India. Bombay became a boom town, with fabulous land values, producing overnight millionaires on a scale unapproached by the gold rushes of Australia and California, with general prices quadrupling in four years.[4] Indian railway construction received a stimulus; British exports to India rose. But drastic crisis

[1] 'The Influence of Increased Education upon the Stock Market', *Econ.*, 7 May 1864, p. 575.

[2] From 1864 onward the best continuous source of data about the course of the British economy is to be found in the Supplement to *The Economist*, '*The Commercial History and Review*'.

[3] See below, Chapter 4, section 3, pp. 127, 128.

[4] 'The Crisis at Bombay', *Econ.*, 10 June 1865, p. 685; 'Effect in Western India of the Sudden European Demand for Cotton', *Econ. C.H. and R. of 1864*, 1865, p. 32; 'The Late Banks of Bombay', *Econ.*, 1868, p. 495.

came in 1865 when the Bombay boom broke in the face of American recovery. Less perceptibly, the buying of Indian cotton meant a drain of the precious metals to the East, lowering the coffers of Western bankers, and eventually affecting their ability to lend. Nor did the drawing upon Indian sources of cotton keep prices stable: heavy speculation, becoming increasingly violent, took place in cotton markets down to the general collapse of May 1866. There were fears that buying from India, a country which, unlike the United States, was not fully integrated with the British pattern of manufactured exports, would provoke a crisis in the exchanges.

On the other hand, though Britain might lose precious metals to the East, she received a good deal of corn and bullion from America, exported from thence because of inflation and currency depreciation. Egypt, too, was profoundly affected by high cotton prices, for from this time stemmed the extravagance of the Khedive, the insolvency of his kingdom, and the appearance therein of European financiers.[1]

Moreover, by the early sixties, with continued prosperity and growing trade, conditions were ripe for a further extension of the railway system. New projects began to appear and to multiply, both at home and abroad. Between 1863 and 1866 some £150 million was authorized by Parliament to be spent on railways in Britain.[2] The contractors and finance companies evolved a new system whereby lines were first built and then offered for sale. Landowners, eager to sell portions of their property for high prices, supported them. All this involved very dangerous financing. Only a small proportion of authorized capital was paid up, so that, when difficulties came, it was necessary to make calls at a time of general embarrassment. Companies were borrowing at rates ranging from 7 to 16 per cent or more.

The potential for economic development abroad centred very largely around projects for opening up and unifying new territories by railway construction. Every corner of the Continent felt the stir of enterprise. There was to be a thrust to join Europe with the Black Sea ports, and eventually, the hope of an Anglo-Indian railway, perhaps continuing to China. Brassey and the rest were laying down new lines abroad at an amazing rate, for though British railway building was much reduced in western Europe, where native initiative had taken over, it was increasingly active in eastern Europe and in India.[3]

---

[1] David S. Landes, *Bankers and Pashas: International Finance and Economic Imperialism in Egypt*, London, 1958.

[2] 'Railway Traffic (United Kingdom) 1866', *Econ. C.H. and R. of 1866*, 1867, p. 31.

[3] Jenks, op. cit., p. 196.

After the Mutiny in 1857, and the subsequent passing of the government of India from the East India Company to the Crown in 1858, the movement of British capital to India was much accelerated.

Shipbuilding, too, felt a strong stimulus. There was an astonishing increase in 1863 and 1864; many new yards were constructed. The ships themselves were better, sailing faster, and carrying better pay loads, with the share of steam tonnage compared with sail steadily mounting. New companies, with little sound finance or experience, started up on a great scale in the high-class iron shipbuilding industry of London. The Suez Canal drastically shortened one of the world's greatest routes after 1869.[1]

Mill construction for textiles, new railway lines, and shipbuilding, all added to the demand for iron. Between 1862 and 1865 production rose from some four million tons to five. In the Cleveland area the extension was most marked, many of the new larger furnaces only coming into production toward the end of 1865. The prices of Scotch pig iron were kept exorbitantly high by a 'rig' or monopoly organized in Glasgow or Liverpool. French and Belgian rivals were becoming increasingly effective.

This very great increase in new projects called into being a frightening number of new companies, under limited liability, in the years 1863 and 1864. The new principle invaded most parts of banking, commerce, transport, and industry. These new flotations sent the rate of interest up, so that it *averaged* 7 per cent in 1864 — a level higher than in any year in English banking memory. Business men were borrowing money for their permanent regular businesses, in order to use their own resources for joint-stock companies or foreign loans. Especially worrying were the new finance houses: would they, in the urge for profits and with their liabilities limited, abandon the conservative tenets of the sound old bankers?

The field in which the actions of the new houses was most feared was that of lending abroad. Down to the sixties foreign loans on any scale had been the virtual monopoly of the greatest merchant bankers, headed by the Rothschilds. But the new companies began to take an active part in lending in areas where money could fetch 15 or 20 per cent. The spread of knowledge of foreign countries among the investing public assisted this development. Thus investment, at home and abroad, in a situation in which the canalizing of savings was becoming

---

[1] Max E. Fletcher, 'The Suez Canal and World Shipping, 1869–1914', *J. Econ. H.*, 1958.

ever more efficient, produced a system in which the element of reserve (gold) was dangerously slight.

In addition to these speculative elements, adverse long-term trends were also at work. In the sixties the rate of expansion of British trade, though still capable of producing a large absolute increase, was falling fairly steeply as Continental and American industrial and trading rivals made their first serious bids to overtake Britain.[1]

Just as there had been premonitory rumblings and fears two years before 1857, so too, before 1866. The year 1864 was full of vague forebodings, not only in London, but in Paris, Frankfurt, Amsterdam, and elsewhere. A considerable number of houses that had barely managed to survive in 1857 had remained unsound ever since and were now precarious.[2] There was a panic in Bombay in 1865 as the short-lived cotton boom expired. The inevitable commercial crisis appeared in May 1866, the most disastrous since 1825.[3] In the previous month there had been much concern about the prospect of war between Prussia and Austria. But the real detonator was the failure of the house of Overend and Gurney, standing second in prestige only to the Bank of England itself. A general collapse followed, led by unsound banks and finance houses, bringing down leading railway contractors like Peto and Betts, and very nearly ruining the great Brassey himself. The railway companies withstood the first shock, but crisis seemed almost to be endemic. In 1867 there was a panic in the railway market. Many companies were found to have sunk the proceeds of temporary loans into permanent works, relying on their ability to borrow again when their short-term debentures fell due. The London, Chatham and Dover Co., the Great Eastern, the North British, the Great Western, the London and Brighton, the Caledonian, and the Midland, were all publicly inquired into, and found to have been mismanaged; several were unable to meet even the interest on their debenture debt. Their discredit was the greater because of the bitter and wasteful war that had been waged over fares and facilities. Yet the great railway interest could act in concert so that under the emasculated Regulation of Railways Act in 1868, intended to remedy abuses, nothing drastic was required of them; downright falsification was still possible.[4] Nerves were shaken much more seriously than in 1847 or 1857 by the amount

[1] Schlote, op. cit., p. 43.

[2] Jenks, op. cit., p. 191.

[3] Robert Baxter, *The Panic of 1866*, London, 1866.

[4] 'The Regulation of Railways Act and the Accounts of Railway Companies', *Econ.* 29 August 1868, p. 992; see below Chapter 4, section 4.

of untrustworthiness revealed in business. For some three years the courts and press were full of inquests on mistaken or corrupt practices. Thames-side shipbuilding received a blow from which it never recovered.[1] The effect upon business incomes was reflected in the fact that in 1867 no less than 1,600 persons discontinued their private carriages.[2]

The business community was depressed and pessimistic for over three years, in strong contrast to its reaction after 1857. The middle classes were indeed sorely tried, more so than ever before. Worries about war among the German states, Fenian troubles, cholera, cattle plague, and dismal weather destroying much of the European corn crop, all came together. With a fine perversity gentlemen turned from drab trousering to fancy, leaving the Huddersfield wool men with large stocks of the more conservative material. So low was the confidence of investors that the young Goschen anonymously invented the phrase 'capital is on strike'.[3] The rate of interest, from its level of 7 per cent, fell and stuck at 2 per cent.

The Paris Exhibition of 1867 provided further ground for disquiet. The government's observer reported that other countries, uninhibited by trade union resistance, and by business complacency, were pressing ahead with improved methods.[4] In woollens, engineering, steel production, and other industries comparisons adverse to Britain were made, and the rise of the great firms of Krupp and Essen was stressed. Britain's lead in many directions was rapidly shrinking. It was clear that in many lines foreigners had made astonishing advances in skill, organization, science, and capital provision. In the woollen industry the company mania of 1866 caused the postponement of the adoption of limited liability; many partnerships, obstacles to efficient reorganization, survived because of fear of the new principle.

But even this severe condition of mind could not last indefinitely. For the railways had by no means exhausted their demands for capital, nor were they the only claimants. After the American Civil War, and the success of the army of the North German Confederation over Austria in 1866, there dawned a new realization of the importance of railways for the mustering and movement of vast armies. American

---

[1] See below, Chapter 4, section 5, p. 147.

[2] 'General Results of the Commercial and Financial History of 1867', *Econ. C.H. and R. of 1867*, 1868, p. 2 n.

[3] 'Two Per Cent', *E.R.*, 1868, p. 260.

[4] 'General Results of the Commercial and Financial History of 1867', *Econ. C.H. and R. of 1867*, 1868, p. 3.

expansion had been long arrested by the waste and confusion of the Civil War, but by 1868 new railway projects, encouraged by low prices, were being offered.[1] The American cotton crop could now earn more abroad, and by 1870–1 was of immense size, greatly aiding the recovery of the United States, though damaging the commerce of India and Egypt. Activity soon became more brisk, and though American iron output increased rapidly, by the spring of 1871 Britain was receiving very large orders for iron. The task was resumed of completing the world's telegraph system, including China, Australia, and both sides of South America.

Foreign government financing mounted to a crescendo between 1870 and 1873. Germany, Russia, Austria, and other countries, at last in a position to engage fully in the industrial race, were making increasing haste to equip themselves, straining every nerve. Eastern and central Europe were swept by a promotion mania. New foreign borrowers appeared in London and found acceptance. Many half-barbarous states pressed eagerly for funds, and spent them with no display of wisdom, sometimes to compound, at extortionate rates, old debts, sometimes in hopeful response to the blandishments of projectors. Among these casual states were Turkey, Egypt, Honduras, Roumania, and a swarm of South American countries. 'We are to them', said Walter Bagehot, 'what the London money dealers are to students at Oxford and Cambridge.' For promoters busied themselves stimulating in such countries a sense of the need to borrow, and its bewitching simplicity. The British colonies of Canada and Australia also made their claims. Under the new pressure of transactions the Stock Exchange in the ten years down to 1872 at least doubled its size.[2]

Before the sudden accession of orders in 1871 British iron-producing capacity had extended slowly if at all, with the result that output could not immediately increase to meet the new demand. There were other difficulties, including strikes, as the iron workers pressed for their share of prosperity. Prices shot up, reaching by 1873 the highest level since the introduction of the hot blast some forty-five years earlier. Between 1868 and 1872 British iron exports increased from some two million tons to some three and one-third million, with the price per ton rising from £7 10s. to over £10.[3] During this phase of very high prices

[1] D. R. Adler, *British Investment in American Railways, 1834–1898*, Cambridge Ph.D. thesis, I.T.A., 1958/9, no. 658.
[2] 'Why the Commercial Depression is so Protracted', *Econ.*, 5 May 1877, p. 505.
[3] 'Iron', *Econ. C.H. and R. of 1873*, 1874, p. 26.

railway companies in Britain postponed their own capital expenditure. Since the fifties the Cleveland area had forged rapidly ahead until, by the mid-seventies, it was responsible for one-third of the output of the kingdom. A spirit of aggressive extension produced ever bigger furnaces and ever higher chimneys. Scotland remained a formidable second, but south Staffordshire and South Wales failed to hold the pace.

Coal, the 'instrumental twin' of iron, followed its upward course. Down to 1871 the industry had been more or less starved of profits; thereafter they began to appear on an exciting scale. The price of coal in London reached an extraordinary level, costing over fifty shillings per ton for the first time since the coming of the railways. So dear was coal that a reaction appeared in shipbuilding, for owners, bitterly complaining that the earnings of their ships 'go up the chimney', began to place orders for sailing ships once more. The railways found their coal bills mounting so as to threaten dividends, for working expenses moved upward much more quickly than traffic receipts. The rise in coal prices in the boom contributed to raising iron and steel prices, and those of other British exports. Fuel-saving inventions began to multiply in industry, and there was a general search for economies or alternatives. As the miners gained increases in wages far beyond all previous experience (some 50 to 60 per cent between 1871 and 1873) their output per man fell appreciably. The Coal-Mines Regulation Act of 1872 was also blamed for raising coal costs. But for all this, British coal output increased by some fourteen million tons in four years after 1871.

From 1869 onward, shipbuilding too, was feeling a mounting pulse. The world's grain traffic was growing; so too was general carriage as trade expanded. Britain gained an increasing share of this increasing world commerce. America and France, heretofore supplying much of their own carriage, now accepted British terms. The use of steam extended greatly: the Suez Canal put a rising premium upon controlled power. Those engaged in the coasting trade were rapidly discovering the virtues of steam. The compound engine, with its enormous economies, was now available. The navy too, was changing, and Sheffield, with its near-monopoly of armour-plate, gained accordingly. The Clyde was asserting its supremacy over the new shipbuilding of iron and steam. A competent authority estimated that by the mid-seventies, with 40,000 shipbuilding workers, the Clyde had capacity enough to replace the entire British Navy in two years.[1] The average

[1] 'Clyde Shipbuilding in 1876', *Econ. C.H. and R. of 1876*, 1877, p. 24.

size of the ships launched on the Clyde rose in the four years down to 1873 from some 800 tons to 1,300 tons, largely due to the screw steamships, which averaged 1,700 tons. The three rivers of the northeast coast, Tyne, Tees, and Wear, were not far behind, but the Thames and the Mersey lost ground, the former never recovering from the collapse of 1866.

The new prosperous conditions were not fully shared by cotton. In 1871 cotton did well, but its period of great gains was much shorter than that of iron and coal. The cotton-buying world responded to some degree to the activities of the engineers, contractors, ironmasters, and the rest, but not to the extent that the cotton men had hoped for. The paucity of profits in the cotton manufacture was due in the main to the fact that the prices of raw material (because of general inflation aggravated by a poor American crop in 1871) were advancing more rapidly than the manufacturers dared to raise selling prices.[1]

The cotton manufacturers had always been heavily dependent upon exports. As early as 1860 only about one-fifth of their output was consumed at home. By 1872 home consumption was only some 12 per cent of output.[2] In their export markets the cotton manufacturers were turning increasingly toward the East, as industrializing rivals gained strength, so that by 1872 India and China were taking about half of what was sent abroad. These were markets in which demand was highly sensitive to a rise in prices. Manchester manufacturers and merchants, in their anxiety to exploit these markets, made excessive credits available to speculative shipping houses in the East. The woollen and worsted industries did a good deal better in terms of the boom, increasing their mills substantially, and replacing old machinery with self-actors and improved mules.

Further elements appeared to add fuel to the boom. The American situation was approaching mania, with railway projects mounting daily. Since 1862 the American monetary system had been on a paper basis (greenbacks) so that the controls involved in a gold system were absent. There was a great wave of borrowing for municipal and state development of all kinds. Canals, harbours, roads, civic buildings, were begun. A good deal of this was ill-considered because of the demoralization of American politics since the Civil War. The Erie and Tammany political rings, headed by men like the notorious Boss Tweed, brought

---

[1] 'The Unprofitableness of the Cotton Manufacture', *Econ.*, 1 June 1872, p. 672.
[2] 'Year 1872. General Results of its Commercial and Financial History', *Econ. C.H. and R. of 1872*, 1873, p. 4.

corruption on a new scale to set alongside the wild speculation in railways and elsewhere of Fisk and Gould. Other unsettling elements appeared. In the three years after 1871 no less than 135 foreign mining projects were floated in London, some two-thirds of them for gold and silver.

The Franco-Prussian War (August 1870 to April 1871) meant that, at a critical moment, two great states partially suspended production, and drew heavily upon the rest of the world, especially Britain. When the war, with its uncertainties, ended, efforts to build railways and other equipment in central and south-eastern Europe were redoubled. Banks, credit companies, discount companies, all multiplied in Germany, Austria, and Italy. The payment of the indemnity of £200 million by France to Germany added fuel to the speculative fever. British stocks of coal and pig iron were almost exhausted.

Thus developed the most dramatic boom of the nineteenth century. Behind it lay the urge to develop in the Americas and in Europe; associated with it was an astonishing convergence of circumstances helping to extend and sustain it.

In Europe and in America the crisis, which on all past experience might reasonably have been expected, did occur. In May 1873 the speculative mania in eastern and central Europe broke in Vienna, and the panic spread to Germany. On 18 September New York was struck by the failure of Messrs Jay Cooke and Co., long the government's loan contractor, bringing down with it other titans of the financial scene, all more or less involved in railway speculation, with Jay Gould at their head. European difficulties took increasing toll as liquidation spread. In 1875 Baron Strousberg, the Prussian railway contractor, after an amazing display of virtuosity in trying to realize enough assets to keep going, was the centre of a gigantic failure spreading far afield.[1]

But there was no general commercial crisis in Britain (though Bank Rate was 9 per cent in November 1873).[2] The story was one of sagging, punctuated by heavy failures, rather than collapse and panic. The Master Cutler of Sheffield went down; so too did the Aberdare Iron Company, many major firms in the cotton trade, and in the Eastern trade the great houses of Alexander Collie and John Gledstanes, both deeply involved in long and complicated credits, failed, the latter with liabilities of some

[1] 'The Gigantic Failure of Doctor (or Baron) Strousberg', *Econ. C.H. and R. of 1875,* 1876, p. 47.
[2] Jenks, op. cit., p. 292; Sir Robert Giffen, *Economic Inquiries and Studies,* London, 1904, p. 99.

£2 million. The West of England Bank collapsed; so too did the City of Glasgow Bank, amazing Britain with the ineptitude and culpability of its church-going principals. Yet somehow the liquidation of unsound enterprises was spaced out in such a way as to avoid general breakdown in the City.

The three great booms of the third quarter-century, 1857, 1866, and 1873, represented the points of fullest employment and greatest excitement in the long trend of rising prices and general expansion. Each had its peculiar characteristics. But taken together they all pointed to the same problem. How was the rate of flow of the mounting supply of capital into fixed forms — railways, gas and water works, drainage, docks, piers, buildings, machinery, and the rest — to be kept at its appropriate level? Even with a rising trend the problem of investment produced frightening breakdowns, with a strong tendency, when opportunity appeared, as in 1845, 1866, and 1873, for investors to seek to place excessive sums at long term.

This tendency to press too far and too fast with new equipment may have been inherent in the psychology of owners of capital and therefore in the system. When a long-term project was offered promising 20 per cent, it had a stronger draw than, say, a short-term investment in trading, or a less remunerative project in industry. The successive railway booms showed the curious blindness of investors to the prospect that the boom itself, with its multiplication of lines, would greatly reduce the gains of any particular venture. Moreover many investors loaned their money with the intention of selling after a rise. Both of the dominant elements in the investment picture, railways and steamships, showed this volatile character. In boom times shipping freight rates rose very steeply, outrunning even rising costs, so that owners placed new orders undeterred by the purchase price of ships. The same was true, though in a less dramatic form, of millowners, who also lost sight of costs when dazzled by rising selling prices. In each case, railway promoters, shipowners and textile manufacturers, generated a sort of mass optimism among their kind.

To the psychology of the investor must be added the effect of new forms of mobilizing capital.[1] The rise of banks and finance companies meant that the centripetal forces were much strengthened — bringing into the money market all kinds of savings scattered about the country and concentrating them for investment. But the newly created centrifugal forces generated by the new institutions were even stronger — seek-

---

[1] See below, Chapter, 6, p 203.

ing means of placing capital with increasing skill. The net effect was a closer mobilization of capital for the more hectic exploitation of a boom, both at home and abroad. Indeed, the capitalist economy in full flood contained within itself elements which served to hasten and even to rush the laying down of new equipment well beyond a rate consistent with stability or orderly growth.

8. GROWTH IN ADVERSITY: 1873–90

The long failure to recover after 1873 provoked much discussion.[1] At first there was a strong tendency, especially in the absence of crisis (the financial panics of 1873 and 1884 passed Britain by) to minimize the difficulties. It was pointed out that though earnings from exports were declining because of falling world prices, there was no great fall in the *quantity* of goods sent abroad. The enormous progress before 1873 in so many lines of production was much stressed. It was argued that the workers had made great strides, and that once cured of the extravagances of the boom, would settle down to new feats of output and cost-lowering. It was observed that though the iron industry in general might be in severe difficulties, the newest and most modern area, that of Cleveland, was much less hard hit than some older regions. The same was said of shipbuilding on the Clyde.

But evidence of deep-seated difficulty was accumulating. France, which seemed, in spite of military defeat, to have avoided serious economic trouble, joined the embattled nations in 1877. Prices moved steadily downward without provoking a recovery in demand. Money wages, too, were cut as employers found themselves obliged to reverse the concessions of the previous years. Unworked plant, some of it completed in the last phase of the boom or even later, was to be seen in all areas. Fuel-saving devices introduced to meet high prices militated against a recovery in demand for coal. Cotton manufacturers, now so heavily dependent upon the level of incomes in the East, found purchasing power there lowered by bad harvests and famines, especially in 1873 and 1877.

Fears began to mount in the later seventies that Britain had been, and was increasingly, since 1873, living on her capital. As the excess of

[1] For this inquest, and for aspects of the preceding phase, see W. W. Rostow, *British Economy of the Nineteenth Century*, Oxford, 1945, esp. chapters III and VII; Sir Robert Giffen, op. cit.; Goschen, *Essays and Addresses*, London, 1905. For a comparative study see J. D. Bailey, *Growth and Depression: Contrasts in the Australian and British Economies, 1870–1880*, Canberra, 1956.

imports over exports increased from £60 million to well over twice
that figure between 1873 and 1877, the question arose: was Britain,
because of relatively high wages and other incomes, consuming more
than current output, paying for the difference by the sale of foreign
assets, the fruits of previous exports?[1] This conjecture caused much
misgiving. For there is little doubt that something of the sort was
happening, and the idea of using foreign assets in this way was new and
disturbing. On the other hand, however, a good deal of the imports
took the form of raw materials – cotton, iron, wool, flax, and jute,
which manufacturers were processing and holding for stocks, rather
than have their factories and workers idle.

Since 1865 Europe had been in a state of heavily armed peace, with
military budgets increasing year by year. Though this might create
incomes of a kind it could do nothing to cause trade revival, for a
general sense of precariousness was the result. In addition the heavy
demands of military service on the Continent stimulated emigration,
especially to America.

But by no means all the casualties of the period after 1873 were
industrial. Agriculture, that had made such great progress since the
forties, keeping abreast of industrial prosperity, suddenly found that
the new equipment in the new world was a tremendous threat to food
prices. Wheat had come in great quantities from America in the sixties
– a warning to British agriculture of what was to come.[2] Now it began
to pour in at prices that home agriculture could not rival. Moreover,
bad harvests ruled in Britain for some six years after 1873. From a
standard normal yield of about 29 bushels per acre, the return fell by
1877 to 22 bushels, and by 1879 was only 18, the lowest ever recorded.[3]
Formerly British farmers would have recouped themselves for lower
yields through higher prices. But grain from the immense harvests of
the new world, increased 4½ times since 1850, was rushed to Britain by
the new railways and steamships.[4] The latter were soon engaged in
something of a freight war, causing transport costs to be halved
between 1868 and 1879, further adding to the British farmers' difficul-

[1] A letter from William Rathbone, *Econ.*, 24 November 1877, pp. 1394, 1395, 1396;
'Are we Consuming Our Capitals?' 15 December 1877, pp. 1483, 1484, 1485; 22 December
1877, pp. 1515, 1516, 1517.
[2] W. Trimble, 'Historical Aspects of the Surplus Food Production of the United States,
1862–1902', *Report of the American Historical Association*, 1918, p. 224.
[3] 'Year 1879. General Results of its Commercial Financial History', *Econ. C.H. and R.
of 1879*, 1880, p. 2.
[4] See Morton Rothstein, 'America in the International Rivalry for the British Wheat
Market 1860–1914', *Mississippi Valley His. Rev.*, 1960–61.

ties. Down to 1872 Britain had imported a good deal of European wheat, mostly from north and south Russia and Germany, but European countries, because they were growing richer, were now consuming a higher proportion of their own produce. Thereafter, when British imports of wheat increased enormously the great suppliers were the United States and Canada.

So ended the golden age of British agriculture.[1] It had been able to meet the threat of the repeal of the Corn Laws by a magnificent improvement in husbandry, but this new challenge was too much. Protectionist proposals were once more heard as agriculture entered upon its long decline. Though it was possible to make adjustment in the direction of market gardening, dairying, animal husbandry, and the like, and though farmers on rich land could do much better than those on inferior soil, cereal growing on the old scale had had its day. The relative decline of farming incomes could not fail to aggravate the difficulties of the time, and at last to shake the massive assurance of the landowning class. The value of agricultural land fell drastically; the area under wheat shrank steadily. Cheap food, though it helped to maintain the real incomes of the workers, was clearly not a sufficient condition for industrial recovery. Yet the sea carriage of this vast new supply of grain helped to stimulate British shipbuilding.

Incomes of the rentier group were dealt a heavy blow by the default of foreign governments. The City of London had sold an enormous range of foreign issues to persons who often failed to realize that not all governments were as sound as their own. Many states failed altogether to pay interest: Spain, Mexico, San Domingo, Honduras, Costa Rica, Paraguay; others partly defaulted – Egypt, Columbia, Ecuador, Guatemala, Peru.[2] There were defaults also in America. But the collapse of Turkish finance in 1876 was the most serious blow. All this meant that incomes were reduced, and an almost panic adoption of economy measures followed; carriages were laid up, establishments of servants reduced, housing activity curtailed, and building operations stopped at resorts and watering places. This damage to rentier incomes was felt until about 1882.[3]

The recovery of general activity in 1879, by its mildness and brevity, served only to emphasize the general difficulties. The business world,

---

[1] See below, Chapter 6, pp. 186, 187; also Chapter 8, section 2, p. 288.
[2] William H. Wynne, *State Insolvency and Foreign Bondholders*, Oxford, 1951, vol. II.
[3] Edwin Goadby and William Watt, *The Present Depression in Trade: its Causes and Remedies*, London, 1885, p. 12.

frustrated for six years, seized with joy upon the signs of hope. After 1873 migration to the United States had fallen off; so too had internal migration to the middle west. But momentum was slowly regained, as the frontier once more exerted its attraction. The virgin soil, men said, had only to be tickled with a plough to laugh in a harvest. Fertility, when flagging, seemed to be restored by taking a crop of Indian corn. In four years down to 1880 the United States received well over a million newcomers. The American currency was restored to a specie basis, thus encouraging business men to believe that the monetary system was once more sound. Railways were again the rage, not only in America, but in Europe and Asia. Some 65,000 miles of line were under construction or projected in 1879; it was calculated that in the early eighties, one-half of the savings of mankind was being absorbed in railways.[1] The sales of American wheat in Britain, so damaging to British agriculture, prepared the way for recovery in British iron and coal, for they made American purchases possible. The thirst for iron was like a gold fever. Other British industries felt the stimulus, including chemicals and shipping. The cotton industry, with half the manufacturers of Lancashire near the end of their resources, was vastly relieved. But for all its hectic beginning, the recovery could not gather real momentum. A particularly bleak year for the British harvest had a dampening effect.

As real recovery was more and more delayed, attention was drawn to the operation of long-term factors, centred upon the previous investment boom abroad, and its consequences. A sense of structural derangement began to spread, together with a querying of the whole basis of Britain's industrial growth and supremacy. The tone of the eighties was set: it was a time of questioning and misgiving.[2]

In a few short years two great rivals had so increased their stature as to threaten Britain. America and Germany had shown the degree to which British expansion had depended upon this very creation of rivals. By 1873 most of the profitable projects had been taken up both in America and Europe, and the new equipment was all that was needed for more than half a decade. It was disturbing to discover that the Americans were often much more efficient in using it, getting bigger returns from the Bessemer converter, by means of better mechanical devices. Serious misgivings were expressed that Britain was about to

[1] 'Possible Railway Development at Home and Abroad in 1880', *The Times*, 20 November 1879; M. G. Mulhall, *History of Prices since the year 1850*, London, 1885, p. 48.
[2] See below, Chapter 10, section 9.

pay a heavy price for her neglect of technical education and of the study of chemistry, in competition with Germany, a country which had attached higher priority to such things.[1]

As in the years after 1815, monetary factors offered an alternative kind of explanation of falling prices and less than full employment. In the great expansive phase from the forties to the seventies new supplies of gold from California and Australia had been pumped into the world's circulating medium. New silver, too, had played an important part. But in the seventies the situation changed. The addition to the volume of precious metals fell off after 1868 as the gold discoveries were worked out. In addition countries which had formerly been on a bi-metallist basis began to adopt gold only. Germany in 1872 began to absorb gold in great quantities for this purpose, following the law of November 1871. France was also attracting gold in order to resume cash payments suspended when her war with Germany began. The Latin Monetary Union, comprising France, Italy, and Switzerland, so long the regulator of the price of silver by the maintenance of a fixed parity of 1 to $15\frac{1}{2}$ between gold and silver, closed its mints to silver in 1873.[2] The United States followed suit the same year. Thus was silver effectively demonetized, leaving the world's monetary system dependent upon gold only. Between 1873 and 1876 the value of silver relative to gold slumped by 20 per cent, to the horror of all those dealing with India and the East, especially in Lancashire. The finances of India seemed to present an insoluble problem as the budgetary deficit rose. So began the long and somewhat barren bi-metallist controversy that was to rage for the rest of the century. With gold production greatly reduced, and silver demonetized, many observers argued that the depression was monetary in origin.[3]

By the mid-eighties it seemed that two factors dominated the situation: low prices and low profits.[4] As prices fell year after year the hope was continually expressed that a bottom must soon be reached, at which the very cheapness of goods would cause demand to recover. There was a good deal of confusion in the discussion. Sometimes the case was put that, from the point of view of the trading world as a whole, low prices would eventually stimulate demand; sometimes it was argued

[1] See below, Chapter 3, section 5, pp. 91, 92.

[2] H. Parker Willis, *A History of the Latin Monetary Union*, Chicago, 1901.

[3] See below, Chapter 10, section 9, p. 421.

[4] H. L. Beales, 'The Great Depression in Industry and Trade', reprinted in E. M. Carus-Wilson, ed., op. cit.; A. E. Musson, 'The Great Depression in Britain, 1873–1896', *J. Econ. H.*, 1959.

that Britain must press her prices lower than those of her competitors and thus achieve a relative advantage.

Capital could be borrowed at very low rates, yet real industrial recovery did not follow. This could not be explained on the basis of the post-1866 situation, for though this time also there had been bad failures, the shock to confidence had been less severe. Further, it was now well in the past. The argument of capital glut was increasingly heard; the view that British and Western society as a whole had laid down, in the years preceding 1873, so extensive a range of capital equipment that little more would be called for until population growth, changes in tastes, new inventions, or the exploitation of new natural resources, altered the picture.

Low prices and low profits taken together suggested that the national income was being redistributed away from capitalists toward the workers. The latter, whose money wages on the whole did not fall so rapidly as prices, were (except for those who lost their employment) better off. Some contemporaries were prepared to applaud the gains of the workers, hoping for a new social solidarity through more equal incomes, but in doing so they inevitably encountered the question: with profits persistently low for so long, was real growth precluded?

Many elements in the situation suggested that the answer was no. In coal mining, in the ten years down to 1885, U.K. output rose from 133 million tons to 159 million. South Wales was beginning its great phase of exporting steam coal for use in ships. Iron output, at 6·5 million tons in 1873, stood at 8·5 million tons in 1883. Yet these were the very industries in which complaints of low prices and low profits were greatest. The shipbuilding industry, too, increased very rapidly after the low point of 1876, and by 1883 was producing 1·25 million tons per year. The demand for ships' plates and angles postponed the impending eclipse of the iron puddler. In cotton and wool there was no great extension of output, but new capital was flowing in, especially to cotton, where a remarkable increase in new and much larger mills occurred on the joint-stock and cooperative principles. By 1882 the price of raw cotton was the lowest since the early fifties. Increasingly the industry was contracting in north Lancashire and concentrating in the south of the county, as the extinction of weaker firms went on. In 1874 there had been some 42 million cotton spindles; by 1885 there were 44·3. Power looms increased from 460 to 560 thousand.[1] But

---

[1] 'The Textile Industries of the United Kingdom', *Econ.*, 19 December 1885, p. 1536 and 26 December 1885, p. 1568.

Scotland, accused by Scotsmen of complacency, failed to get her share. In the Yorkshire woollen industry never before had there been such ingenuity, taste, and skill displayed. Yet, though there was real growth in the output of these staples, it was clear by the eighties that, in spite of dramatic absolute increases, the proportionate rate of growth had fallen off sharply since the great days of the fifties and sixties.

Steel was rapidly replacing iron; indeed the age of steel began with the Great Depression of the seventies. New capital could often be found for steel projects even when there was idle capacity, for each new venture was justified on the plea that some local advantage, or a cheaper method, would defeat competitors. Between 1877 and 1882 the British capacity for the production of steel rails rose from ·88 million tons to 2·75 million, mostly using the Bessemer process.[1] High-grade iron imports flowed in, especially from Spain, to feed the new steel mills. The price of steel rails approximated to that of iron, with the result that steel, with its longer life, was preferred.

In explanation of the extension of iron and steel output in the face of bitter complaint of the lack of profits, two circumstances were adduced. Output was pushed toward capacity because if the high fixed costs of producers could be spread over a greater output, average cost fell sharply. But prices were kept near cost by two circumstances: if above cost by any great margin, the workers would restrict their work in an attempt to raise wages; if below, the masters curtailed output in order to push up prices.

Lloyds in 1877 sanctioned the use of steel for merchant vessels, and the Admiralty contracted for some steel vessels of war. Steel very greatly increased the dead-weight carrying capacity of ships, and greater safety was secured by its extraordinary ductility. Palmers of Jarrow, though their output was halved between 1883 and 1884, pressed forward with the manufacture of their own steel plates. The triple expansion engine appeared, with other innovations. The merchant navy grew more rapidly in the period 1880-83 than at any previous time, embodying great improvements in design. By 1886 even the oil tanker had made its appearance.

Many branches of engineering were learning new techniques; new factories were opened and old ones extended, in an atmosphere of keen competition.[2] Britain was still the world's greatest exporter of machinery, and was to hold this position for the rest of the century.

[1] 'Metals', *Econ. C.H. and R. of 1882*, 1883, p. 32.
[2] See below, Chapter 4, sections 6 and 7, pp. 149, 150.

Though home demand for agricultural machinery was poor, exports were quite well maintained. Heavy engineering made great strides; the contract for the Forth Bridge, far surpassing in magnitude and daring any bridge in the world, was placed in 1882. There was a growing demand for equipment for refrigeration and ventilation, chiefly for the new frozen-meat trade.

The jute industry was still in its youth; oil cloth and linoleum, and plate glass, were new, and soap and spirits had taken on new life. Canned foodstuffs began to make headway against the tide of conservatism and adverse publicity. Electricity was upon the scene, and new companies were quickly formed to exploit power and light, raising some £7 million of capital in 1882. Tramcars began to appear on the streets. All this meant both investment and diversity, and an eager search for a means whereby electric power might be generated and transmitted.

The fact that stocks of primary products — tea, sugar, coffee, cotton, and wool — were not growing to any great extent was evidence that, in spite of vastly increased output, consumption was increasing in step with production. Indeed, from the mid-seventies onward, with the predominance of the steamship and the extension of the electric telegraph, a great reduction in stocks of commodities held became a permanent feature of the world economy.

The output of what contemporaries knew as 'incorporeal products'— transport, banking, retail trade, and services generally—all rose, as the distribution of goods became increasingly efficient.[1] Indeed there was much complaint that the fall in wholesale prices was not sufficiently passed on to consumers because of the rise of an excessive number of retailers. Old products were taking new forms, and new ones were appearing; in most of such cases the rate of growth was that appropriate to a new start.

Nor must we forget the changes in the pattern of incomes as they affected the demand for the products of industry. The number of 'solid fortunes' built up since the forties meant that representatives of an important element in society enjoyed incomes, which, for all the troubles of the times, allowed them to avoid serious contraction of their spending. Those who possessed land, houses, or obligations of states or companies, were now a much more substantial element in society in Britain, Europe, and America. Some might lose by state defalcation or commercial failure, but others gained from the fall in prices, just as the employed workers were doing. Their spending,

[1] See below, Chapter 7, section 1, p. 218.

though variable with good or bad times, constituted a buffer of real, if not dominant, importance. But on the other hand there were some signs that the income position of many of these fortunate people was such that they now sought not merely to consume at a fairly high level, but also to place their resources in bonds and the safer shares, in the interest of the prestige and continuity of the family. Calculations based upon income tax returns suggested that though the national income was continuing to grow, the rate of increase 1875–85 was considerably less than that of the previous decade.[1]

The trading pattern too was changing, and, as always, those men of business who responded to the new needs could do better than the rest. The old markets, involving the greater nations, though still important, had now less potential than those of the smaller countries and the colonial possessions. For the greater states had developed their own industries, and had, moreover, become protectionist. Exports and capital could be better placed in Canada, Australia, New Zealand, the African colonies, and India, and in China, the Pacific Islands, and South America; colonies formerly paying 6 per cent for loans could borrow at 4 per cent. British capital investment went increasingly to the colonies. Because of this the worker was invited to share the imperial enthusiasm of many of the middle class.

A further element of encouragement lay in the productivity of the worker. In the boom years this had tended to fall in many trades, due to negligence and lack of application. In coal mining, between the mid-sixties and the mid-seventies, output per man fell by more than one-fifth, from 21 cwt per day to $17\frac{1}{2}$.[2] Now, with employment harder to come by, there was real encouragement to work harder. Though the strikes that accompanied many wage reductions resulted in loss of working time, this was probably more than compensated for by harder work and improved discipline. There were increasing signs, also, from the seventies onward, of the growth of the concept of scientific management, together with a changing attitude on the part of employers toward the wages bargain.[3]

Nor must we overlook the falling cost of imports. The previous booms had prepared a situation in which the production and marketing of primary produce could greatly expand, and this meant real gains for Britain as the cost of her imports fell. This, of course, contributed

[1] Giffen, op. cit., p. 100.
[2] 'Metals', *Econ. C.H. and R. of 1874*, 1875, p. 19.
[3] See below, Chapter 10, section 7, pp. 415, 416.

to the general world fall in prices as the cost lowering effects of investment in such countries showed themselves. Down to 1896, when the general price fall ceased, British wage-earners gained from the fall in prices of imports.

There was, in all the foregoing, much scope for new initiative, even within a general situation of low profits affecting the basic trades. In fact, many men were prepared to argue that these were the best conditions for the improvement of efficiency, both of business men and workers. For a surprising length of time producers were willing and able to extend output in the attempt to compensate for low prices. But there were limits to the power of the economy to move forward without significant profits in its important sectors.

9. THE NEW DEMOGRAPHIC REVOLUTION

The population boom of the seventies reflected the new structure of society.[1] Urbanization was now far advanced. It was in the town areas that the great increase in birth rates appeared. But it was also under urban conditions that death rates had remained highest.

The general picture was now due to undergo radical change. From its peak of 1876 the crude birth rate of England and Wales began its long decline.[2] From over 36 per thousand it had reached 28·5 by 1901 and was still dropping. The early phase of this decline might be explained on a number of grounds not involving fundamental change in society. The increase down to 1876 could hardly have continued, for it depended upon ever younger marriages. The change in economic conditions was also bound to have its effect. Employment became less easy to obtain and to keep. Those born in the boom of the sixties, were, from the mid-seventies onward, seeking jobs. But these men and women could not multiply as their parents had done, and for the first time in a century the absolute number of births ceased to grow, remaining constant for some seven years. But though 1876 was the year of downturn in birth rate for the country as a whole there were wide variations between towns: the date for Blackburn was 1851,

---

[1] For general data see *Statistical Memoranda and Charts . . . relating to Public Health and Social Conditions*, Cmd. 4671, 1909.

[2] See E. Cannan, 'The Changed Outlook in regard to Population, 1831–1931', E. J., 1931, p. 533, for a table of births and their ratio to marriages, 1851–1930; A. Newsholme and T. H. C. Stevenson, 'The Decline of Human Fertility in the U.K. and other Countries as shown by Corrected Birth-rates', *J.R.S.S.*, 1906.

Glasgow 1857, Liverpool 1864 or earlier, Preston 1866 or 1867, Huddersfield about 1872.[1]

Even such improvement in real incomes as was enjoyed by those in employment might fail to affect marriages and births. For if better real wages are brought by falling prices rather than by rising money incomes, they may not prompt an increase in births. When prices fell in the seventies and eighties, the workers in employment, whose real income (with roughly constant money wages) was increasing, did not respond by increased marriages and births, but continued at something like the old level.

The first phase of the long decline in birth rates is thus explicable in terms of well-known demographic and economic factors. But its persistence from the mid-eighties onward, in the face of a recovery in the marriage rate, points to a new situation. Only the advent of a change of attitude toward the size of the family can explain this new phase.[2] This was reinforced by the renewed vigour of the birth control movement. The responsiveness of the birth rate to boom conditions, observable in the sixties and early seventies, was much reduced — higher wages were no longer capable of inducing a national marriage and child-bearing spree.

Accompanying the revolution in births came one in deaths. The death rate, more or less stable for so long, also began to fall; from an average of 22 per thousand in 1875, it had reached 16·9 by 1901. At last the factors favourable to survival had become dominant over those that were adverse, even in the cities. After the eighties the urban death rate, though continuing above that of the countryside, dropped more rapidly, and so the gap in favour of the rural areas was reduced. But the general birth rate moved downward even faster than the death rate, so that the rate of natural increase declined from a peak 14·5 in the years 1876–80 to 11·6 in the last five years of the century.

There was a renewal of the urge to emigrate. Indeed, so far as England and Wales were concerned, apart from the fairly modest losses of the fifties, the first mass migration ever experienced came in the later seventies and early eighties. Statistics of the movement of peoples are so imperfect that indirect evidence must be used, namely the number by which the actual increase in population was less than the excess of births over deaths. In the decade 1871–80 England and Wales lost by

---

[1] G. H. Wood, Discussion on G. U. Yule's Paper, *J.R.S.S.*, 1906, p. 137.
[2] J. A. Banks, *Prosperity and Parenthood; A Study of Family Planning among the Victorian Middle Classes*, London, 1954.

net emigration thus calculated 164,000 people.[1] But the number of persons of English and Welsh origin going abroad was much larger than this, for of the 661,000 leaving Ireland a great many entered England to replace departed denizens; moreover there was an inward balance of aliens to the United Kingdom of 62,000, thus reducing the figure of net loss. Indeed it would seem that the gross loss (English and Welsh emigrants) was something like three-quarters of a million. In the eighties English and Welsh net loss was very much higher, at 601,000, with Ireland exporting 738,000, Scotland 218,000, and a net inward balance of aliens of 95,000. The gross loss for England and Wales must have been between one and one and a half million. It is also worthy of note that over half a million foreigners sailed from United Kingdom ports in the seventies, and no less than a million did so in the eighties.[2]

In this great phase of migration from Britain it would appear that though elements of expulsion were present in the form of a decline in the rate of economic growth at home, the element of attraction to new countries was now much stronger than ever before. Railway building once more dominated the American scene; men were drawn westward by the earnings offered and by the prospect of land made newly accessible. But it remained true in the later phase as in the earlier, that Ireland sent mainly unskilled labourers, whereas England sent a substantial element of skill and business enterprise; the Welsh and the Scots sent even more of such eligible men.

10. THE EXTERNAL CONTEXT

Britain was experiencing fundamental changes in her trading pattern also.

During the first half of the nineteenth century, down indeed to 1857, the net barter terms of trade (the prices received for British exports compared with the prices paid for imports) moved against her. Increasing quantities of manufactured textile exports had to be provided in exchange for imports. This was due to the enormous economies made possible in the production of such goods, both in the form of more effective methods of growing and processing cotton in America, and in the form of the factory system in Britain. Foreign buyers responded by ordering ever larger quantities. This was the basis upon

---

[1] *Statistical Abstract for the United Kingdom*, Cmd. 3084, 1928.
[2] *Statistical Tables Relating to Emigration and Immigration from and into the United Kingdom*, 1899.

which the village textile industry could be destroyed in India, Brazil, and elsewhere, but it was also the basis upon which the world's supply of clothing was vastly increased. This, of course, was not necessarily bad business for Britain, for to use the new methods, with their great economies, was a condition of expansion, not only for Britain, but the world. But the prices of Britain's imports were also falling, though at a lesser rate. In the export of manufactured goods in the thirties, textiles predominated to the extent of about 75 per cent.[1]

In the second half of the century prices in general began to rise, but British export prices moved up faster than those of her imports. In part this was due to the increase in other lines of exports to challenge the long ascendancy of textiles—machinery, hardware, coal, and iron and steel, in all of which the economies of new production methods, though real, were less dramatic. The great age of railway building and indus-trialization in the United States, Germany, Russia, and elsewhere, gave a great stimulus to British metal production, machine making, and engineering. Not that foreign countries relied exclusively upon Britain for such goods; indeed they were building up their own resources of this kind at least from the forties, with the result that, increasingly, such countries called upon British production only when their own plant has been brought into full operation, that is to say, when prosperity was well under way. In part, also, the movement in the net barter terms of trade in favour of Britain after the fifties was due to the development of the new primary producing countries, adding to the flow of food and raw materials under conditions of falling costs. This meant that, with the exception of a brief period from 1873 to 1879, Britain enjoyed favourable terms for the second half of the nineteenth century.[2] But this was not a dramatic reversal, for the degree to which the terms favoured Britain was quite modest.

Not only did the terms of trade alter; so too did the composition both of imports and exports, though this change came some twenty years later. Previous to the seventies the order of size of imports had been raw materials, foodstuffs, and manufactured goods.[3] With the cotton and woollen industries making their enormous demands for materials, and with British agriculture responding vigorously to the needs of the economy and thus obviating the necessity for great

---

[1] Schlote, op. cit., p. 71.
[2] Imlah, op. cit., p. 102; see also R. E. Baldwin, 'Britain's Foreign Balance and Terms of Trade, *Explorations in Entrepreneurial History*, 1953.
[3] Schlote, op. cit., p. 53.

foreign purchases of food except when harvests were short, it was raw materials that dominated the import picture. Food imports had, in fact, been increasing from the forties, but it was animal products, especially meat, that accounted for most of it. In 1842 the prohibition against importing livestock ended; this was followed by a very great increase in imports.

But from the mid-seventies came the great influx of foreign cereals, together with a decline in the relative importance of textiles in the economy; food imports began to outrun those of materials. Foreign manufactured imports were also increasing steadily from the sixties, reflecting both the industrial progress of other countries, and the rising incomes of Britain. Down to the nineties it was mainly consumer manufactures that were imported; thereafter chemicals, machinery, and other heavy imports gained. That is to say, the growth of heavier industries in Britain had made their import needs as serious as those of textiles by the last decade of the century.

From the seventies onward, not only was Britain importing more manufactured goods, she was also altering her ways by exporting more by way of raw materials and goods that were only semi-manufactured: coal, textile yarns, and iron plates increased in importance.

The classic pattern then, of the first half of the nineteenth century, was much altered by the eighties, and the trends then apparent were accelerating. No longer did Britain merely export manufactures for raw materials and foodstuffs. Even before the mid-century invisible earnings accruing from banking, insurance, and shipping services were of very great importance, and increased rapidly as Britain's role in the expansion of other countries, especially primary producers, grew. But industrial rivals were appearing in Germany, the United States, France, and elsewhere. Though Britain might continue to seek new primary areas to develop, especially within her expanding empire, and though great efforts were made to increase invisible earnings, the trend was too strong. Britain was bound to accept the more involved role forced upon her by the rise of industrial rivals, together with the urge in primary countries, as in Canada, India, Brazil, and elsewhere, to alter the balance of their economies, and the unwillingness of other nations to see Britain continue to pre-empt so much of the world's invisible earnings. With these developments the complexities of Britain's problems mounted, so that it became difficult to assert in simple terms where her interest really lay, and what policies were appropriate to it.

From 1872 onward these changes showed themselves increasingly in the balance of payments. Britain's deficit on merchandise account with the world at large rose steeply, averaging over £100 million annually from 1876 to 1885, when it fell slightly.[1] A free trade nation, she became an enormous receiver of other countries' exports, while her own sales abroad were curtailed by the rising protectionism of other states. But by accepting this role Britain greatly assisted the smooth operation of the world's economy.

Four principal circumstances enabled her to play this beneficent part. First, her past accumulation of foreign balances was now such that the dividends upon these were very great, and were rapidly increasing; interest and dividends of earlier investment, plus the invisible earnings of insurance, shipping, trading services, and profits, could more than compensate for the deficit on merchandise account. In 1850 the grand total of her foreign holdings was some £225 million; by 1875 it was in the region of £1,000; by 1885 it was half as much again.[2] These investments gave her a large and growing stake in the prosperity of other countries, including the primary producers.[3] Secondly, her invisible earnings, increasing steadily through the century, were now very large — payments for shipping services, insurance, brokerage, and banking, now earned for Britain immense sums that did not appear in her trading balance. For from the sixties onward a much more involved pattern of multilateral settlements developed between trading nations as capital movements extended, and as Britain was joined by other industrial nations as purchasers in the world's markets for primary products. London formed the centre of this pattern; the great ascendancy of the London Money Market was a further reflection of Britain's gains from her early predominance. Though the rise of rivals might damage her as an industrial competitor, it added to her power to provide these profitable services. Thirdly, there were the adjustments going on in British markets. Increasingly she came to rely upon the Empire, and especially upon India. By the eighties Britain was heavily dependent upon her surplus with India to settle her deficits in Europe and the United States.[4] Britain had a near monopoly of exports to

---

[1] S. B. Saul, 'Britain and World Trade, 1870–1914', *Econ. H.R.*, 1954, p. 57.

[2] Albert H. Imlah, 'British Balance of Payments and Export of Capital, 1816–1913', *Econ. H.R.*, 1952, pp. 231, 232, 235, 237.

[3] L. H. Jenks, 'British Experience with Foreign Investments', *J. Econ. H.*, Supplement, 1944, p. 75.

[4] S. B. Saul, *Studies in British Overseas Trade, 1870–1914*, Liverpool, 1960, esp. chapter VIII.

India; India was able to supply Europe with primary products. In this way India served as a kind of by-pass round the new protectionism, contributing to the ability of Britain to maintain free trade. Finally, British agriculture, like India, made its contribution. Agricultural protection, in spite of the plight of the cereal farmer, was not revived, so that the decline of arable incomes in Britain, and the reduction of arable acreage may be regarded as sacrifices to the multilateral system.

II. THE EIGHTIES AS WATERSHED

Though British traders were learning to make adjustments that were very helpful in the long run, by 1884, and increasingly in the following two years, signs of difficulty were accumulating in much of the business world. Values of capital assets fell drastically, as the Stock Exchange and the land market revised them downward. The volume of trade, the maintenance of which (in contrast to the value of trade) had encouraged observers since the mid-seventies, at last showed signs of falling. Iron and coal output, so long continued under difficulties, contracted with extreme depression in Lancashire and the north. The iron masters sought scapegoats, and found them in the 'three Rs' – rents, royalties, and railway rates, all of which were condemned as too high. Continental producers had applied the new Gilchrist-Thomas basic steel process much more vigorously than had been done in Britain. By 1886 the output of the shipyards had dropped to about a third of the figure of three years earlier (·47 million tons against 1·25) with desperate conditions on the Wear and the Tyne; for the nation as a whole 20 per cent of shipyard workers were idle.[1] Shipowners were very hard pressed, with many vessels laid up around the coasts.

Labour troubles increased as downward pressure was put upon wages. Agitators found a ready hearing; riots broke out in London and some of the provincial cities. The number of paupers was rising. The Oldham cotton strike of 1885 was the longest on record. It began to look as though the long ideological and political truce observed by labour, more or less since the forties, was about to end.

Talk about the possibility that the system had entered upon definitive crisis spread from the Continent to Britain. Volume I of Marx's *Capital*, with its statement of the doom of capitalism, was first published in English in 1887. Joseph Chamberlain's celebrated radical programme, sounding, at last to all the world, the 'death-knell of

[1] 'Shipbuilding', *Econ. C.H. and R. of 1886*, 1887, p. 34.

the *laissez-faire* system', was launched in 1885.[1] Internationally the fear of the outbreak of a great European war between the nations, now heavily armed with the weapons of the new age, retarded business.

The difficulties of the time inspired many men of business to seek to create their own conditions of stability through monopolistic arrangements. In 1884 and the years immediately following there was a rush of attempted combinations. British and Continental steel railmakers sought agreement to curtail output, but two years later their syndicate broke up. The British Iron Trade Association agreed upon the wisdom of a general reduction of output by 25 per cent, but failed to settle a plan. Various groups of shipowners sought to form rings, but with the same result. The Cleveland ironmakers were rather more successful. In chemicals there was a similar attempt, especially with a view to maintaining the price of bleaching powder. In the cotton trade efforts were made to form combinations of employers. The brewers joined together in order to push down malt and hop prices. In spite of the relative lack of success of all these attempts, they are significant as showing how the business mind was reacting to its difficulties, seeking safety, as in earlier times, in regulation and agreement.[2] The general sense of excess capacity added to the gloom.

There was a considerable revival of protectionist arguments in Britain. Russia, Germany, France, and Italy in the late seventies and early eighties had raised their tariffs; the United States, increasingly defensive since the Civil War, had, by the eighties, the steepest protection in the world. By the nineties only Britain, Holland, and Belgium remained faithful to the older free trade creed. Not surprisingly, strong pleas were made that Britain should behave like the majority. As early as 1872 the much discussed Commercial Treaty with France, agreed in 1860, was not renewed, partly because of the argument that it represented special treatment for France over other countries, and partly because it might stand in the way of British protective measures should they be desirable. Bitter complaints were made about French, Italian, and German shipping subsidies. In the matter of colonies, the British government was criticized for its passivity, with France active in Indo-China, Belgium in the Congo, and Germany in southern Africa and New Guinea; indeed 1884 was a colonial crisis year.

[1] Anon, *The Radical Programme*, London, 1885, p. 13; See below, Chapter 9, section 10, pp. 378, 379.
[2] Clapham, op. cit., vol. II, pp. 151–3; W. Page, *Commerce and Industry*, London, 1919, p. 320.

To add to Britain's difficulties there were disturbing signs, as in Brazil, of a lessening assent by primary producers to their role. It would thus appear that the mid-eighties were a real watershed. Between the breakdown of 1873 and the mounting difficulties of 1884, business men, deprived of foreign outlets, had bent their energies to the adoption of new methods and new products at home. Most industries tried to find a way out through application and ingenuity. The Royal Commission on Depression heard much evidence to this effect. But this policy, in itself, could not redeem Britain's situation. It was necessary for real development to be resumed abroad.

Foreign investment did eventually revive to relieve the strain. It had fallen in 1873 and for three years down to 1878 was negative. At home capital expenditure was reasonably well sustained until 1883 or 1884. Thereafter it fell. Less was done in terms of local government projects, public services, railway developments, shipbuilding, textiles, coal and iron. But from 1885 onward mounting investment abroad, encouraged by the continued low yield at home, provided relief to the British economy, reviving exports of capital to accompany exports of men.[1] On the whole home investment was somewhat more stable than foreign, and had to sustain a serious lag before foreign investment recovered. Together they caused the national income per head to expand once more after 1884.

House building was a most important element in home investment.[2] It had risen steeply with the boom of the seventies, remaining high until 1876 or 1877. Thereafter it was more or less in the doldrums for some nineteen years. This meant that housing boomed with domestic prosperity in the seventies, fell away with collapse, and then recovered a little as lending abroad extended after the mid-eighties. It should also be remembered that in the eighties, though the supply of houses was increasing only slowly, the additions to the population were less than

---

[1] Herbert Feis, *Europe, the World's Banker, 1870-1914*, New Haven, 1931; A. K. Cairncross, op. cit., esp. chapters I and VIII, *passim*; Brinley Thomas, op. cit., esp. chapter VII; Jeffrey G. Williamson, 'The Long Swing Comparisons between British and American Balance of Payments, 1820-1913', *J. Econ. H.*, 1962. For an earlier opinion C. K. Hobson, *The Export of Capital*, London, 1914.

[2] B. Weber, 'A New Index of Residential Construction, 1838-1950', *S.J.P.E.*, 1955; E. W. Cooney, 'Capital Exports and Investment in Building in Britain and the U.S.A., 1856-1914', *Econa.*, 1949; 'Long Waves in Building in the British Economy of the Nineteenth Century', *Econ. H.R.*, 1960; K. Maiwald, 'An Index of Building Costs in the U.K., 1845-1938', *Econ. H.R.*, 1954; G. T. Jones, *Increasing Return*, Cambridge, 1933, part II; *Royal Commission on the Housing of the Working Classes*, 1885; H. Richards and P. Lewis, 'House Building in the South Wales Coalfield, 1851-1913', *M.S.*, 1956.

in the previous decade, due to emigration. Large-scale emigration was in fact a concomitant of capital exports, for both men and money went abroad seeking better opportunities; this weighed against housing construction at home. This factor was especially important for some of the great cities, like Glasgow, where the building trade came almost to a standstill when emigration was high as in the eighties. It would appear also that the increase in the number of larger houses was less than that of the more modest, suggesting the relatively more severe impact of the times upon those who were better off. The new houses for ordinary people were slowly improving in size and amenity.

Some towns ceased to gain population, and in fact began to lose, especially in the northern textile and industrial areas as migration abroad increased in the eighties. Many active men in the countryside, and even in the towns, rated the opportunities of the new world more highly than those in home industry. By the nineties the southern areas were gaining over the northern in a more positive sense; with emigration abroad once more at a low level, it appeared as if, within Britain, the south was beginning to reassert its ancient predominance.[1]

General improvement in prices, profits, and employment eventually came in 1886, stimulated by both home and foreign investment culminating in full employment in about four years. In some sense those who had agreed with the Royal Commission of 1886 that the situation must be left to correct itself had proved right, for correction seemed to have come. But stability had not, for in 1890 crisis returned, this time complete with an old-fashioned commercial panic in the City, precipitated by the embarrassment of the great house of Baring, heavily involved in lending abroad, especially to the Argentine. Nor did agriculture enjoy prosperity, even in the boom.

The worrying question remained: how long and how strong a recovery was possible?[2] To place against the misgiving expressed by so many was the knowledge that, even with falling profits and prices, average real incomes could at least be maintained.[3] New products and

---

[1] Cairncross, op. cit., p. 71.
[2] Giffen, op. cit., p. 99 *et seq.*
[3] C. H. Feinstein, 'Income and Investment in the United Kingdom, 1856–1914', *E.J.*, 1961; Phyllis Deane, 'Contemporary Estimates of National Income in the Second Half of the Nineteenth Century', *Econ. H.R.*, 1957, p. 460; A. R. Prest, 'National Income of the United Kingdom, 1870–1946', *E.H.*, 1948; J. C. Stamp, *British Incomes and Property*, London, 1922. James B. Jeffreys and Dorothy Walters, *National Income and Expenditure of the United Kingdom, 1870–1952*, Income and Wealth, Series V, 1956.

enterprises could appear.[1] Moreover, though Britain might be losing ground relative to other national competitors, the very rise of such rivals signalled a new age of expansion for the world economy as a whole, thus providing the greatest single condition necessary for British prosperity.

[1] For capital formation in the last third of the century, see J. H. Lenfant, 'Investment in the United Kingdom 1865–1914', *Econa.*, 1951; Paul Douglas, 'An Estimate of the Growth of Capital in the U.K., 1865–1909', *Journal of Economic and Business History*, 1930; Cooney, op. cit. For the long cycle see E. H. Phelps Brown and S. J. Handfield-Jones, 'The Climacteric of the 1890s', *O.E.P.*, 1952; D. J. Coppock, 'The Climacteric of the Nineties', *M.S.*, 1956; W. A. Lewis and P. J. O'Leary, 'Secular Swings in Production and Trade, 1870–1913', *M.S.*, 1955.

PART TWO

The Sources of Initiative

THE economy as a whole grew, diversified, and pulsated. The manner in which it did these things depended upon the initiatives taken by the men and women concerned, both at critical times and in their daily lives.

In a competitive economy, as in nineteenth-century Britain, there are three principal kinds of economic initiative: invention, organization, and the provision of the means for gaining control of resources. The first is the task of the scientist and inventor. The second is the role of the man of trade, industry, or agriculture, who perceives the possibilities inherent in the new knowledge and is prepared to attempt to exploit them. The third is the function of the capitalist and banker, the man who is willing and able to place real capital at the disposal of the business man. For an economy to grow it must somehow be able to invoke, by appropriate incentives, an adequate supply of each.

CHAPTER THREE

The Men of Invention

I. THE AGE OF THE EMPIRICS

INVENTIVE activity is closely related to the general condition of society; whether its members are of an inquiring frame of mind, whether it has the means and the will to retain and disseminate knowledge, and whether it permits the adoption of new modes of production, with the inevitable threat to established positions and incomes.[1]

In Britain the new phase of industrial invention was greatly accelerated by a demand factor: the urgent and irrepressible desire developing in Europe, greatly stimulated by the cargoes of the East India Company, to wear lighter clothes. Britain was dramatically confronted with a product cheaper and more versatile than her own light woollen fabrics. The traditional response to such a situation had been to import enterprising practitioners of the superior techniques. But India was too far away, both geographically and culturally; Britain had to meet the challenge from her own resources of ingenuity. The result was to place a premium on the adaptation and elaboration of textile processes.

It was mechanical improvement on the basis of untheoretical perception that gave rise to this first great increase in output. The early engineers were men who extended and applied their own skill and learning, and who only incidentally sought aid from textbooks. The machine makers knew the behaviour of levers and the like through experience; they did not need to know the mathematics involved, for they could judge stresses and safety limits by eye. It was impossible to bring together at this stage the brilliant generalized mechanics of d'Alembert and Lagrange and the experience of the artisan. The

[1] Charles Singer *et al.*, eds. *A History of Technology*, vol. IV: *The Industrial Revolution, c. 1750 to c. 1850*, Oxford, 1958; vol V: *The Late Nineteenth Century, c. 1850 to c. 1900*, Oxford, 1958. For a shorter account see T. K. Derry and Trevor I. Williams, *A Short History of Technology: From Earliest Times to A.D. 1900*, Oxford, 1960. For biography see J. G. Crowther, *British Scientists of the Nineteenth Century*, London, 1935.

staples of Cambridge teaching, Newton's *Principia*, Simpson's *Algebra*, his *Conic Sections*, and his *Fluxions*, were all beyond the untutored mechanic, and far removed from practice. Mushet, very early in his iron experiments, abandoned as useless the celebrated textbook of Fourcroy. The real need, as felt by Stephenson and others, was for a simple statement of known practice, with fairly straightforward explanation: the nearest approach was James Ferguson's *Lectures* (1805 and 1823), written by a self-taught natural mechanic, son of a farm labourer.[1]

Though the new age of observation and experiment inspired by Bacon, Boyle, and Newton, did not contribute much that was of great use to the early inventors, it was more significant indirectly, in its effect in turning even humble men who had never heard of its specific scientific results, away from traditional acceptance toward a life of querying. This is not to deny that the mechanical inventors were scientific in the sense of deliberately construing and solving their problems, nor to deny that they drew on the observations and experience of others. But it is to say that the engineering progress of the day was made by men with little philosophical knowledge, working with what contemporaries called 'mechanical instinct'. Though they certainly did not scorn scientific knowledge, the early engineers believed that the eyes and the fingers were the best means of learning; as Nasmyth put it, 'The nature and propensities of the material comes in through the finger ends.'

In three cases scientific knowledge and the empirics did come into contact: the invention of the steam-engine, medicine, and the development of chemistry. In the case of the first, James Watt was undoubtedly conversant with scientific speculation in the University of Glasgow, with consequent effects on his thinking. But the practical insight of the instrument-maker was at least as important as the idea of latent heat. His immediate challenge was the puzzling fact that the tiny Newcomen engine used for teaching purposes at the university did not work because of its diminutive scale; the search for the answer led to the separate condenser. So was achieved that great advance in steam-engine construction that brought the harnessing of inanimate energy. The new age was to be one in which the engine and the machine conjointly worked miracles, both in production and in the furtherance of new discoveries.[2]

[1] For this 'natural' see *A Short Account of the Life of James Ferguson, F.R.S., Written by Himself*, London, 1826.

[2] A. R. Ubbelohde, *Man and Energy*, London, 1954.

Among the doctors were to be found some of the most scientific of men, trying by observation and experiment to understand body processes.[1] Jenner and others took a great slash at the death rate in the late eighteenth century with the vaccination technique, at least in some regions. But controlled experiment was very difficult, and there was much conservatism. The first major surgical operation under ether anaesthesia in England was performed in 1846 by Robert Liston; at long last the tension in the operating theatre, where everything depended upon the strength, dexterity and presence of mind of the surgeon, could be relaxed. Lister's development of the antiseptic method at Glasgow in the sixties further diminished the casualties. At last, surgery, that had hardly changed from Biblical times to the days of Liston and Lister, could make great strides.

Medicine and chemistry were closely allied; in fact in the Scottish universities, where most progress was taking place, it was the medicos who were most active in chemical inquiry. Chemistry also united the scholar and the practical man. It was with chemistry in mind that Davy announced in 1801 that the man of science and the manufacturers are daily becoming more assimilated to each other. Producers of chemicals in Scotland consulted the professors in their universities, and attempted their own research.[2] Early in the century, Josiah Gamble, having attended chemistry lectures at Glasgow University, abandoned his ministerial calling to set up a chemical works. William Woollaston produced his sliding rule of chemical equivalents, of great practical value to the manufacturer.

But though manufacturers and scientists worked on the same range of problems, cooperation did not lead very far, for there was not, even between them, sufficient knowledge of chemical behaviour to make much progress. The early nineteenth-century chemists, like Davy, were only on the threshold of systematic knowledge, though he in 1807–08 had isolated the alkali and alkaline earth metals.[3] The great work of Dalton on the atomic theory of matter in the first decade of the century could contribute little to immediate progress. Yet the intelligent manufacturer kept his eye on the chemist, and could gain thereby. He could read with profit the works of Lavoisier and Berthollet, and especially Nicholson's *Dictionary of Chemistry*, and the texts of

[1] Fielding H. Garrison, *An Introduction to the History of Medicine*, 4th ed., Philadelphia and London, 1929, chapter 12; T. K. Munro, *The Physician as Man of Letters, Science, and Action*, Edinburgh and London, 1951.

[2] A. Kent, ed., *An Eighteenth Century Lectureship in Chemistry*, Glasgow, 1950.

[3] Anne Treneer, *The Mercurial Chemist: a Life of Sir Humphrey Davy*, London, 1963.

Thomas Thomson.[1] But an air of alchemy still hung over the subject, with its overtones of the mystical and the magic. Chemists still proceeded by trying everything available in combination; the difficulties of controlled experiment and accurate recording made for unpredictability. There was a considerable miscellany of processes practised commercially, as with Dundonald's preparation of tar and in the manufacture of paint, but the commercial crux of the chemical industry was the production of alkalis, mainly from kelp, and of sulphuric acid. From these came bleaching powders, improved dyestuffs, and soap, all closely linked with textiles. Lord Dundonald with his *Connection between Agriculture and Chemistry* of 1795 and Davy with his *Elements of Agricultural Chemistry* of 1813 made an attempt to awaken farmers' interest in the new chemistry, but with no immediate success, though eventually the influence of Davy's book both in Britain and the U.S.A. was very great.

In a sense, then, in spite of the genius they embodied, the inventions of the first half of the nineteenth century were extensions of intuitive 'rule of thumb' and 'scowl of brow' methods. The somewhat acrimonious debate over the priority of invention of the miners' safety lamp showed how Stephenson's powerful native wit could make him a worthy rival to Humphrey Davy, with his laboratory knowledge of the behaviour of explosive mixtures. Each steam-engine produced was peculiar and had to be separately nurtured and adjusted. The early engineers spent most of their time, not in the consideration of principles, but with this engine or that, in Cornish mines, Shropshire ironworks, Lancashire cotton mills, or even Peruvian silver mines, spending weeks coaxing it into motion. Murdoch, Trevithick and Stephenson were all men of notable strength and agility well able to try conclusions with their workmen. They needed to be.

At Oxford and Cambridge mechanics was taught as a branch of mathematics, and was bound up with an intensely competitive examination system, with non-experimental lectures of a repetitive kind. The elder Rennie, though well placed, and a great respecter of theoretical power, refused in 1809 to allow his son to enter either university, because, he argued, after three or four years of such an atmosphere he would have great difficulty in turning to the practical parts of engineering.[2] Yet though there was some justification in the thirties for charging

---

[1] William Nicholson, *A Dictionary of Chemistry*, London, 1795; A. F. de Fourcroy, *Elements of Chemistry, a General System of Chemical Knowledge*, London, 1804, 11 vols. trans. by W. Nicholson; Thomas Thomson, *The Elements of Chemistry*, Edinburgh, 1800.

[2] *Autobiography of Sir John Rennie*, London, 1875, p. 4.

the English universities with being too apt to undervalue the science of the factory, it is hardly fair to criticize them for failure to take the lead in a task which sprang so directly, in this early stage, from the practice of the artisan and craftsman.[1]

Many streams of experience ran together to produce the great flowering of British engineering in the first half of the nineteenth century, all of them of a highly practical nature. London was the traditional centre of craftsmen, artisans, and smiths in the precious metals. Though the level of skill in producing automatic toys attained by continental craftsmen like Vaucanson did not appear in London, yet the handicrafts of the City were of a high degree of excellence. But though certain processes, like the perforation of metals by presses, were carried to a high pitch, the London artisan, at least in the earlier part of the century, saw little of the new standardized, mechanized age. The manufacturers of Birmingham and Sheffield on the other hand, had made great progress with the cutting and shaping of metals on a large scale.[2]

In contrast, the trade of the millwright, the greatest single source of engineering training, was concerned with complex and highly specific engineering jobs. The northern millwright was an aristocrat among artisans, an itinerant engineer, with a fair knowledge of elementary arithmetic and geometry, levelling and mensuration and practical mechanics, able to calculate velocities, to draw and plan in section, and to construct buildings, conduits, and watercourses. The engineers trained by millwrights make a formidable list: Brinley, Rennie, Cubitt, Fairbairn, Donkin, John Wilkinson, and a great host of lesser men. Similar in initiative and outlook were the millwrights' rivals — the builders and installers of the new steam-engines, like Murdoch and Trevithick.[3] Nor must we overlook the ironmasters as a source of engineering inventiveness. Military engineering was also important; Maudslay first acquired his skill in the arsenal smithy at Woolwich. But there was nothing in Britain approaching the mechanical education provided for its officers by the Prussian army.

Most tool builders set up their own engineering works, first doing what jobs they could get. The making of tools called for men who were close to production problems. But this was not of course to say that

[1] For the impact of scientific thought on the British universities see Sir Eric Ashby, *Technology and the Academics*, London, 1958.

[2] See below, Chapter 4, section 6, p. 148.

[3] Francis Trevithick, *Life of Richard Trevithick*, London, 1872.

such men could not introduce new tools or machines capable of general development; indeed they were in the best possible position to do so. So prolific had they become in invention by the thirties that obsolescence was very rapid. A steam-engine might serve a generation or more, but, according to Babbage, in order for textile machinery to be profitable, it was customary for the manufacturer to reckon to recover his money from the machine in five years, and to scrap it in ten.[1] Birmingham, London, Manchester, and Leeds became the new centres of this engineering activity.[2] Bramah trained Joseph Clements, Arthur Woolf, and Henry Maudslay; the latter in turn employed Nasmyth, Roberts, Napier, and Whitworth, who in succession produced their large progeny. In the works of such men the line between inventor and organizer was difficult to draw: Rennie was known to his men as 'Almighty Rennie' partly for his engineering skill, but more for his ability to deal with critical and complicated situations.

Samuel Smiles, in assigning inventive priority, took the view that the palm should go not to the man who had a vision of a new method or product, but to him who first executed the idea in fully practical terms.[3] These early engineers were both percipients and executants within the field of mechanics. This was true both technically and commercially. Their engineering enterprises were on a considerable scale, by the thirties, the larger with perhaps one hundred to one hundred and fifty employees in good times. Like other men of business, they paid the penalty when costs outran selling prices, or when assumptions or partners went wrong.

There were difficulties in judging both technical and commercial viability. The men of science sometimes acted as temperers of the enthusiasm of the practical men whose confidence, fed on success, knew no bounds. But often the engineers, in their untutored state, were right. Sir Humphrey Davy, so advanced in his theoretical knowledge of gases, ridiculed the proposals to illuminate the streets with gas; Sir Joseph Banks, great scientist and traveller though he was, believed that the steam-engine could not be used for the propulsion of ships because it required a firm platform on which to work. What are we to think of a mind that could conceive a channel railway tunnel

---

[1] Charles Babbage, *On the Economy of Machinery and Manufactures*, London, 1832, p. 231.

[2] See Chapter 4, section 4, below, p. 134.

[3] It was Smiles who brought the great inventors to the attention of Victorian England, especially in his *Lives of the Engineers*, London, 1861–62, *Industrial Biography*, London, 1863, and *Men of Invention and Industry*, London, 1884.

and yet believe that the horse would never be superseded as prime mover by the steam-engine?[1]

There were, of course, many examples of inventors whose ideas outran technical capacity; a great deal of such invention had to take its place in the queue awaiting better and cheaper materials and labour. There were still other ideas that proved wholly incapable of execution because of the state of mechanical knowledge, but which came into productive application with better techniques – the principle of Bramah's hydraulic press was known a century and a half before his time. Indeed, it was the opinion of Babbage in the thirties that of the vast multitude of inventions then coming forward, a large proportion of the failures was due to the imperfect nature of the first trials. There was also the safety factor: Watt's hostility to high-pressure steam, which brought a bitter conflict with Trevithick, was due to his belief that the then state of the iron manufacture made it impracticable and dangerous.

2. NATURAL PHILOSOPHERS, ACADEMIC AND AMATEUR

Though scientific speculation was not a major direct contributor to economic change at this phase, its evolution cannot be ignored, for by the mid-century, it could make large contributions, and by the later decades, it began to seize the initiative.

The early nineteenth century saw the hegemony of the Royal Society challenged by the formation of new, more specialized, scientific bodies. The parent itself was much strengthened in 1800 by the founding of the Royal Institution, its 'workshop'. Natural philosophy was much practised, not only by academics like Herschel and Babbage, but also by noblemen, gentlemen, and rising men of business and commerce who, inspired by Priestley, interested themselves in phlogiston and the like. The medical profession, with its scientific training, provided more than its share of such men as Dr Harland, who, with Sir George Cayley, kept a skilled mechanic continuously at work constructing equipment for experiments in electricity, magnetism and chemistry. Humphrey Davy, apprenticed to a physician, more than once blew his master's bottles from their shelves. This growth of specialization helped to remove the spell of old controversies.

In large measure, until the mid-nineteenth century, natural philosophy

---

[1] Henry Fairbairn, *A Treatise on the Political Economy of Railroads*, London, 1836. For the history of the tunnel project see Thomas Whiteside, *The Tunnel under the Channel*, London, 1962. See below, Chapter 4, section 4.

was regarded in a cultural light. It was a mitigator of philistinism; attendance at the lectures at the Royal Society was an intellectual and social pursuit. When David Brewster, while pursuing his optical researches, invented the kaleidoscope in 1816, so great was the sensation that no less than 200,000 were sold in London and Paris in three months, unfortunately by pirates of his patent. In the early decades of the century the tradition of the aristocratic *virtuosi* was not yet dead. The complete gentleman, even though educated in the classics, still aspired to at least a conversationalist's knowledge of the leading branches of scientific speculation, and generated his stinks and bangs and erected immobile machinery in cellars and outhouses.

It was this aristocratic patronage of science that was in the mind of de Tocqueville in the thirties when he compared England and the United States in this regard.[1] In the older country the principle of aristocracy prevailed, providing the men of culture and leisure who could promote speculation; in the younger, with the democratic principle dominant, all men could hasten to exploit nature in the competition for fortune, with the result that pure science languished, but a marvellous adeptness in application and development could occur.[2]

In the United States Yankee ingenuity was indeed a major element in economic growth in terms of the design of products, standardization of parts, and organization of the production process; in Britain aristocratic science made no equivalent contribution. But aristocratic pragmatism did: the contribution of the great agricultural improvers was very real. Animal breeders and seed selectors were seeking improved strains, working by trial and error toward an understanding of genetics, so that man could operate upon the animal and vegetable world, not merely by varying the context within which growth takes place, but by consciously controlling the selective process. Most aristocrats were not speculators in natural philosophy; with their landed background they were of practical bent, with an engrossing interest in animals. Lord Spencer, the least theoretical of men, even at the height of his political career, found stock-breeding his most compelling concern; thus something of an alliance between science and practice became possible in practical genetics, under aristocratic aegis.

The failure to bring together science and the artisan was not to be

---

[1] Alexis de Tocqueville, *Democracy in America*, Phillips Bradley, ed., New York, 1945, vol. II, chapter X.

[2] See H. J. Habakkuk, *American and British Technology in the Nineteenth Century*, Cambridge, 1962.

serious for Britain until well after the mid-century. There was plenty
of scope for the application of the 'mechanical instinct'. George
Stephenson described in striking terms how his own mind worked: he
would construct in it a kind of notional workshop, picturing the parts
of the machine he was working on, their shapes and dimensions and
their relationships to each other as they went through their movements.
This the historian Taine compared with the method of Foucault
the French physicist.[1] The latter arrived at his idea of a gyroscopic
governor as he was exploring the theoretical implications of a forgotten proposition of Huygens and Lagrange. But Stephenson could,
on this fairly rudimentary basis, achieve very great things. Cayley, on
the other hand, a genius in ordered creative thought, has been, until
recently, almost forgotten.[2] He was able to state the challenge of
powered flight in scientific terms long before the terms could be met:
'the whole problem is confined within these limits — to make a surface
support a given weight by the application of power to the resistance of
the air'. To this end he carried out a mathematical investigation of the
powers of steam machinery when compared with the muscular force
of birds.

3. PROGRESS BY THE PRACTICAL MEN

The progress made by the practical men in the first half of the nineteenth century was very great.

The first generation of machine tool makers — Bramah, Maudslay,
Murray, Clement, and Fox, had by the twenties made it possible to
make machines to make machines. Expressed in terms of the total
national product, their value was small. But they were the necessary
prelude to the great extension of engineering activity in industry.[3]
These men executed highly specific machines for particular jobs as
ordered by their customers, or machines embodying new principles
capable of general application. Bramah's hydraulic press provided a
new means of concentrated power for compression and lifting; his
pupil Maudslay introduced his slide rest, and his screw-cutting machine,
the prelude to standardization of this essential component. Richard
Roberts developed the 'mortising engine' of Brunel and Bentham into

[1] Taine, op. cit., p. 242.
[2] J. Laurence Pritchard, *Sir George Cayley: The Inventor of the Aeroplane*, London, 1961;
C. H. Gibbs-Smith, *Sir George Cayley's Aeronautics, 1796–1855*, London, 1963.
[3] Charles Wilson and William Reader, *Men and Machines: A History of D. Napier and
Son, Engineers Ltd., 1808–1958*, London, 1958.

a metal slotting and paring machine; his metal plane was available by 1821; by 1825 he had introduced standard templets for reproducing parts.[1] At the same time old machines were being improved.

The metallurgists were making materials cheaper. Cort's puddling and rolling processes had been in use to some extent since 1783, and were spreading in England in the twenties. But puddling, though a great step forward, was very costly in manpower, and constituted a standing challenge to the metallurgist and engineer. Neilson, with his discovery in 1828 of the hot-air blast, driven by steam power, though he could not supersede the puddler who retained his position until the time of Bessemer and Siemens in the mid-century, did enormously extend the range of ores and coals usable in iron manufacture. In the coal mines, though there was less to dazzle the eye and stimulate the imagination, the great managers, like John Buddle on the north-east coast, were learning to drive deeper and further.

Ingenuity of a new kind was flowing into the construction of the mills into which the new machinery was to be fitted, and into their layout and powering.[2] It is not surprising that the initiative came from two millwrights, who sensed the widened horizons of their craft – Rennie and Fairbairn.[3] As long ago as the 1780s the older Rennie, in the construction of the Albion Mills near Blackfriars Bridge, had made a new beginning in flour mill construction, based upon the steam-engine. By his death in 1821 he had a considerable practice of this sort among his many other projects. But the new cotton mills offered almost a virgin field. Fairbairn and Lillie of Manchester in the twenties turned their attention to cotton mill architecture and were soon prepared to contract for the complete construction of mills. Rennie had been confronted by general hostility in the flour trade; Fairbairn found a much more enterprising attitude among the big cotton men, for it was clear that their future lay with the planned enterprise. Many of the new mills were fireproof, for this was a true economy. Iron could support a higher payload in buildings just as in ships. Soundness of construction was increasingly called for; not infrequently overtaxed mills crashed to the ground. There were, of course, errors of excessive ingenuity. The attempt was made to use the hollow iron pillars support-

---

[1] H. W. Dickinson, 'Richard Roberts', *T.N.S.*, 1945–47, p. 123.

[2] J. M. Richards, *The Functional Tradition in Early Industrial Building*, London, 1959; S. B. Hamilton, 'The Use of Cast Iron in Building', *T.N.S.*, 1943, p. 139; T. Bannister, 'The First Iron-Framed Buildings', *Architectural Review*, 1950.

[3] Sir Wm. Pole, ed. *The Life of Sir William Fairbairn, Partly Written by Himself*, London, 1877.

ing the floors to circulate steam for heating purposes, but the expansive effect was dangerous to the structure, an extraordinary example of failure to use existing knowledge.[1] Fairbairn and Lillie, in reconstructing old mills, were capable of getting an increase in power of some 20 per cent. The engine, too, was used to operate lifts to improve movement in the factory. When Nasmyth visited Manchester in 1830 he was astonished and delighted at the sight of whole buildings that worked as one grand and perfectly constructed machine.

In chemicals the road to progress was represented by the production of alkalis from common salt rather than the expensive kelp. Leblanc had shown how this could be done; the heavy salt tax was removed in 1823, creating the opportunity.[2]

Well before the railway locomotive demonstrated its powers at the Rainhill trials in 1829, the civil engineers had done much to alter the face of Britain with their improvements to transport and communications. In so doing their inventive faculties were constantly evoked. Telford did splendid work on roads, bridges, harbours, and canals just before the railways made their great challenge.[3] The younger Rennie, sent to Aberdeenshire in 1815 to obtain granite blocks of the unprecedented weight of twenty-five tons, to support the 6,000 tons of iron of the new Southwark Bridge, had to create his own handling methods. Similarly in the twenties the two Brunels, working on the Thames Tunnel, showed a fine combination of ingenuity and physical courage in running their enormous cylinder through the Thames mud. Before the rise of the great engineers and contractors the largest construction establishments were the Royal Dockyards. These were largely unplanned – their elements occupying the ground like a dropped pack of cards. It was natural that recourse should be had to the engineers to improve them.

The railway from the thirties onward brought a redoubling of constructive energy and ingenuity. A considerable number of inventors had struggled to improve a locomotive engine to the point of making it commercially feasible, including Murdoch, Trevithick, Blenkinsop, Murray and Blackett. Stephenson's most original contribution was to use the steam discharged from the piston to improve the draught of the chimney; this steam blast made a strong and continuous supply of power possible. The production challenge involved

[1] Andrew Ure, *The Philosophy of Manufactures*, London, 1835, p. 25.
[2] See below, Chapter 4, section 10, p. 170.
[3] See L. T. C. Rolt, *Thomas Telford*, London, 1958.

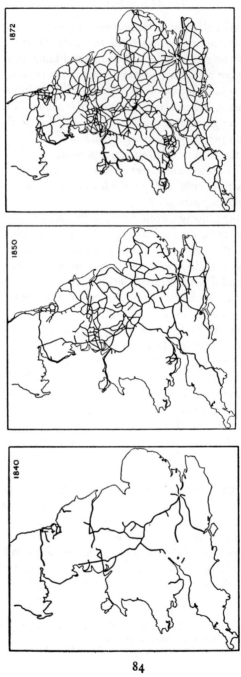

THE GROWTH OF THE RAILWAYS

Source: Wilfred Smith, An Economic Geography of Great Britain, London, 1949, p. 160; Report of S.C. on Railway Amalgamations, 1872

in the engine was formidable: Stephenson founded his own locomotive works in 1824 largely because it was the only way to provide skill and tools better than those available in the collieries. But the greater challenge was that of building the permanent way, considered in close relation to the engine. Telford in the construction of his great canals had not needed to know a great deal about mechanics, but the railway builder had to merge knowledge of the prime mover with the insight of the civil engineer. Engine and road were, as Stephenson put it, like man and wife. He found that he was called upon not merely to demonstrate the feasibility of his engine, but to become, almost incidentally, a civil engineer on the grandest scale; it is not surprising that the first attempt at a Bill for the Liverpool and Manchester Railway failed and Stephenson was told that he had no science to apply. But if he had no science he certainly had insight, as he demonstrated in his concept of building the line over Chat Moss on the principle of an elongated raft. So powerful was his confidence in his own perception that, with no knowledge of the chemical nature of the problems of combustion, he carried his new miners' safety lamp into the strongest pocket of fire damp available.

Watt had sought to solve the problem of accuracy of execution by training and keeping sets of workmen to specialize on particular types of work. Sons followed fathers in the same tasks. But the limits of precision of hand labour were being approached; progress was very difficult when each bolt had to be fitted with its specially made nut. By the thirties and forties it was clear that, to some extent at least, the available skill, which had contained such potential a generation or so earlier, was now a limiting factor. It was necessary for the engineer to design improved methods of manipulating metal.

There was a crying need for standardization of parts and of design. The latter was especially obvious in the case of the railways. The catalogue of the Great Exhibition reflects a great revelling in virtuosity in the handling of metals. The great machine makers had always delighted in precise workmanship; they were, perhaps, the first men to feel the aesthetic thrill of almost perfect mechanical execution. But the time had come to alter the emphasis toward uniformity in workmanship so that the skill of the élite might be multiplied. Both Bramah and Maudslay had been preoccupied with the problem of precision. Maudslay's pupil, Joseph Whitworth, building on the work of his master, made the greatest contribution. He developed his series of standard gauges for obtaining great accuracy and uniformity in the moving parts of machinery. His method of producing a true plane was available by the late forties. The metal planing machine, embodying his ideas, gave not only much greater accuracy than older methods, but made one of the greatest economies in engineering production, reducing the labour cost of truing an iron surface from twelve shillings per square foot to less than a penny.[1] Yet it was by no means easy to secure adoption for the new method. Whitworth, in his missionary zeal, encountered much opposition, partly because chipping and filing had become one of the principal means of training and testing skilled mechanics and was one of the leading aspects of their skill. With these changes came a new emphasis on mechanical drawing.

The whole scale of machines was altering. By 1839 the old tilt-hammer had reached the limit in terms of the size of forging it could effectively strike. Nasmyth's steam hammer, available after 1842, was the response: capable of cracking the top of an egg in a wineglass at one blow, and of shaking the parish at the next. The new hammer meant that shafts and other parts of unprecedented size and exactness could be forged. Further applications soon followed in the form of the pile driver, placing new power in the hands of the civil engineers: a pile requiring twelve hours to drive by the older method now required four and a half minutes. Even more important, the steam hammer revolutionized the process of 'shingling' before the iron from the puddling furnace was rolled into bars. The steam-engine itself was posing new problems: high-pressure steam was the obvious answer to the new demands made upon it. Fairbairn and Hetherington patented their revolutionary double-tube boiler in 1844. With these great extensions of scale and power came the need for ever greater precision of execution.

For these engineers and inventors the standardization of parts and methods was closely related to the question of the labour supply. As their enterprises grew, and as the demands upon them extended, their dependence upon skilled men increased. Bottlenecks could easily develop that could disrupt the flow of work; even more important was the danger of withdrawal of labour. The engineers, in their crucial pioneering position, depended greatly upon their ability to locate talent quickly, and to raise it to positions of responsibility: 'free trade in ability', as Nasmyth called it.[2] But this quick promotion of newcomers was sometimes contrary to the sense of equity of the

[1] Fairbairn, *Life*, p. 46.
[2] *James Nasmyth, Engineer: an Autobiography*, London, 1883, pp. 218, 226.

men. Moreover, trade-union organization was spreading. It was very tempting, in times of good trade, to strike for higher wages. The engineers met these problems partly by direct resistance to the unions, and partly by perfecting methods to reduce their dependence upon labour. Fairbairn, who as an aspiring young man had been excluded from employment by the Millwrights' Society of London, developed his riveting machine in 1837 to defeat a strike. Roberts's self-acting mule was produced under the same conditions; so too were new methods of shaping and welding tubes. The whole development of self-acting machine tools underwent the same impetus. Nasmyth went so far as to argue in 1868 that on balance strikes and lockouts or 'kicks from behind', rather than a leading forward by profit incentive, produced a greater net gain for society. But it must be remembered that this was an engineer's view: his industry lent itself uniquely to quick responses in labour-saving changes.

War and preparations for war must also take their place among the factors bringing forward new inventions.[1] The need to make naval ships more manoeuvrable so that they might 'break the line' inspired the experiments of the Edinburgh banker, Patrick Miller, in 1788, the first ever made with an actual steamboat. In order to improve the manufacture of ships' blocks, Henry Maudslay, working for six years upon plans and suggestions by the elder Brunel and Sir Samuel Bentham, produced a range of no less than forty-four magnificent machines. The labour power required for the task was reduced to less than one-tenth. Even more important, the new machines embodied the prototypes of nearly all the modern engineering tools. The potency of naval demand was illustrated in the case of iron tanks for ships: these produced in small quantities were enormously expensive; Maudslay, given an order for two thousand, made special tools, reducing the cost of rivet holes from seven shillings to ninepence.[2]

Those inventors who could carry their own ideas into commercial execution, managing their own enterprises, could do well: this was the case of many, perhaps most, of the great engineers. But those whose inventions were in metallurgy too often failed to gain. From Cort onward there was a sad procession of men who did great things as inventors, but ended in poverty. Mushet, Crane, Neilson, Heath, and Rogers spent the latter part of their lives in 'a long contention with

---

[1] See Theodore Ropp, *War in the Modern World*, Durham, N.C., 1959; Cyril Falls, *A Hundred Years of War*, London, 1954.

[2] Babbage, op. cit., p. 100.

pirates', unable to obtain redress by law.[1] Their defeat was due largely to the fact that their ideas could only be commercially implemented on a large scale using much capital; they were unable to fill the role of giant entrepreneur.

4. PRACTICE AND THEORY DRAWING TOGETHER

Well before the mid-century there were many points at which the theorist and the practical man were drawing together. A notable example was the invention of photography.[2] From chemistry came the possibility of producing plates so sensitive to light that a picture could be recorded and the image fixed. From optics came the lens, so long used for the study of the immense and the minute. Now it could aid in the production of permanent pictorial records; the photographing of history begins with the Crimean War.[3] There followed the possibility of a new kind of scientific record, and, perhaps paradoxically, the possibility of a new view of nature in the romantic vein, for the camera was capable, at the will of its user, either of a new scientific objectivity, or of a new impressionism.[4]

The new emphasis upon precision further helped to close the gap between the engineers and the philosophers. It was by the performance of the steam-engine that the laws of thermodynamics were suggested. Progress in theory had been delayed by inexactness of workmanship; while pistons fitted to cylinders with very large tolerance the question of the strict relationship between potential power output and a given fuel input was not seriously entertained. As the engine itself gained in efficiency it slowly became clear that there must be relationships of a perfectly general kind between heat and power.

Increased scale, speed, and pressure, all posed new problems of measurement and mathematics. There had long been an adage which reflected the individualism of the old school of craftsmen and engineers who relied upon their own instinctive insights or long established prejudices: 'Two of a trade can never agree.' When Seppings, the brilliant shipwright, sought after 1810 to get his new principles for the stressing of ships adopted, he was met with blind conservatism. In bridge construction the theory available in the early nineteenth

---

[1] *Mining Journal*, 19 February 1859.

[2] Helmut and Alison Gernsheim, *L. J. M. Daguerre: The History of the Diorama and the Daguerreotype*, London, 1956; *The History of Photography*, Oxford, 1955.

[3] *The Radio Times Hulton Picture Library* is perhaps the greatest collection.

[4] See below, Chapter 8, section 8, p. 316.

century was purely geometrical, with little or no regard to the performance of materials. When Telford and Douglas in 1801 proposed a new London Bridge a long debate was started which revealed the chaos of opinion on the theory involved; indeed it was often unpredictable, when the supports of a new bridge were removed, whether it would stand or fall. This attitude could not survive much longer. Bridges and other great works were now of unheard of size, and ever more daring constructions were being adopted. It was Rennie, whose millwright education had been reinforced by study at Edinburgh University, who laid the basis of true theory in bridge building. Stephenson, too, knew the importance of getting at principles. It was for this reason that he so ardently supported the new Mechanics' Institutes. In designing the great Britannia Bridge in the later forties, models were constructed and tested and the best form for the bridge itself deduced mathematically. Yet it remained true that much depended upon straightforward construction and testing as practised by Fairbairn for Stephenson. I. K. Brunel, the most daring engineer of his day, attempting all manner of gigantic constructions on the basis of instinctive judgement, could often succeed, but sometimes failed.

The philosophically minded were making an increasing contribution, turning to the theory of hydraulics and other branches of physics. In ship construction the laws of resistance of hulls to higher speeds, of increasing importance since the elder Rennie introduced the paddle-wheel steamer into the navy in 1819, inspired General Bentham in the late twenties and the thirties to conduct systematic experiments with models in a testing tank. In the forties, David Elder carried the work further.

Not only was engineering being drawn closer to philosophy; through metallurgy it was contacting chemistry, and through it the main stream of science. As industry grew, the limitations of its basic material, iron, became increasingly apparent. Metallurgy, which had always concerned the engineer, assumed a new significance. The railways had begun with cast-iron rails in three foot lengths, an obvious obstacle to rapid growth. In 1835 the British Association commissioned Fairbairn and Hodgkinson to investigate the properties of cast iron, especially as affected by Neilson's hot blast. In 1844, the same body sent Playfair and Bunsen to find out what went on inside a blast furnace; proprietor and professors vied in ingenuity to penetrate the infernal mysteries. Among other observations, they established that over 80 per cent of the fuel was wasted; it was to this challenge that Bessemer and Siemens responded.

The warnings of Babbage about the dangers of stultification of industrial progress by failure to adapt science to industry helped to provoke the founding to the British Association in 1831. It became both meeting ground and tactical weapon for those seeking to bring science and industry together. In addition, the Royal Institution continued throughout the century to act as a sort of university of research, supported by private munificence. The Royal Society of Arts, founded in 1755, was busy encouraging 'Arts, Manufactures and Commerce', sponsoring the Great Exhibition of 1851 to this end.[1] The Prince Consort gave science his ardent support. The English universities were producing a new generation of scholars making it easier for them to find common ground with the practical men. Charles Babbage, Lucasian Professor of Mathematics at Cambridge, in his *Economy of Manufactures* of 1832, aspired to present industrial change in a philosophic light, after visiting the leading manufacturers of England and the Continent. So too did Andrew Ure in 1835.[2] By this time there were men like William Whewell teaching science at Cambridge, and James Smithson at Pembroke College, Oxford, founder of the Smithsonian Institution. By the mid-century the scientists of the schools, like Playfair and Kelvin, were increasingly called into consultation by the engineers. On their side, academics like Hopkins (the teacher of Kelvin at Cambridge) in his experiments on the behaviour of matter under pressure, in an attempt to understand the cooling process of the globe, called in the engineers in constructing apparatus and in giving practical advice. In the forties, University College, London, occupied a unique position in the scientific world, but the older universities were soon to feel the stimulus of the removal of religious tests.

Some of the engineers, in their turn, especially in their established phase, were attracted by natural philosophy; the old aesthetic delight in elegance of design and precision of execution was now joined with the fascination of learning to express their engineering experience in terms of general laws. Fairbairn and other engineers published many works in the interest of generalized knowledge. By the sixties, though there was no formal alliance as in Germany, the engineers and the academics were drawing closer together, at least in terms of the outstanding men of both groups.

[1] See D. Hudson and K. W. Luckhurst, *The Royal Society of Arts, 1754–1954*, London, 1954.
[2] Andrew Ure, op. cit., 1835.

5. THE DEBATE ON SCIENTIFIC AND TECHNICAL EDUCATION

But not everyone was confident that the renaissance of academic science was going fast and far enough. By the fifties, condemnations of the 'monastic teaching' of the universities were increasingly heard.[1] Not a single government department was prepared to collaborate with the commissioners of the Great Exhibition. George Birkbeck's 'Mechanics' class' at the Andersonian University had become in 1823 the Glasgow Mechanics' Institution, giving intelligent artisans the opportunity of learning the new science and technology.[2] But the Mechanics' Institutes elsewhere could not maintain this objective.[3] Damaging comparisons between the theoretical grasp of French and German workmen with those of England were made after the Paris Universal Exhibition of 1855. There were even more pointed remarks after 1867. As Lord Ashburton put it, Britain had gained her initial advantage from instinctive genius; this was no longer enough. In 1855, in a lecture before the Prince Consort, Faraday stated that the training of the mind by the older classical education still left men wholly incapable of understanding simple questions in chemistry or mechanics, and indeed may have impaired their powers. The electric telegraph, the steam-engine and the railroad, he said, had been developed by men outside traditional education. Gladstone was perhaps the most striking example of this widening breach between men raised in the classics and those whose minds and characters were formed by the new science, for Gladstone, endlessly cogitating theological refinements, willing to expatiate on any aspect of church and state, would have nothing whatever to do with the new knowledge.[4]

The resistance to the introduction of an effective system of technical education, though passive, was powerful.[5] The traditional landed interest saw no great need to nurture a new set of institutions that would further hasten their own eclipse. The universities continued to do little in sponsoring research programmes; the President of the Chemical Society in 1872 blamed the universities for the fact that

[1] For a survey of the state of English industry, science, and education at this time see J. Vargas Eyre, *Henry Edward Armstrong 1848–1937: The Doyen of British Chemists and Pioneer of Technical Education*, London, 1958.

[2] Thomas Kelly, *George Birkbeck, Pioneer of Adult Education*, Liverpool, 1957.

[3] See below, Chapter 8, section 7, p. 310.

[4] See Martin Lowther Clarke, *Classical Education in Britain, 1500–1900*, Cambridge, 1959.

[5] J. Blanchet, *Public Opinion and Technical Education in England, 1867–1906*, Oxford D.Phil. thesis, I.T.A., 1953–54, no. 552.

there was less chemical research in England in the early seventies than there had been in the fifties.[1] The industrialists, or rather an innovating minority among them, had provided both the initiative and the resources for the inventions that underlay economic growth, and continued to do so pretty well to the end of the century, but it is hardly surprising that those who had made no great contribution through new ideas should feel no great need for new educational development. Even those with a strong innovating bent often felt that it would do no good to try to institutionalize teaching and research dealing with industry. But Faraday and Huxley found some allies among businessmen who were becoming frightened of German rivalry.[2]

The question of how science and technology should be financed disturbed the liberal minded: *The Economist* took up the duty to 'expose the ambition and greediness of those who beg in the name of science'.[3] The upshot was the effective denial of state assistance on the scale required in the critical period from the fifties to the eighties, when the new basis for the meeting of the challenge from newly industrialized rivals ought to have been laid.[4] Yet, though belated and slight, there was some progress. The School of Mines and the Royal College of Chemistry were founded in 1845, helped into being by the tour of England made by the German chemist, Liebig, in 1842. Men like William Lyon Playfair campaigned for an increase in technical instruction, demonstrating what could be done and pressing for greater support.

6. THE NEW TECHNOLOGISTS

Though Faraday was conducting important researches into alloys, it was Bessemer and Siemens who made iron and steel available on a new scale. Their careers were in significant contrast to those of the mechanical engineers who had held the palm until the mid-century. Bessemer and Siemens were experimentalists first, who became industrialists in order to demonstrate the commercial soundness of their

[1] Dr Frankland, Presidential Address, *Trans. of the Chemical Society*, 1872, p. 20.

[2] George Haines, 'German Influence upon Scientific Instruction in England, 1867–1887', *V.S.*, 1958.

[3] 'Beggars for Science', *Econ.*, 21 February 1857, p. 196.

[4] See D. S. L. Cardwell, *The Organization of Science in England*, London, 1957; S. T. Cotgrove, *Technical Education and Social Change*, London, 1958; *Report on Technical Education, Schools Inquiry Commissioners*, 1867. P.P., vol. XXVI, p. 261; David H. Thomas, *The Development of Technical Education, 1851–1889, with Special Reference to Economic Factors*, London Ph.D. thesis, 1940.

processes. Their discoveries did not come to them in the ordinary course of business, as they sought to meet this demand or that of their customers. They rather looked to the needs of the economy as a whole, sought by study and experiment to provide a means of meeting them, and then, as the final development, became producers themselves.

William Siemens had been trained in the technical colleges and universities of Germany, with the explicit aim of making advances in the enormous new field of science and technology.[1] His chronometric governor, to regulate the action of machinery more effectively than Watt's had done, brought him in his twenties to the attention of English scientific circles, over which Faraday presided. By 1844, the inventiveness of William and his elder brother Werner had almost run riot, with schemes for the improvement of paper making, new modes of ship propulsion, winged rockets and flying machines, new kinds of locomotives, railway equipment of one kind or another. In addition, because of their successes, they were approached by others who wanted their own inventions taken up. It was time to call a halt.

William concentrated on the immense range of problems concerned with the efficient application of heat. The most obvious field was that of iron production. Here the challenge was: how could the method of processing iron be improved in order to eliminate the great waste of fuel? Both Bessemer and Siemens were seeking to improve and control the combustion process. In a sense both were rule-of-thumb innovators, standing, in spite of technological training, on the threshold of the new age in which the chemist rather than the heat engineer becomes dominant in metallurgy. Bessemer's answer to the problem was to drive a powerful blast of air through the molten metal from below.[2] This gave much higher temperatures, correspondingly more effective in burning out impurities, and resulting either in pure malleable iron or steel of various qualities. There were of course difficulties in altering the construction of iron works. Moreover Bessemer's process still left unsolved the great problems of the presence of occluded oxygen and of phosphorus, making the metal industrially useless. Robert Mushet, metallurgist, quickly saw how to deal with the former difficulty, thus taking a step beyond Bessemer the inventor. The oxygen was

[1] See Sir William Pole, *The Life of Sir William Siemens*, London, 1888; also J. Munro, 'Sir William Siemens', *Journal of the Society of Telegraph Engineers and Electricians*, 1884, p. 447.
[2] Sir Henry Bessemer, *Autobiography*, London, 1905.

removed by causing it to combine with Mushet's special compound of iron, manganese, and carbon.[1] Using appropriate, non-phosphoric, ores the success of the Bessemer-Mushet process was enormous, both in profits for the inventor and in lower costs.[2] Nasmyth at the British Association in 1856 held aloft a piece of Bessemer's iron, announcing, 'Here is a true British nugget'. Siemens's open hearth process consisted of a means of leading back to the furnace the gases formerly wasted, so that much more complete combustion was obtained, thus purifying the pig and scrap contained in a large bath in the furnace. Again great economies were achieved. Bessemer published his results in 1856; by the early sixties he himself and John Brown had adopted the process. Siemens's method was first used industrially in 1861. In this way the sixties and seventies were provided with a cheaper source of iron and steel, though the puddler still turned out over two million tons well into the eighties. Even more important, steel was now available as an effective general rival to iron, though its great victory did not come until after the seventies. The Siemens-Martin product had the especial merit of uniformity; it was subjected in the sixties to Whitworth's process of compression by hydraulic machinery, producing a steel of unique quality and toughness – once more shafting and constructional parts could assume a new scale, and precision of workmanship too could express itself anew. Siemens steel was also especially suitable for tinplating. The last great leap, the Gilchrist and Thomas method of removing phosphorus by lining the furnace with limestone, was made in 1879; lime, so important in the general history of chemistry, now made possible one of the greatest of metallurgical breakthroughs.[3] The formation of the Iron and Steel Institute in 1869 provided a valuable forum for the exchange of ideas.

In the third quarter of the century the new engineering and the new metallurgy combined to produce an extraordinary burst of construction on the largest scale. The younger Brunel set the pace with his steamships, of which the Great Eastern was the most fabulous and most disastrous to its backers.[4] But less dramatic change was going forward. Railways were extending at an extraordinary rate, both at home and abroad; steel rails were available from 1860 onward. Ship construction

---

[1] J. C. Carr and W. Taplin, *History of the British Steel Industry*, Oxford, 1962, p. 22.

[2] The troubled relations between inventor and metallurgist are described in Fred M. Osborn, *The Story of the Mushets*, London, 1952, part II.

[3] R. W. Burnie, *Memoir and Letters of Sidney Gilchrist Thomas*, London, 1891; Lillian G. Thompson, *Sidney Gilchrist Thomas*, London, 1940.

[4] James Dugan, *The Great Iron Ship*, London, 1953.

was changing to iron, and by the late sixties, even to steel. Armour plate and ever more formidable ships of war appeared. Indeed, in this age of peace, much was owed to preparations for war. Siemens noted with concern: 'We have scarcely recovered our wonder at the terrific destruction dealt by the Armstrong gun, the Whitworth bolt, or the steel barrel, consolidated under Krupp's gigantic steam hammer, when we hear of a shield of such solidity and toughness to bid defiance to them all.' He solaced himself with the reflection that the civilized nations of the earth could not work these engines of offence and defence for long 'without effecting the total exhaustion of their treasures'.[1] The power of ships leaped upward with compound engines, coming into use in the late sixties and seventies. The constructors of docks and bridges produced a continuous stream of marvels, though engineers were chastened by the Tay Bridge disaster of 1879.

As the number of names of inventors grows it is impossible for the historian to specify them; invention was in the air. Herbert Spencer, better known as the philosopher of individualism, offered an ungrateful world a flow of salt-shakers, candle-extinguishers, jugs, invalid chairs, watches, and other ingenuities. Inventions affecting railways alone defy recounting. With such plenitude a curious air of anonymity spreads over invention, for there are too many discoveries to mention, let alone remember. The age which saw the engineer as hero is seen by posterity in this curious impersonal light, so greatly had heroes multiplied.

Each successive invention in engineering was becoming less potent in its effect on the general picture. Even before the seventies the age of the Stephensons, Rennies, Brunels, and Fairbairns was ending. The renovation by the engineers of transport and the mechanization of the traditional processes, chiefly in textiles, had both been more or less carried through by 1880. Thereafter the role of the engineer changed. He lost the initiative in the sense of originating new concepts capable of changing society, but yet grew to ever greater stature as his monuments became ever larger. New revolutionary ideas must now come not from the manipulators and empirics, but from the higher reaches of mathematics, physics, chemistry, and electricity.

But in spite of very great achievements, the impression grows of a failure of ingenuity in British industry from perhaps the seventies onward. The provision of scientific and technological education had

[1] William Siemens, 'Presidential Address to Mechanical Sciences Section, *R.B.A.*' 1867.

been much less effective than many had hoped. Huxley was still thundering in 1884: 'The whole theory on which our present educational system is based is wrong from top to bottom'; there were many who shared his view. Even the scientists and engineers who were forthcoming could often not be effective in the situation of the time because industry could not locate and use them. The incentives to choose such a career were not great, for social prestige still went very much to the traditional occupations of aristocracy – the church, the universities, the armed forces, law, diplomacy and the foreign service.[1] Britain had stood in so advantageous a relationship with so many primary producing countries for so long that her response to the challenge of industrial initiative could be delayed until her rivals were very formidable. Finally, with expansion checked by the difficulties of the seventies and eighties, there was less scope for innovation, just when it was most needed.

Yet Britain, though failing to make an adequate response to the need for scientists and technologists, could draw upon the ingenuity and vigour of foreigners who, disgusted or discouraged with the political and social situation in their own countries, converged on Britain. Graduates of the German polytechnics, wishing to try their arm free of authoritarian government, and Swiss engineers and technologists, seeking a wider area of operation, moved to Britain. The security afforded by the developed state of the English patent law was a further attraction. Siemens was perhaps the greatest of these immigrés, but there were others of importance, including Hans Renold the Swiss and Henry Simon from Silesia, who converged on Manchester, Renold to perfect and produce the precision chain for the transmission of power, and Simon to displace the traditional grinding stones in British mills with rollers and so to revolutionize mill design, and to build the first by-product coke oven.[2] These men made a notable general contribution to education and culture, and in particular fought for the improvement of British scientific education.

7. THE NEW SCIENTISTS

The intuitive engineer had one further victory with the steam-engine before the theory of the subject was revolutionized. John Elder, in

---

[1] See below, Chapter 8, section 3, p. 295.

[2] Basil H. Tripp, *Renold Chains*, London, 1956; Anthony Simon, *The Simon Engineering Group*, Stockport, 1953.

1854–56, following Hornblower and Woolf, flying in the face of accepted 'caloric' theory, which precluded the conception of heat as a form of motion, produced his compound or combined high and low pressure engine.[1] The result was a fuel economy of from 30 to 40 per cent, making possible further revolutionary developments in ship design and performance. With the availability of the laws of thermodynamics, made explicit by the efforts of Joule and Kelvin, a new phase of engineering history began to be realized.

It was Joule, the son of a wealthy Manchester brewer, who had come under the spell of John Dalton, who showed experimentally between 1843 and 1878 that it was possible to express quantitatively the mechanical equivalent of heat. He had found that if water was churned with a paddle its temperature rose; he set about seeking the relationship between the amount of work done and heat generated, and arrived at a new fundamental theorem, the identity of heat and energy. Joule occupies a position of great interest in the history of technology. He was not a 'profit-seeker', rummaging in science as so many had been, for he was already wealthy and had no active interest in the family business of brewing. Yet he applied his liberated energies to a point of inquiry of crucial economic significance, for the potential of the engine now depended on improved theoretical grasp.

The new theory did not, however, react quickly upon engineering practice. Indeed, engine construction proceeded much as before, though the compound engine could be treated with greater confidence. The laws of thermodynamics found their real economic application from the eighties, with the internal combustion and turbine engines, and especially with refrigeration machines. There were soon many types of the latter available, with their great potential for preserving foods over time and space.

In chemicals the stage was now set for very rapid expansion.[2] Knowledge still rested mainly upon the researches and initiative of producers. Indeed down to the end of the century chemistry continued to be largely empirical and commercial, concerned with the purification of natural substances and their analysis, and without any viable understanding of the structure of the atom.[3] There was the further limitation that the necessary mathematics (particularly in Series) was slow in development, a serious impediment to the expansion of heavy

[1] W. J. Macquorn Rankine, *Memoir of John Elder*, Edinburgh, 1871, p. 18.
[2] See below, Chapter 4, section 10.
[3] J. D. Bernal, *Science and Industry in the Nineteenth Century*, London, 1953, p. 14.

chemical production, bound up as it was with factors of time and quantities, a set of problems not mathematically treated until the eighties. But in spite of these difficulties producers could make great extensions of output. William Gossage, apprenticed to a chemist and druggist, had in 1823 begun his long series of patents which are in a sense a summary of the alkali trade. He was hampered by engineering difficulties, for it was becoming increasingly apparent that chemical discovery depended very much, as did mechanical success, upon efficient execution. Moreover, when laboratory results are reproduced on a production scale, unexpected things may happen: the manufacturing chemist could not operate as did Wollaston, who, when asked to show a visitor his laboratory, produced all he required on a small tray. In 1836 Gossage patented his condensing towers which disposed of effluvient by absorption in coke, and made available the valuable by-product of muriatic acid. In 1853 he perfected his process for producing caustic soda, which revolutionized soap making. By the fifties heavy chemical production could go forward on a new scale. Widnes became the centre of great production and adaptive invention. Oldbury from the forties onward was the centre of phosphorus manufacture for matches. The chemical producer, as the millowner had been a generation earlier, was now confronted with the dangers of obsolescence. But it was a danger associated with adaptation rather than a change in the principal process: the Leblanc method held its own against that of Solvay, discovered in 1863, down to the later seventies.

But though the expansion of chemical output owed little to the scientist down to the seventies, the beginnings were appearing much earlier, as in engineering, of a new situation. By the mid-century chemists were beginning to appreciate the importance of Dalton's theory that matter consists of atoms combined in molecules in mathematically precise and constant proportion, and to consider how to give substance to such conjectures. Liebig, the son of a Darmstadt druggist, had created a great stir in the thirties and forties with his stress on chemical law, education, and research, and the importance of organic chemistry for industry and agriculture. Though Liebig's own patent fertilizer was a failure, it was the beginning of a great new branch of the chemical industry.

William Perkin built on Liebig's ideas.[1] At London University Chemical School, in trying in 1856 to produce quinine by altering the

[1] R. Brightman, 'Perkin and the Dyestuffs Industry in Britain', *Nature*, 1956, pp. 815–21.

atomic structure of certain coal tar derivatives, he discovered mauve, thus revolutionizing the dyeing industry. Perkin was especially fortunate that purple, the widowed Queen's favourite colour, had defeated the traditional dyers. He also gave a great impetus to the new synthetic organic chemistry. James Young, with his pioneer work in the manufacture of mineral oils, both contributed to the elaboration and interplay of the elements of which the chemicals industry was composed, and liberated those of mankind who had no access to gas from the great expense of using natural oils as illuminants.[1] Yet there was a long way to go before the importance of chemistry as a profession was realized: as late as 1872 a society formed to promote academic research in chemistry failed.[2]

But changes were afoot. In 1840 William Whewell had used for the first time the word scientist. To him it seemed right that, as practitioners of the arts had long been known as artists, it was now proper that those whose concern was with science should be called scientists, rather than savants, virtuosi, or 'natural philosophers'. By the sixties Thomas Huxley was stating in the strongest terms the view that the understanding of the natural world, sought after as an end in itself, would also yield the greatest uncovenanted benefits in terms of new wealth.[3]

8. INDUSTRY FROM THE LABORATORY

Electricity is unique in having sprung direct from the laboratory, without previous roots in traditional practice as had been the case in engineering, textiles, and chemistry. In 1831, Faraday demonstrated the principle of electrical induction.[4] But electricity was by no means new to scientific men. As early as 1800 Alessandro Volta had communicated to the Royal Society his invention of the 'pile' — a discovery basic to further research, for, unlike Leyden's jar which gave off its whole charge instantaneously, the voltaic pile could provide continuous current. With it the use of electricity in chemical experimentation began — for Davy it was the 'golden branch' giving access to new vistas.

[1] See John Butt, *James Young, Scottish Industrialist and Philanthropist*, Glasgow Ph.D., 1964.

[2] Sir Henry Roscoe, *Autobiography*, London, 1906, p. 43. See also Chapman A. Chaston, *The Growth of the Profession of Chemistry during the Past Half-Century, 1877–1927*, 1927; Wemyss Reid, *Memoirs and Correspondence of Lyon Playfair 1818–1898*, London, 1899.

[3] For Huxley's views on many matters see Cyril Bibby, *T. H. Huxley: Scientist, Humanist, and Educator*, London, 1959.

[4] Thomas Martin, *Faraday's Discovery of Electro-Magnetic Induction*, London, 1949.

Nicholson and Carlisle in 1800 had accidentally decomposed water into its elements; there followed a fury of electro-chemical experiments. In 1819 the Dane, Oersted, discovered that a wire carrying an electric current created a magnetic field; Ampère set about the development of electromagnetic theory. Finally Faraday, the blacksmith's son who became the prince of experimental philosophers, made possible the large-scale supplying of electricity, by demonstrating the generation of an electrical voltage by the rotation of a coil in a magnetic field. All these men were *savants*; with electricity, science for the first time held the initiative in social change.

The first commercial uses of electricity were in the deposition of metals and the electric telegraph.[1] By 1838 the Elkingtons of Birmingham had deposited both gold and silver on ornaments, and had taken out a patent for electro-plating copper and brass with zinc. With the financial aid of Josiah Mason, the pen maker, the firm flourished. As early as 1842 Robert Davidson had shown the first electric locomotive built by the Edinburgh and Glasgow Railway Company.[2] But, though it hauled six tons at 4 m.p.h., its battery system was uneconomic. Electricity could not challenge steam in transport until cheap current was made generally available. But the electric telegraph was of much greater importance. The promontories of Europe had for generations supported towers with long wooden signalling arms. Now the hectic operation of this visual system (brought to an amazing level of manipulative virtuosity) gave way to mysterious invisible impulses. Military men grasped the importance of the new system at once.

Out of the electric telegraph came the electrical industry proper.[3] The production of wire and cable, insulating devices and control gear, all prepared the way for the general provision of electricity. Communications were revolutionized throughout the world by the telegraphic cable in the decade down to 1875.[4] The pioneer phase of cable laying in the fifties and sixties provided a unique example of cooperation between projector, manufacturer, government, and scientist, for though the beginning was made on a curiously optimistic and simple-minded basis, it soon became clear that operations of this kind gave a new significance to the laws of electricity. The discovery, after the failure of

[1] See below, Chapter 4, section 11.
[2] F. J. G. Hant, 'The Early History of the Electric Locomotive', *T.N.S.*, 1949-50 and 1950-51, p. 153.
[3] See Willoughby Smith, 'A Résumé of the Early Days of the Electric Telegraph', *Journal of Telegraphic Engineers*, 1881, p. 312.
[4] G. R. M. Garratt, *One Hundred Years of Submarine Cables*, London, 1950.

the first Atlantic cable, that the purity of the copper used affected the conductivity of the cable, linked the electrical industry with chemistry and metallurgy. It was cables that posed for the first time, in a way that officialdom could not ignore, the problem of the role of government in the sponsoring of science, and the interrelationships of its practitioners.[1] Joseph Swann demonstrated the first incandescent electric lamps in 1878.

The output of the electrical industry grew substantially, but it did not rise to the first rank until the last decade of the century. Though there had been experiments as early as the fifties in lighthouses, there were many difficulties before the economic provision of electric light was possible: indeed the President of the Society of Telegraphic Engineers remarked in 1881 that its progress 'has not been a very brilliant one'.[2] Even further removed was the use of electricity as a source of power. But once well begun, in the eighties, the growth of the electrical industry was to proceed at a rate inconceivable by the standards of fifty years earlier.

In the phase 1815–75, industrial practice grew from below, in a step-by-step attempt to meet new needs and to reduce costs. The expense of invention and development could be met on the same principle. For they were still fairly moderate, with much still to be learned, especially in chemistry, biology, metallurgy and electricity, by the use of modest equipment on a laboratory bench; the new age of massive apparatus was not yet. In toolmaking the costs were relatively small, and were usually embodied in the price charged for machines sold. In civil engineering there was still scope for insight without vast development costs; moreover, as with machine making, such costs could be embodied in contracts with the sponsors of the work. In chemicals, the very large gains of the Leblanc process could pay for much research; chemical study was largely endowed by this one great discovery.

But the effort thus sustained tended to run in a groove – the fundamental nature of matter was left to the amateur. In metallurgy the enormous prizes awaiting the men who could solve the problems of the day served both as landmarks and incentives; moreover, iron producers were prepared to help by the loan of facilities. But electricity revealed the shape of the future, springing as it did from science. The

---

[1] *Report on the Construction of Submarine Telegraph Cables*, P.P., vol. LXII, 1860, p. 591.
[2] *Journal of the Society of Telegraph Engineers and Electricians*, 1881, p. 160. See also, Henry Schroeder, *History of Electric Light*, Washington, 1923.

new generation after 1875, borne along by the new physics, chemistry, and electricity, could no longer safely rely, as in 1815, upon the producer-inventor, important though his sense of immediance continued to be, but was obliged to think increasingly of positive sponsorship of scientific investigation.[1] Should this come from the very large productive units, with monopolistic tendencies, that were inherent in the new modes of production, or should it come from the state?

Formerly the problem of inventors had been how to improve upon the performance of animal or simple natural power in doing simple manipulative and carrying tasks. Now the natural world was to be exploited in new ways. Science was to become increasingly specialized and remote from public comprehension, with its own highly trained custodians. It was to move, also, with the work of men like Maxwell, away from engineering concepts toward new modes of thinking of a non-visual kind.[2] Mankind, with Britain in the lead, was approaching the end of an immemorially long phase of occupying empty territory and bringing into use unworked natural resources, especially agricultural land. Now a new era in the manipulation of matter was dawning.

Such new, highly specialized, modes of thought, were not confined to the scientists. Increasing sophistication of thought about the natural world, eagerly applied to productive methods, created problems that depended for their solution upon growth of specialized knowledge in fields other than natural science. Man's powers of business, social, and political organization were challenged. To achieve new arrangements meant new and more viable social theory. So too with respect to the individual. The effort to help him in his new situation made it obvious that the older simple psychology of the eighteenth century was not enough. The age of the powered machine was inducing the age of the specialist in all aspects of human life.

[1] For a statement of the continuing scope for the individualist-inventor see John Jewkes, *The Sources of Invention*, London, 1958.
[2] J. G. Crowther, op. cit., p. 309.

CHAPTER FOUR

The Men of Business

I. THE BUSINESS COMMUNITY

It is probably not far from the truth to say that the period from 1815 to 1885 in Britain represents that range of human experience in which individual economic initiative had its greatest opportunity to operate upon men and things, and in so doing to re-make an ancient society. Before 1815 the technical capacity for massive manipulation of the factors of production by the individual was lacking; after 1885 the individual was merged in the joint-stock undertaking, with an incipient trend toward monopoly and official intervention.

The business man is at once the most important and most enigmatic character in a free enterprise economy of this kind. It was with him that the principal initiative lay: his decisions to embark or to remain passive, and his ability to estimate and manipulate, were the proximate determinants of growth and change.

Down to 1815 only a few men had been confronted with the new kind of commitment demanded by industry. For centuries there had been great traders, often with immense resources, together with many petty master craftsmen, in quasi-paternal relationship with their employees. But now it was necessary for those seizing the new industrial initiative to revolutionize their outlook. Instead of dealing largely in liquid resources, they had to create very great elements of fixed capital, and re-embody these in the product over future time. Instead of hiring the crews of ships and a few clerks, it was now necessary to mobilize a new labour force, and subject it to a discipline utterly alien to the immemorial experience of man. A whole new psychology of business, which embraced a psychology of labour, was demanded.

The two elements commingled, for the attitude to labour was to an important degree derived from the attitude to investment. There was a natural tendency for the new man of business to be very greatly concerned about his capital — his own stake in the firm, and to see labour as an adjunct to be hired and manipulated according to the needs of the enterprise.

Both the initiatives of the competitive business man and his omissions might be mistakes, and often were; indeed it is hardly surprising that some contemporary theorists after the mid-century, like Herbert Spencer, could fit business enterprise into a Darwinian system in which the fittest were selected and the less fit rejected. Yet Darwinism is by no means a complete key to business history even in Britain in the nineteenth century, for it was not easy for the business man to estimate the trend of circumstances and to adjust to it. It is certainly true that men frequently were able to alter the business organisms over which they presided, in the light of the experience of other men, including both the successful and the unsuccessful. The flat refusal of Thomson of Clitheroe in the forties to move away from the exclusive production of short runs of his continentally famous expensive prints ended in insolvency, and thus provided a notorious warning to the trade of the need to follow market changes.[1] Others learned much from the dramatic engineering failures of the younger Brunel, for they were public knowledge. But it might more often occur that business men did not know the experience of others and could not therefore adjust their own behaviour. Indeed one of the characteristics of a free enterprise system is to keep business policies secret.

Hence it became necessary for the business community to find some compromise between the need for secrecy and the need for data. This was done in many ways: a body of what might be called 'community experience' grew up, partly on a quasi-personal basis, partly within groups that somehow had discovered some kind of common identity, partly through institutions like the banks and insurance companies, and partly through long periods of business conversation, spent exploring the mind of others with minimum disclosure from oneself. Bankruptcy proceedings too might be very revealing. The formation of professional and trading associations like the Iron and Steel Institution (1869) led to some degree of pooling of information that was commercially useful, and to the formation of attitudes. Recruitment of management was specially significant in this regard, as when the employees of the greater iron and steel masters, like Siemens and Bessemer, sought positions of responsibility elsewhere. The coming of limited liability gave rise to the question of governmental responsibility for business

---

[1] Sir. T. W. Reid, *Memoirs and Correspondence of Lyon Playfair, First Lord Playfair of St. Andrews*, London, 1899, p. 53.

behaviour. How far should the state, in sanctioning the new forms of organization, enforce disclosure of company affairs through the publication of balance sheets? On this issue *The Economist* equivocated for almost a quarter of a century.

As enterprises and transactions extended, the man of business found that he had to think harder about the concepts used in calculating profits; soon a separate profession of accountants had established itself.[1] With the increasing necessity for business men to act together in the interests of a particular region, Chambers of Commerce came into being, with, in 1860, a national Chamber.[2] With the need of the trader and industrialist to be able to delegate risks, insurance developed into a major profession with many branches.[3] The companies evolved an attitude of collective responsibility towards their customers.[4] All these were aids to the business man in assessing his position.

Partial knowledge could be dangerous. Shipowners were notorious in following one another in bursts of optimism and pessimism, for if they were to share in a rising freight market they had to place orders for ships well in advance so that they might be in a position to compete to the best effect in the new situation. 'As you know', remarked a former President of the Chamber of Shipping, 'shipowners are very gregarious; when they see other people building, they build too.'[5] This was true, to greater or lesser degree, of all heavy capital users. Estimates of the great crops could be very far wrong, especially in cotton and wheat. Again, much of the data gained about the experience of others was of a kind not to be easily transferred to new situations. The first attempts to use the Bessemer process in Scotland were not successful; this provoked an excessive conservatism which seriously and unjustifiably delayed progress well after the initial difficulties had been removed.[6]

---

[1] A. C. Littleton and B. S. Yamey, eds. *Studies in the History of Accounting*, London, 1956; Nicholas A. H. Stacey, *English Accountancy 1800–1954*, London, 1954.

[2] G. H. Wright, *Chronicles of the Birmingham Chamber of Commerce*, 1910; M. W. Beresford, *The Leeds Chamber of Commerce*, Leeds, 1951; Arthur Redford, *Manchester Merchants and Foreign Trade 1850–1939*, Manchester, 1934, 1956; A. R. Ilersic and P. F. B. Liddle, *Parliament of Commerce: The Story of the Association of British Chambers of Commerce, 1860–1960*, London, 1960.

[3] D. E. W. Gibb, *Lloyds of London: A Study in Individualism*, London, 1957; George Blake, *Lloyds Register of Shipping, 1760–1960*, London, 1960; P. G. M. Dickson, *The Sun Insurance Office, 1710–1960*, London, 1960. There are no less than forty titles under 'Insurance' in the list of histories published in 1959 by the Business Archives Council.

[4] Though this did not come at once: see W. Webster, 'The Moral of the Albert Life Assurance Company', *C.R.*, 1869.

[5] *Royal Commission on the Depression*, 1886, Evidence of John Williamson, Q. 11, 219.

[6] I. F. Gibson, 'The Establishment of the Scottish Steel Industry', *S.J.P.E.*, 1958.

Nor must we imply that all men of business, even in this period, were imbued with a restless and assertive spirit of innovation. Even at times of great change, many important examples can be found of men who presided over large and flourishing enterprises, but without exercising any great daring.

If business men can arrange a sufficiently large combination among themselves they can affect events by mutually sustaining each other in corporate error or in organized distortion of the economy, so that the losses are passed to others, often in such a diffused way that the community may not be aware that they have occurred. In fact, both the drive towards a competitive society and the attempt to organize positions of mutual safety are always present.[1] But it is clear, by and large, that though signs were apparent by the eighties of a reversion to the much older and more typical system of combination and control, the years from 1815 to 1885 were a period in which individual economic initiative was more important than the urge to combine.

The relations between business men were not confined to the gathering of data and the driving of bargains. They were obliged to come together for the provision of common services then thought not to be a governmental concern, especially in the extension of harbour and dock facilities.[2] Even more important was the need to evolve a code of behaviour. A system of self-regulation, perhaps deriving from the ancient guild principle, had been evolving for a long time in the great markets. The Stock Exchange, the Baltic Shipping Exchange, and the vast commodity markets (in London in Mark Lane dealing in corn, and in Mincing Lane, the centre for sugar, tea, coffee, indigo; in Liverpool the cotton and corn markets) were all supervised by the traders, without governmental intervention.[3] The Bank of England, too, was an example of the business community learning, after long trial, to regulate a crucial aspect of its life.[4] This ability of British business men to manage their great market-places, free of state intervention, contributed much to the expansion of the nineteenth century.

Because of their freedom, allowing the full operation both of judgement and speculation, the great markets were the barometers of the business mind as it surveyed the international scene, and as it

---

[1] See Clapham, op. cit., vol. II, pp. 198–205 for various combinations and lobbies.

[2] *Mersey Docks and Harbour Board: Business in Great Waters 1858–1958*, London, 1958.

[3] C. Duguid, *The Story of the Stock Exchange*, London, 1901; J. G. Smith, *Organized Produce Markets*, London, 1922; Thomas Ellison, *The Cotton Trade of Great Britain*, London, 1886, part II.

[4] See below, Chapter 6, pp. 201, 202.

sought to estimate domestic and foreign trends. This did not preclude the development of elaborate schemes based not upon the realities of the situation, but upon guesses about the guesses of others. The holding of stocks became one of the recognized modes of holding assets, so that commodity prices might rise or fall because of attitudes toward other investment outlets. In this way the money market, the commodity markets, and indeed all other markets were linked.

Down to the sixties business was mainly conducted either by the individual owner-manager, or by simple partnerships. Though the latter might well involve passive provision of capital, on the whole the connection between ownership and active management was close. But this system could be a serious limitation to the growth of business. It made for discontinuity, for the English partnership ended when any of its members died. By limiting the raising of capital to a relatively few men it could be most cumbersome. Finally, with each partner liable for the debts of the partnership to the full extent of his resources, the new extensions of business could be very risky. The solution was the creation of impersonal joint-stock corporations, in perpetuity, appealing to the public for capital, with a committee of directors in charge of policy, though responsible to the shareholders, and with the liability of each limited to the value of his subscribed shares. The new system was made possible by the Limited Liability Acts of 1855 and 1862.[1]

There was a good deal of resistance by those who feared that the creation of impersonal perpetual corporations would lead to the evasion of responsibility; they were overborne by those who advocated escape from old limitations. New companies began to come into being, at a sharply accelerating rate, contributing to the booms ending in 1866 and 1873, involving both home and foreign investment. Though the advocates of the old principles of individual responsibility and initiative had been obliged to come to terms with new conditions, the old individualism had not been greatly encroached upon by 1880. But depression was demonstrating the need for greater staying power in companies than the old basis afforded.

[1] See B. Carleton Hunt, *The Development of the Business Corporation in England 1800–1867*, Cambridge, Mass., 1936; George Herberton Evans, Jr. *British Corporation Finance 1775–1850; a Study of Preference Shares*, Baltimore, 1936; H. A. Shannon, 'The Coming of General Limited Liability' and 'The Limited Companies of 1866–1883', and J. B. Jefferys, 'The Denomination and Character of Shares', all in E. M. Carus-Wilson, ed., *Essays in Economic History*, vol. I, London, 1954; J. Saville, 'Sleeping Partnership and Limited Liability 1850–1856', *Econ., H.R.*, 1956; G. Todd, 'Some Aspects of Joint-Stock Companies 1844–1900', *Econ. H.R.*, 1932.

The new business group emerging in Britain in the early nineteenth century was the result of a far higher order of social and economic flexibility than was present in the rest of Europe. The parts of Britain in which the new business men appeared in greatest number (the north and the midlands, and the Clyde area), combined the advantages of an old society, with much political experience, great stability, and developed economic institutions, with the initiative of a new one. Industrial and trading opportunities appeared in the pastoral, iron and coal-bearing north, where guild control of industry, chartered company monopoly of trading, and the landed, aristocratic tradition derived from intensive arable agriculture, had all been least developed, and could have little serious inhibiting effect. In a sense this first of industrial revolutions took place in a frontier area, long inferior to the south in point of wealth and population, inhabited by men little accustomed to debilitating influences.

But business men by their own actions had created a new situation by the seventies and eighties. By this time a hereditary element had developed in the industrial sector which, though bringing continuity and experience, also brought an insensitivity to new challenges. The fourth quarter of the century called for a new response no less vigorous than that of the first; in many firms it was not forthcoming. In general the established industrialist was not very sensitive to the need for scientific and technical education. Nor was there a very lively appreciation of the threats to British industrial predominance developing in countries where the urge for assertion had not been impaired by success. Finally, the availability of large invisible earnings abroad obscured the decline of manufactured exports, and the hopes of new gains from imperial expansion could sustain the idea of Britain continuing an advantageous trade with client societies.

2. THE TRADERS

The great drive for the extension of world trade came from the merchants; it was they who, for good or ill, took the risks, performed the calculations, and selected the men for the ventures which carried the surplus products of Britain abroad, and in so doing promoted world expansion of output. They not only ruled the produce markets, but stood between the manufacturers and their customers, at home and abroad.

Their position in society was of long standing. The great merchants of London had from medieval times held a most important place in the nation, as the great possessors of liquid wealth and the sponsors of new projects abroad. They had traded, in the main, with the traditional markets of Europe, with the West Indies, and to the narrow but high value-content markets of the East. After Waterloo a new generation of merchants arose, impatient of the restraints that encumbered traditional trade. New men and new partnerships could do better by trading to new markets and by the aggressive pushing of their goods. It was to Liverpool that the great merchanting initiative largely passed. But though Liverpool had far eclipsed London in exports, it was essential for the merchant, once a certain stature was attained, to have a London house. Moreover there were early signs of the concentration of trade in large concerns, making firms anxious to figure among the swallowers rather than the swallowed. It was the import of the leading industrial raw material, cotton, and the export of manufactured goods, both from the textile areas of the north, and the metal and engineering areas of the north and the midlands, that gave the ascendancy to the new generation of merchants.[1]

In the cotton manufacture, the leading merchant converters, like Sir Thomas Potter, were performing the traditional role of the merchant clothier on a new scale.[2] His Manchester warehouse, purposely removed from the old, crowded, commercial area, became the centre of a great trade, both home and foreign, with, by the sixties, some five to six thousand retailers as customers. The warehouse, built in 1836, represented a revolution in merchandizing architecture; as the centre of carriage between all the elements of the industry (spinners, weavers, bleachers, dyers, printers) held together by the converter, and as the focus for retail and wholesale trade, it was a hive of transhipment. Scientific layout, steam hoists, and vans were all in evidence, so that coal-black cloth for Italian clerics, and gay fabrics decorated with Chinese designs, found their destinations with certainty and speed. In the sixties and seventies stupendous fortunes were made in this kind of merchanting, for the manufacturers were dependent upon the merchants to keep their machinery running; no less important, the retailers had no other source of supply, and were frequently financed by the greater traders.

[1] See Sheila Marriner, *Rathbones of Liverpool 1845–73*, Liverpool, 1961; Vincent Otto Nolte, *The Memoirs of Vincent Nolte* (first published 1854) ed. of 1934, B. Rascoe, ed., New York, p. 316 *passim*.

[2] H. R. Fox Bourne, *English Merchants, Memoirs in Illustration of the Progress of British Commerce*, London, 1866.

Records often tended to fall badly into arrears, resulting in endless confusion, especially when litigation occurred. This was especially so at the foreign end of business, as in China. For there, conscientious clerks were not easily come by, and were very costly. At home, in the early decades of the century the great army of office workers, with its modest remuneration, careful conventionality, and slender ambition, was only slowly beginning to mobilize.[1] The methods of keeping records was often poor, especially before the Limited Liability Acts began to affect the trading world in the seventies. More or less loose partnerships were the rule, and many traders were involved in a multiplicity of such engagements, which made it exceedingly difficult even for themselves to understand their position. From the mid-century, the number of clerks was growing fast; in Liverpool in the sixties they were receiving salaries of from £60 to £80 per year. Among these toilers of the pen were not a few sons of wealthy business men and landed gentry, hopeful of a partnership.

The export trade was largely in the hands of commission agents who consigned manufacturers' goods to foreign buyers; the manufacturer drew bills on the agent for a substantial proportion of the sale price, and discounted these with his banker. The agent allowed credit at three to six months to the foreign buyer. The trade to America grew rapidly after the war, in spite of the misgiving of those who perceived very early that British machinery and skill would produce rivals in the new area.[2] It was the cotton trade which dominated the Anglo-American nexus, at least down to the mid-century, or perhaps to the Civil War. America increasingly attracted attention, with traders closely studying the political kaleidoscope with one eye, and with the other watching eagerly for an opportunity to adopt new methods of transport to both North and South America. The American trade was especially treacherous, appearing so simple and easy to the uninitiated that it was frequently overdone, often by men of slender capital. But heavy speculation could occur in almost any direction; invariably it did so when a new market was opened as in South America in the twenties, the East in the thirties, and in Australia in the fifties.

The conditions of trade to India in the years between the opening of the trade in 1813, by modification of the charter of the East India

[1] B. G. Orchard, *The Clerks of Liverpool, being ten chapters on their Numbers, . . . salaries, grievancies, Marriages, etc.*, Liverpool, 1871.

[2] N. S. Buck, *The Development of the Organization of the Anglo-American Trade 1800–1850*, New Haven, 1925; F. Thistlethwaite, *The Anglo-American Connection in the Early Nineteenth Century*, Philadelphia, 1959.

Company, and the seventies, were peculiar.[1] An entrance into a merchant's counting house was equivalent to an appointment in the government service. So bright were the prospects that clerks worked without payment for their princely employers, but looked forward to principalships, solacing the interval with private transactions on their own account. Men dealt on their own capital in the main, thus avoiding excessive speculation. But profits did not flow evenly. Over long periods merchants dealing in Indian cotton would merely keep going — awaiting the marvellous gains of a few hectic years. The Indian crop went mainly to China and to Europe, entering England only when great prosperity or an American crop failure occurred. When this happened, the price of Indian cotton shot up in all its markets, and a fine killing was made. But this possibility of heavy gains was gravely damaged first by the weakening of the Usury Laws in 1833, which by making it possible to bid for loans by offering higher rates of interest, caused a rush of new traders in speculative times, and by the electric telegraph which gave world notice of impending shortage, and dampened the price rise in Indian cotton.

Even earlier, by 1850, most of the Agency Houses that had developed out of the private trading of the East India Company's servants, had succumbed to commercial pressures, especially in 1836 and 1847, when many had increased their own difficulties by granting mortgages on estates in India. They were replaced by new firms based on Liverpool, Glasgow, and London, the new wave of merchant princes. In the trade to the East generally, the ability to profit by movements in exchange rates between currencies was almost as important as trade itself.

The new trading initiative caused not merely English, Scots, and Welsh to react; men in other parts of the world did so also. There was the quite literally fabulous family of the Sassoons.[2] The young David Sassoon, part of the Jewish community of Baghdad, learned to traffic between Persia and India to such effect that in 1832 he removed to Bombay. Here economic expansion, following the extension of British control, was going forward; Bombay had become the gateway to the new India. Sassoon distributed the products of British factories throughout the sub-continent. Many other Oriental (sephardi) Jews were drawn by the same new opportunities. In the fifties and

---

[1] Arthur Crump, *A New Departure in the Domain of Political Economy*, London, 1878, part I, p. 2.

[2] Cecil Roth, *The Sassoon Dynasty*, London, 1941.

sixties railways into the hinterland began to appear. Sassoon carried British textiles from India to Persia and Mesopotamia, bringing back native products and oriental cloths for sale to the British traders who marketed them in Britain. He moved from this to an even wider field. He realized that great transactions could be conducted between India and southern China. Bombay yarn, Lancashire piece-goods, and opium were sent from India to China; back, for British use to a considerable extent, came silk, tea, and other expensive goods. Branches of the house appeared in the leading Chinese ports. Eventually in 1858 a London office was formed, with Liverpool and Manchester branches soon following. Thus the evolution was complete; the Jewish aristocrats of Baghdad had, in the persons of David's sons, become British merchants, in Britain. Soon they were moving in the highest society, intermarrying with their co-religionists, the Rothschilds.

But, by the seventies, the sun was setting for the merchant princes. Many manufacturers began to develop their own direct marketing. The new screw steamships had brought great profits for a decade or so, but the speeding up of shipping, so enthusiastically hailed by the merchants, in the longer run worked against them. For with greater speed and efficiency, fewer ships were needed and smaller stocks were held. Services provided by the merchant were now the special function of new banking concerns, especially the discounting of bills and dealings in exchange. As S. G. Rathbone, who left a perhaps unique merchant's memorandum on the situation, dated 1878, put it, 'We have had processes in operation which have economized distributing power in a ratio disproportionate to the increase in the Trade of the world and which has therefore for a time disturbed the balance between the work to be done and the power available for doing it.'[1] Though great trading houses like that of Jardine Matheson and Co. did not cease to exist, the share which they corporately took of the total product was greatly diminished after the seventies. Survival depended increasingly on closer specialization upon a narrower range of activities and markets. Some firms turned to shipowning; others to some specialized group of products, some became simply warehousemen. A few like James Finlay and Co. found their solution in the internal development of the countries to which they traded, in their case sponsoring plantations and jute mills in India.[2] But the older type of universal merchant was now more

---

[1] *Rathbone Papers*, Harold Cohen Library, University of Liverpool.
[2] Anon., *James Finlay and Company Limited, Manufacturers and East India Merchants, 1750–1950*, Glasgow, 1951.

or less obsolete – in cotton importing because specialization had eliminated him, in the produce markets (except for tea) for similar reasons, in the exchanges (once so important an aspect of trading) because of new and more efficient banking and transfer facilities, and in exports because of the new skill in direct selling abroad developed by manufacturers.

The supersession of the merchants altered once more the balance between the outports and London. From the late eighteenth century onward the vigorous traders of Liverpool and Glasgow had seized and held the initiative. Now the concentrative power of London was re-asserted as the commercial challenge of the outports fell away as their merchant princes were eclipsed or centred themselves on the capital. The passing of the general merchant did not mean the end of British invisible earnings abroad. On the contrary, British insurance services, banking, shipping, all continued to grow, and to make an immense contribution to the balance of payments. But these were highly specialized services.

As the international traders were reconciling themselves to a relatively declining role in the economy, a new portent was appearing. With rising real incomes among the artisans and to some extent the labourers, it was becoming clear that the old aristocratic bias of trade, disdaining to display its wares, was inappropriate. Trade began to move away from the provision of a narrow range of goods to a new diversity and a new cheapness.[1] As early as 1856, David Lewis in Liverpool had begun his preparations to meet the new situation.[2] There were several keys to the new mode of trade, all of which Lewis adopted: large-scale sales at small profit margins, no credit to purchasers, the exchange of un-satisfactory goods, marked prices, and mass advertising. Much of this had been anticipated. The Rochdale Pioneers and the cooperative retail trade movement they started in 1844 had insisted upon cash payment because it reduced risks, sponsored independence, and destroyed the degrading bond of indebtedness between customer and shopkeeper.[3]

Advertising had shown something of its potential a good deal earlier.[4] By the forties the magistrates of London were struggling to stem the rising tide of handbills and wall posters. The sandwich man had arrived together with vehicular advertisements, including the

---

[1] J. B. Jefferys, *Retail Trading in Britain, 1850–1950*, Cambridge, 1954, especially chapter I.

[2] Asa Briggs, *Lewis's of Liverpool. Friends of the People*, London, 1956, p. 29.

[3] See below, Chapter 9, section 8, p. 264.

[4] For a provocative account see E. V. Turner, *The Shocking History of Advertising*, London, 1952; Blanche B. Elliott, *A History of English Advertising*, London, 1962.

'perambulating hat', seven feet high, which had so outraged Carlyle. The French historian, Taine, complained after a visit to England in the sixties, 'All are clamouring . . . Everyone tries to devise some refinement to flatter a desire, a whim, a mania.'[1]

Explicit prices had been a Quaker principle as early as the seventeenth century; it was rediscovered from time to time by others. Early in the century Thomas Chalmers had denounced exploiting shopkeepers from his Glasgow pulpit. Among his hearers was James Campbell, father of a future Prime Minister, who thereupon set exact prices upon the goods in his shop, and subsequently prospered greatly.[2] Even the technique of the 'sale' was known, for at least as early as the thirties traders in drapery had bought job lots or 'remainders' at great discounts and sold them far below ruling prices, discovering in the process the responsiveness of consumer demand. The youthful Mr Morrison practised this system in London and the provinces to great effect, justifying his election among the luminaries of the Political Economy Club.[3] Mass sales at low profits as a general policy was a genuinely new principle, wholly alien to a generation of shopkeepers who lived above the premises who could keep contact by means of a speaking tube with the clerks below. David Lewis made a new beginning in gathering up these ideas and in new initiatives in advertising. By the seventies and eighties the new method was beginning to show its power.[4]

These beginnings, in the fifties, of the new age of retailing, were the symptoms of many changes. The merchants of the older generation had held an elevated view of their calling, conceiving themselves in a quasi-aristocratic light. They could hardly conceive that Thomas Lipton beginning with his little shop in Finnieston, Glasgow, in the early seventies, would own an empire of such stores and receive a knighthood. The cultural development of Liverpool and Glasgow strongly reflected the merchant prince idea, with Greek and Italian renaissance models much in mind. Their status rested upon mass dealings in basic commodities – sugar, coffee, cotton, iron; they were far removed both from physical contact with goods and from ultimate consumers and did nothing whatever to stimulate demand.

[1] Hippolyte Taine, *Notes on England*, E. Hyams, ed., London, 1958, p. 190.

[2] J. S. Jeans, *Western Worthies*, Glasgow, 1872.

[3] Anon., *Political Economy Club*, vol. VI, London, 1921, pp. 255, 263.

[4] H. Pasdermadjian, *The Department Store, its Origins, Evolution and Economics*, London, 1954; R. S. Lambert, *The Universal Provider, a Study of William Whiteley and the Rise of the London Department Store*, London, 1938.

But the extraordinary improvement in consumption standards meant that great gains could accrue to the men who were prepared to study, to meet, and to manipulate consumers' taste. The great exponent of advertising method was the celebrated Barnum; it was from him that men like Lewis took their cue. It is significant that advertising found its first great development in America, where traders had never sought to model themselves on an aristocracy, and that the first men to adopt such methods in Britain were the humble but ambitious retailers. When the young Lipton returned from a visit to America in the early seventies he brought back the useful couplet:

> The man who on his trade relies
> Must either bust or advertise.[1]

3. THE TEXTILE MANUFACTURERS

The marvellous economies in the mechanical manipulation of natural fibres provided the first great opportunity for the new age of machines and factory owners.[2]

In the woollen industry men of some capital had for generations put work out among the spinster-housewives and the farmer-weavers of the countryside. With the development of new machines, and the increased need for steady output, came a new emphasis upon investment and organization. Those who had been successful as weavers in the later eighteenth century could now sometimes acquire four to six looms, and some put work out on their own account. Many of such men quickly rose to positions of considerable substance. In fact, so many entered the field that, in bad times, the resulting competition could greatly lower cloth prices — an early warning of the problems of scale and specialization. This trend to factory organization was even more dramatic in the cotton industry.[3] But factor masters did not really oust piece masters — those who continued to put work out to cottagers — until the enormous gain in efficiency of the power-loom became obvious after 1830. Yet much earlier, even with hand-powered looms, factories were yielding higher wages than the cottages.

[1] Sir Thomas Lipton, *Leaves from the Lipton Logs*, London, n.d., p. 96.
[2] For an account of nineteenth-century cotton technology see Andrew Ure, *The Cotton Manufacture of Great Britain*, London, 1836, 2nd ed., London, 1861.
[3] The cotton industry is analysed in sociologist's terms in Neil J. Smelser, *Social Change in the Industrial Revolution: An Application of Theory to the Lancashire Cotton Industry 1770–1840*, London, 1959.

By 1815 Crompton's mule was overwhelmingly the most popular method of spinning. The carriage of the machine still required to be pushed back by hand, and the winding of the yarn was still a separate operation. Roberts's self-acting mule of 1825 and 1830 removed both limitations. But it was costly, and less appropriate to fine spinning. It conquered the coarser counts in the fifties, but the finer held out until the eighties.

In weaving, the power-loom, invented by Cartwright as long ago as 1785, was not really made commercially effective until the thirties. Once launched, however, it could not only remove the weaving bottle-neck, but also brought spinning and weaving together again; this time under the spinners' aegis, for weaving sheds were added to spinning mills. Yet this synthesis was mainly confined to the larger firms and was not to persist very generally even among them. Though a considerable number continued on an integrated basis, the advantages of specialization reasserted themselves and spinning and weaving once more separated. By the last quarter of the nineteenth century this distinction was even visible in terms of regions, with spinning concentrated in south Lancashire and north Cheshire, and weaving centred on the northern boroughs.

In the early decades of the century cotton men could start modestly, often financed by cotton dealers, buying a few mules, hiring space and power, and employing half a dozen workers or less. In 1834 there were some 300,000 cotton looms; of these the hand-looms still outnumbered the powered by two to one. The whole nature of the industry placed a premium upon initiative – on the selection of the correct combination of alternatives, first on a small scale, then followed, if the producer were up to it, by a move into a bigger enterprise. He could then benefit through extended operations in well-planned factories using the best principles of steam engineering. The cotton manufacturer, in fact, was the classic case of the entrepreneur who hires his factors of production from others, whose function most obviously consisted in making combinations of these factors, and who may rise from petty production to the ownership of a large firm.[1]

Robert Owen's principles of management were a revelation. Many of the larger men took up Owen's mode of factory management just as they had adopted Arkwright's plan of construction; the more readily as a painful soreness of conscience and a vulnerability to social criticism

[1] For the early general history of the cotton industry, see E. Baines, Jr., *History of the Cotton Manufacture in Great Britain*, London, 1835.

was nagging many of them. For it was often the larger and more successful, making their way into higher social circles with wider horizons, aspiring to the House of Commons, who found that personal conscience and hostile criticism made it increasingly difficult to blink the problems of the workers' condition. But Owen and conscience and criticism were not the only goads. Foreign competition was felt with increasing strength as the war receded. At home the conditions of entry to the industry were not difficult, and the new men found that their challenge to those already established was often aided by the flow of new inventions.

The phases of development in the woollen (using the shorter, more easily matted fibres) and the worsted (using the long staple wools) showed important differences, both in comparison with cotton and between themselves.[1] Wool was much more difficult than cotton to work with high speed machinery; it is not surprising that the success of the new inventive urge was first seen in cotton. The woollen manufacturers had behind them a long tradition of profitability based upon older methods, and were not particularly amenable to change. The supply of raw wool had not grown at anything like the same rate as that of cotton; indeed it was not until the advent of the Australian supply after 1830 that the supply of raw wool allowed of real expansion. The lag of Yorkshire behind Lancashire in fact was such that when machinery was finally adopted on any scale in the woollen trades, the steam-engine was well established, allowing the wool men little credit for its development.

It was the worsted manufacturers who more quickly followed the lead of the cotton men. They were the younger branch of the industry, less bound by the past, more efficient, and in consequence, more highly capitalized. They enjoyed a good market for their yarn outside Yorkshire, in the worsted and hosiery trades of East Anglia. The longer silkier fibres were more amenable to machines, and the products of the industry, light, plain 'dress goods', were much closer competitors to the new cotton fabrics than were the older broadcloths and other heavy goods, not to speak of blankets and flannels. Indeed, worsted could be

---

[1] For general accounts see R. M. Hartwell, *The Yorkshire Woollen and Worsted Industries 1800–1850*, Oxford D.Phil. thesis, I.T.A., 1955–56, no. 740; J. Bischoff, *Comprehensive History of the Woollen and Worsted Manufactures*, 1842, W. B. Crump, *The Leeds Woollen Industry 1780–1820*, Leeds, 1931; Eric Sigsworth, *Black Dyke Mills*, Liverpool, 1958; F. J. Glover, *Dewsbury Mills: A History of Messrs. Wormalds and Walker Ltd., Blanket Manufacturers of Dewsbury; with an Economic Survey of the Yorkshire Woollen Cloth Industry in the Nineteenth Century*, Leeds Ph.D. thesis, I.T.A., 1959–60, no. 666.

allied with cotton, for after 1840 women's dress fabrics were often made with cotton warps. Spinning for worsted manufacture had completed its transition to a factory basis by the thirties. But in the older woollen industry, machine spinning did not begin its advance until later.

Yet in spite of these delays, the technical gap between cotton and wool generally was narrowing rapidly by the thirties. Gig mills (for raising nap) and shearing frames (for cropping it) were appearing rapidly in Yorkshire after 1815. The adoption of the power loom, as we have seen, was delayed even in cotton, so that the thirties saw its almost simultaneous appearance and extension in both industries. Yet the greater initiative of the worsted men over the woollen was still apparent. Woollen cloth manufacture as late as the mid-fifties was still half domestic, with hand-loom weavers surviving, like the father of Philip Snowden, until the sixties and even later. The hand-loom was especially persistent in the fancy woollen trade of Huddersfield. For all this, the Yorkshire wool men's slowness to change was as nothing compared to the attitude of their rivals in the West Country and in East Anglia.[1] Both of the latter areas, once so thriving, suffered virtual eclipse by the forties. In part this was due to the successful resistance to new methods by the workers.

The most intransigent and persistent bottleneck in the handling of wool was at the combing stage. For fifty years after Cartwright's machine of 1785, inventors and business men struggled to reap the great profits to be gained in superseding this wasteful manual process. But hand-combing held the field until Samuel Cunliffe-Lister's invention appeared in the late forties; thereafter this last stronghold of the domestic worker was quickly reduced and the flow of combed wool much increased.

Some big worsted mills were in operation in the Bradford district by the first decade of the century, including those of Gott and Wormald.[2] There was rapid growth both in their size and number, culminating in a notable burst of activity in the thirties. Worsted spinning called for the larger establishments, employing some fifty to sixty people; wool-weaving factories perhaps contained some forty workers. The worsted spinner was a specialist, with a single line of output; in the wool establishments the factory proper, in the thirties, was merely the centre

---

[1] K. G. Ponting, *A History of the West of England Cloth Industry*, London, 1957; M. F. Lloyd, 'The Decline of Norwich', *Econ. H.R.*, 1951.

[2] See H. Heaton, 'Benjamin Gott and the Industrial Revolution in Yorkshire', *Econ. H.R.*, 1931, p. 61.

of operations containing the 'in' workers, often greatly outnumbered by the 'out' workers, both spinners and weavers, as well as those supplying ancillary services.

Outside the factory system altogether were the domestic clothiers. There were some three thousand of these in Yorkshire after Waterloo, producing their goods on a petty scale in the traditional way, weaving in the home, and selling in the Cloth Halls of the West Riding. Though diminishing, they were still a substantial element of the industry in the forties, and even later. But these smaller clothiers were becoming less numerous – some removed by failure or retirement, some by success, moving upward into the ranks of the larger employers as, with the spread of power and machinery, the industry left the home and concentrated in the factory.

Though there was a more or less continuous trend towards larger establishments, its speed must not be exaggerated, nor must it be forgotten that it was a good deal slower in wool than in worsted; the census of 1851 revealed that more than one-half of the one thousand woollen cloth manufacturers had less than ten employees, and only eighty-two had more than one hundred. Yet there were heroic figures in wool as in cotton. Cunliffe-Lister, applying his combing machine, became one of the great figures of Bradford, not averse to expounding his own merits, publicly doubting that there was a man in England who had worked as hard as he had.[1] Sir Titus Salt, exploiting alpaca wool, founded the model village of Saltaire. Edward Akroyd of Halifax was another such man of decision who also sought to cherish the 4,000 workers in his charge.

Clearly, not all textile men were of the same background. Not many were capable of seeing the factory itself in terms of a machine. There were not many managers who could be safely placed in charge of a mill. But there were some who could 'think big' and sustain their thoughts by persistence, courage, and a sense of how to consult and how to delegate. They were the men alive to new ideas, as demonstrated and expressed by the Arkwrights, the Strutts, Oldknow, and Owen.[2]

In cotton, worsted, and to a lesser degree, wool, the result was almost two industries, in each case, one large-scale and the other small, both

[1] Samuel Smiles, *Thrift*, London, 1875, p. 190. Samuel Cunliffe-Lister, Baron Masham, *Lord Masham's Inventions Written by Himself*, Bradford and London, 1905.

[2] R. S. Fitton and A. P. Wadsworth, *The Strutts and the Arkwrights, 1758–1830: a Study of the Early Factory System*, Manchester, 1958; G. Unwin, *Samuel Oldknow and the Arkwrights*, Manchester, 1924; Robert Owen, *The Life of Robert Owen by Himself*, 1857, new ed. M. Beer, ed., London, 1920.

producing the same products, yet with very different outlooks, practices, and labour relations. In the thirties, Charles Babbage, the Cambridge professor, and Andrew Ure, professor in the Andersonian Institution, Glasgow, the rather bland apologist for the millowners, visited the great establishments, and presented what they saw as though it were typical. But the reformers, Fielden, Oastler, and Ashley, and the Commissioners and Inspectors, sought their evidence in the smaller, and usually more degraded units. Further, the distinction in size, to a considerable degree, was related also to situation, especially in cotton: the town spinners, in general, worked on a smaller scale than those in the villages.

There is a considerable list of names of cotton manufacturers who operated on a very large scale by the thirties.[1] Owners who replied to an inquiry of 1833 reported as follows: there were seven mills employing over 1,000, twenty-three over 500 and less than 1,000, and thirty-six between 250 and 500.[2] Horrocks, Miller and Co. of Preston, had some 700 spinners in four mills, with surrounding hand-loom weavers numbering some 7,000, as early as 1816. The second Richard Arkwright had almost 2,000 spinners at Cromford. At New Lanark, Dale and Owen employed some 1,600. There were many others: the seven Ashton Brothers of Hyde, the Grant Brothers of Ramsbottom, the Ashworths of Egerton and New Egerton and New Eagley, the Strutts of Derby, the Orrells of Stockport, Greg and Sons of Manchester, James Finlay and Co. in Scotland.[3]

Many arguments from the literature of the thirties and forties showed the advantages of the large concerns. It was their owners who called in Fairbairn and Lillie and others to show them how to get optimum size, shape, and layout. It was they who called for ever more powerful and efficient steam-engines. They could gain the great benefits from the thirties onward, of combining spinning and weaving in the same concern. Their goods were more acceptable to the merchant than those of the small man, and were better pressed in home and foreign markets. Their reserves were usually greater in bad times; in good times they had more ready access to finance. Great advantages could accrue in labour relations. Loyalties could be created through housing schemes and the like. The best foremen rallied to them. In

[1] A. J. Taylor, 'Concentration and Specialization in the Lancashire Cotton Industry, 1825–1850', *Econ. H.R.*, 1949.

[2] See Ure, op. cit., vol. I, p. 390.

[3] See Samuel Smiles, op. cit., 1875, pp. 186–9. Henry Ashworth communicated papers to the Statistical Society on cotton: see *J.S.S.*, 1842, p. 255, and March 1860, p. 7.

fact, the large owners could occasionally take over a good deal of the patriarchal role of the landowner, thus stabilizing their labour supply. Even the powered machine could be less repugnant under their control, for in the large well-planned concerns arbitrary speeding up, either by foremen or masters, was less likely.

It is worth noting that though these influences operated to promote size, the cotton entrepreneur seldom integrated vertically. He was flanked on the one side by the cotton broker and on the other by the textile merchant; within the industry proper there was a great deal of specialization on one or two operations. Fine spinners, for example, were distinct from coarse ones. As always, in so diverse an industry, there were exceptions. Ashtons had their own coal pit, and embodied all processes from spinning to merchanting. James Finlay and Company themselves marketed the products of their Catrine and Deanston Mills, in India and China, and so effectively as to become leading general merchants, assuming once more the role of the first Finlay. It was clear to engineers like William Fairbairn that ideally, a mill 'should be so constructed and furnished as to enable him [the master] to open his bales of raw cotton at one end of the factory, and receive it in the state of yarn and cloth at the other'. In the early phase of the adoption of the power-loom this was approached by some, but for most producers, short of the very largest, it was a doctrine of perfection; most, especially from the fifties onward, busied themselves with a single process, relying upon the merchants for integration. In the upshot, Lancashire produced a group of single-minded entrepreneurs with a sharply defined focus of interest, in contrast to the engineers of ironmasters whose products found their way into so many aspects of the national economy.

The larger owners were the men who so excited the philanthropic observers of the day, for it appeared that they were demonstrating in the most practical terms that scale was the cure for squalor. But textile manufacturing was also the kind of game in which a player, threatened with eclipse, or seeking a disproportionate share of the gains without real ability, might seek to compensate for inferiority by illicit means. Such a man could neither find relief in coercing the suppliers of raw materials, the cotton market, or the suppliers of machines, the strongly placed engineers, or the merchants. Stimulated by the prospect of defeat, or a desire to gain without effort, he found himself in a position of great strength against the labour force he had summoned to his factory, so many of whom were women and children, confronted by the rapid extension of the power-loom, the number of which almost doubled in the three years after 1830.

Relative competitive failure did not cause the neat eclipse of such a man but rather tempted him to play outside the rules. Especially was this so with the smaller man, who could not reap the economies of scale and planning available to the larger; yet he was small simply because he was less successful at the game. Even large firms, which might positively gain by improved management, too often, when the men at the top were lax, or the overseers had preserved their power and independence, fell back upon the labour force by requiring longer hours, thus receiving the fierce criticism of Sadler and Ashley.

The situation had arisen by the thirties in which the best employers, by and large, carried their work-people with them into higher incomes and higher living standards, on a quasi-paternalistic basis. But the less successful organizers could only turn upon their work-folk, either actively, by attacking conditions and increasing the pace of work, or passively, by losing interest and delegating the insoluble situation to petty managers. When bad times came the differences between good and bad employers became stark, for the larger men, sustained both by ideals and financial resources, often attempted to maintain employment. Further, it was in their interest to try to keep their elaborate equipment in production.

The battles over the Factory Acts in the thirties and forties came at a time when abuses were near their height.[1] They were especially bad in the first of the great standardized processes — spinning. Yet they were approaching, if not their end, at least a great falling away. For they were, in large part, the last hectic attempt of those firms that were now obsolescent because they were too small, and of the less efficient of the larger firms which had fallen back on the cushion of labour exploitation, to maintain themselves.

Yet many of the larger and more efficient factory owners did not rally to Fielden and the rest of the promoters of the Factory Acts because their minds were dominated by other considerations. They knew that the terms of their own emergence and their own scale of operation were that the state, which had only very lately vacated their field, should continue to be excluded. Hence Factory Acts, the effective beginning of a new industrial code, were often disapproved. Yet with time this attitude changed. The bigger men became more confident in accepting the competitive economy; moreover, it became manifest that

[1] See below, Chapter 7, section 7, pp. 246-8.

the protests being engendered in the minds of the workers in the smaller mills were likely to become generally infectious, leading to labour combination and extended wage claims. Also, the time had come to concede factory regulation to the social reformers lest a general reaction against competition set in.[1]

But there was another, more subtle and more worrying ground for misgiving. It arose from the relationship between hours of work and the size of investment in plant and machinery. Millowners, in fact, seem to have been obsessed by the question of fixed capital. Partly, no doubt, this was due to the extent of the obligations contracted by many of them, and their dependence upon short-term borrowings through bills of exchange. It should be remembered that they were, in their day, a unique historical phenomenon in terms of the extent of their obligations. Fixed or overhead costs were a high proportion of total costs, though not perhaps so high as some seemed to argue. Bolton spinners claimed in the early forties that they had some four-fifths of their capital tied up in buildings and machinery.[2] On the other hand, their attitude to labour must also be considered: in general many manufacturers had, in the early decades of the century, little confidence in the responsiveness of the workers to good treatment — they thought almost wholly in terms of man-hours extracted by close supervision, by under-managers, whose own interest was involved in output. They assumed that output varied directly as the length of the time worked, and stressed that the cost of plant and machinery per unit of output fell as output was extended. The small man, whose general conditions of lighting, ventilation, cleanliness, and safety were often inferior, often sought to place an even greater pressure of hours upon his workers in order to remain competitive. How was a man of good conscience to free himself from this law of degeneration?

Fluctuations in trade aggravated the position.[3] When a boom occurred, and orders flowed in rapidly, there was a tendency to seek to fill them on the basis of existing plant, partly because of the delays involved in extension, and partly because of the cost of new capital. In consequence working hours were lengthened. The millowner who tried to resist this trend found that his profits were less, and his capacity for

[1] See C. H. Driver, *Tory Radical: the Life of Richard Oastler*, New York, 1946; also below, Chapter 7, section 7.

[2] Henry Ashworth, 'Condition of the Manufacturing Population at Bolton, 1836–1842', *J.S.S.*, 1842, p. 74.

[3] For an attempt to understand these see Herbert Heaton, 'An Early Victorian Business Forecaster', *E.H.*, 1933.

subsequent growth diminished. If the boom continued new plant would eventually be built. When the slump came, and orders fell off, the position of the humanitarian millowner was even more difficult. The owners who ran for very long hours for large outputs again had lower costs, sometimes with the extraordinary result that some of the mills worked at great pressure and the rest were almost idle. Even the humanitarians felt themselves obliged to follow the example of their rivals and increase hours. These ideas were held during the post-1825 slump, which persisted down to the early thirties, during the prosperity of 1834–36, and during the somewhat ambiguous years to the early forties.

But with the great difficulties of 1841–42, with both cotton and woollen trades embarrassed with enormous new growth, there was some tendency to lose enthusiasm for the principle of long hours. The old obsession concerning the fixed link between men and machines was at the same time becoming weaker. Owen and others had long argued that a reduction of hours produced such a response among the workers that there were many fewer thread breakages and mechanical breakdowns, the time wasted at the beginning and ending of the day was reduced, and the rate of turning of the spindles could be increased under generally better working conditions. Moreover the economies gained in mill and machinery construction were very real in the second and third quarters of the century, thus keeping down the cost of capital equipment per unit of output.[1]

But not all grounds for misgiving were removed. Two kinds of difficulty were argued by those who remained hostile, and who got very close to applauding the millowners in their evasive struggle in the fifties and later against the Factory Inspectors. There were very great differences between the productive problems of different mills, concerning both the need for flexible hours (especially where water power, with its erratic behaviour, was still relied upon), and questions of machine-guarding and ventilation. Indeed there were suggestions in the fifties that those manufacturers whose factories were most adaptable to the new code were disposed to use it against those less favourably placed. Even more important was the fear that the protean ability of industrial initiative to shoot out in new directions would be permanently impaired by a code arising from a particular set of experiences and abuses. Those who held such views were especially incensed

[1] M. Blaug, 'The Productivity of Capital in the Lancashire Cotton Industry during the Nineteenth Century', *Econ. H.R.*, 1961, p. 369.

when the undoubted improvement in the condition of the workers, from 1842 onwards, was ascribed to the operation of the Factory Acts, without mention of the enormous rise in productivity that had come with new methods and new mills, brought into being by the employers' initiative.

In cotton, in this great phase of growth, it was thought impossible to abridge competition. Certain groups like the spinners of Preston were strongly enough organized to bring their workers to heel. The spinners as a whole might combine to wage war against the raw cotton suppliers, through short-time working, as in 1839, when an attempt was made to corner the supply in America and thus hold the British industry to ransom. But in general the conditions under which the entire industry would agree to a common rule had to be so extraordinary, that it was in practice unthinkable. Especially powerful in this respect was the growing fear of the foreigner, for he would be placed in a position to sweep the board if British producers kept up prices by agreement among themselves. In the later twenties and early thirties France, Austria, Switzerland, and the U.S.A. had all been busy constructing mills, especially for spinning the coarser counts. The upshot was that the cotton industry was exposed to the full vicissitudes of prices and profits engendered by high fixed costs.

In spite of a growing sense of precariousness about raw material supplies, the first half of the fifties saw a great revival in factory investment, on a scale that astonished the Factory Inspectors. But the later fifties produced ominous comments about 'maturity'. Thus production pursued its erratic course, dominated in the sixties by the almost fatal failure of American supplies — representing some 80 per cent of British cotton imports.

The same problems of growth and investment were confronting the wool men.[1] The worsted industry ran well ahead of the woollen in this respect, due largely to its greater suitability for mechanization. Between 1835 and 1838 the number of woollen mills in the West Riding rose by 30 per cent; worsted showed an increase of 65 per cent. The number of workers grew on a similar scale. Such growth was extraordinary. It showed a burst of confidence in the industry that was almost unlimited. Yet when slump came, reaching its worst point in 1842, those who had rushed to build new mills loudly complained of excess capacity. Indeed the master spinners of the West Riding, worried

---

1 See K. V. Pankhurst, 'Investment in the West Riding Wool Textile Industry in the Nineteenth Century', *Y.B.*, 1955, p. 93.

about glutted markets, petitioned the government for additional legal limitations upon hours, using arguments of morality and education.

A fully developed liberal competitive system involved so many unknowns and so much instability that the larger entrepreneurs almost inevitably felt that the workers ought to regard them with the same loyal and passive feelings formerly held toward their patriarchal landed masters. This was especially true in the country districts where the employers had become accustomed to a much greater control over their workers than was the case in the manufacturing towns. Though they themselves were engaged in breaking free of the tutelage of state, church, and landed aristocracy, they saw no need to universalize the drive to self-expression, but wished to be accepted by the workers as the custodians of local welfare, a role to which some of them gave much thought. The smaller, less scrupulous, and more hard-pressed masters held no such doctrine of trusteeship; in fact most were devoid of social philosophy altogether.

In cotton, by the mid-thirties, the tiny mills made by knocking the partition walls out of rows of cottages, and powered by small fourteen horse-power engines, were almost extinct. In the Glasgow area, where the trend to greater size had been very pronounced by 1838, it cost somewhere between £40,000 and £80,000 to establish a cotton factory. Even in Lancashire a minimum of some £5,000 was required. The Factory Acts, especially with the difficulties involved in the shift system required to meet the restrictions on children's hours after 1833, gave a further advantage to the larger employer, as indeed did the general regulations improving conditions.

Yet, for all this, in the forties plenty of factories of modest dimensions had survived. Even more remarkable, there seems to have been, contrary to what might have been expected, a slowing down, if not an ending of the trend toward greater size. The average number of workers per cotton factory in the seventies was not much greater than it had been some thirty years earlier. But each worker managed more machines. The amount of machinery in the average mill, measured by the number of spindles, had gone up very greatly; similarly in terms of looms, though not to the same extent. With the enormous increase in world demand, and consequently in total output, there was room for the smaller as well as the larger man. Once the merits of specialization in spinning or weaving had reasserted themselves after the early decades of adoption of the power loom, the integrated concern became less important. The investment boom of the fifties was a good deal more vigorous in the setting up of looms rather than spindles as the industry strove to fill the now fast growing markets of the East. The smaller man could hold his own in the weaving sector as its independence was reasserted. Typically a weaving shed could be built with a good deal less capital than a corresponding spinning mill; moreover it was economical to work on a smaller scale. Increasing difficulties of management had also contributed to the arrestation of the trend to greater size and integration. This, perhaps, was accompanied by a falling away in the initiative of owners as one generation succeeded another. As the reversion to specialization by process continued, it became apparent that different kinds of personal qualities were appropriate to different branches of the industry. But the revival of the smaller enterprise from the fifties did not bring a reversion to the bad labour conditions of earlier times.

By 1860 the British worsted trade was one of the most concentrated in the world, with Bradford its centre.[1] The magnetic pull was also felt by foreigners. German merchants and manufacturers had been strongly represented in Manchester for some time; the growth of Bradford attracted new men, from over the Pennines, the Scottish border, and especially the North Sea.[2] Schusters, Steinthals, Behrens, Rothensteins, Edelsteins, all appeared in Bradford, encouraged by the tariff building of the German Zollverein. By the sixties Bradford had a notable colony of such men, often liberal refugees, living in close propinquity in central Bradford, in 'Little Germany'. The trade owed much to them, especially in its export branches. They combined a generous patronage of culture and charity with a mustard keenness for profits. In Leeds, somewhat later, the immense new demand for clothing, assembled by the new sewing machines, was felt.[3]

The cotton famine of the sixties, caused by the virtual stoppage of American supplies during the Civil War, and the years of boom culminating in the seventies, threw a good deal of light on the attitude of the cotton men at the time, and upon the condition of Lancashire.[4] The

[1] See E. M. Sigsworth, 'The West Riding Wool Textile Industry and the Great Exhibition', *Y.B.*, 1952; J. James, *History of the Worsted Manufacture in England*, London, 1857; C. R. Fay, *Round and About Industrial Britain 1830–1860*, Toronto, 1952, pp. 114–31.

[2] Margaret C. D. Law, *The Story of Bradford*, London, 1913.

[3] Joan Thomas, 'A History of the Leeds Clothing Industry', *Y.B.*, Occasional Papers, No. 1, 1955.

[4] See W. O. Henderson, *The Lancashire Cotton Famine, 1861–1865*, Manchester, 1934; R. A. Arnold, *The History of the Cotton Famine*, London, 1864; John Watts, *The Facts of the Cotton Famine*, London and Manchester, 1866; Barnard Ellinger, 'The Cotton Famine of 1861–4', *E.H.*, 1934.

profits of the trade had been vigorously ploughed back, so that when the supply crisis came two things were starkly visible — the size of the English commitment to cotton, and the dreadful failure to make any attempt to improve the conditions of urban life in the cotton districts. The cotton men reacted to their situation in a way which put critics in a quandary. Many of the smaller less efficient mills, as usual, quickly disappeared. But export prices of cloth rose sharply, creating a burst of optimism among the bigger men. They took little interest in the attempts being made to create employment and improve amenities through public works. Instead, they bent their energies and their capital to investment and improvement in the mills so that, in addition to short-term gains made by efficient processing of cotton that had run the blockade or come from the East or West Indies, and Egypt, they might secure their competitive position when the emergency ended. This confidence in the future, and the foresight it demonstrated, could not fail to find approval, even though many owners, as the crisis lengthened, lost heavily.

Yet conditions in Lancashire were very bad; indeed the propaganda of the Relief Committee frightened the country with its revelations.[1] It was argued that, even without the famine of raw materials, had the growth of industry and population continued in Lancashire without better sanitation and planning a plague would have ensued, involving great damage, if not destruction, for the cotton trade. The government came to the aid of Lancashire with public works loans. Too much of the gains of the past had gone into the creation of new productive capacity, to the deterioration of the workers' social condition, but the realization of what had occurred did not provoke any great misgiving about the social implications involved. The cotton industry not merely had failed to meet its social costs; it was also rendering itself increasingly vulnerable to market crises by its drive toward new investment.

There were some eight years of heavy investment in cotton down to the early and mid-seventies. A further effect of this vast increase in new equipment became apparent in the relationship between cotton prices and those of finished goods. Formerly a significant rise in the price of raw cotton was followed by a rise in prices of manufactured markets. By the seventies this was no longer so obvious. Manufacture prices

---

[1] For the thoughts of a contemporary cotton worker see R. Sharpe France, ed., 'The Diary of John Ward of Clitheroe, Weaver, 1860–64', *Trans. of the Historic Society of Lancashire and Cheshire*, 1953.

might well remain stable – due to the fact that, with demand elastic, a price rise was impossible, and with heavy investments in fixed equipment, output could not easily contract, so that the manufacturers were obliged to absorb much of the rise in the price of materials. They, in turn, sought, through wage reductions, to pass the burden to the worker. The danger of multiplying the number of spindles and looms per worker lay, not merely in simple redundancy, but more subtly in the deterioration and vulnerability of those who remained in employment.

By the seventies the heroic age of textile entrepreneurship was past.[1] As manufacturing technique became more developed, the old hectic preoccupation with new methods and new machines greatly diminished in some sectors. In the primary stages of manufacture technical stagnation had set in well before the seventies; the element of routine became increasingly important in spinning and in the weaving of staple cloths, especially for Eastern markets. The management of such concerns called for little spontaneity or initiative so far as factory organization was concerned. Spinning was traditionally less competitive than weaving, because of the much greater capital required to embark, though it remained necessary to keep a keen eye upon cotton prices and market outlets. But in the weaving of more diverse cloths, demanded by rising incomes from the sixties onward, and in the finishing stages, much more suppleness was called for in response to market changes. This meant that rearrangements of processes were frequently necessary, but revolutionary changes in method were not called for.

By the seventies the great families of the cotton and wool areas were well established socially, making landed marriages, and aspiring to knighthoods, if not quite to peerages. In cotton the Gregs, Birleys, Ashworths, and others were active in politics, and were promoting social amenity of one kind or another. In worsted, the Fosters, the Behrens, and other families enjoyed great regional prestige. In wool, the Cunliffe-Listers and the Ackroyds had assumed some of the paternalism of their landed predecessors. By the seventies Samuel Smiles was extolling the social conscience of such families rather than their innovatory characteristics.[2]

Many family businesses were being impersonalized under limited liability. The first substantial move in this direction in cotton did not come from professional promoters, but from within the industry itself.

[1] R. Smith, *The Lancashire Cotton Industry, 1875–1896*, Birmingham, Ph.D. thesis, I.T.A., 1953–54, no. 679.
[2] See below Chapter 8, sections 4, 6 and 7, pp. 297, 305, 313.

In the Oldham spinning area specialization had gone a long way, and the standardized nature of the operations made flexible management less necessary. The 'Oldham Limiteds' shared some of the characteristics of cooperatives, for the shares were locally held in fairly small blocks. Indeed, the success of cooperation in the grocery, drapery, and other trades, inspired the movement. It was born as long ago as the late fifties, but did not flourish until the boom conditions of 1872–74. The Lancashire machine manufacturers were eager to help, and would erect and equip a mill as a unit. About one-half of the loan capital was provided by workers in the industry, but the operatives took up only a minor portion of the risk bearing, share capital. Yet workmen in other trades, such as engineering, did buy a substantial part of the equity shares. The first of the Oldham 'Limiteds', the Sun Mill, paid 32½ per cent on capital in 1871. But the collapse brought a flood of bankruptcies, and the long persistence of difficulties into the eighties provoked misgivings about the future.[1]

One of the attractive answers to low prices and the increased efficiency of foreign rivals was to by-pass the merchants; the larger firms adopted this device and some did very well through direct marketing. In so doing the manufacturers showed more vigour than had the merchants, and, perhaps even more important, developed a system of stock holding for the quick supply of buyers. Much could be done to meet foreign tariffs by the establishment of mills abroad, within protected areas, a policy in which J. and P. Coats succeeded notably, in North America, and in Russia. But these policies left the principal problem untouched — namely redundant output.

Part of the answer was combination among producers: in the eighties and later nineties there was a burst of textile amalgamations in Britain, both in cotton and wool. Because foreigners were still the chief consumers, without whose purchases the industries would have to contract very drastically, it is not surprising that textile men, especially in cotton, were not, in general, sympathetic to protectionist solutions. But so far as the Empire was concerned, particularly India, they were adamant in resisting the plea that India should have the right to protect its own manufactures by means of a tariff.[2] The cotton men, no longer capable of the astonishing initiatives of the early decades of the century,

---

[1] For near-contemporary discussion see E. Helm, 'The Alleged Decline of the British Cotton Industry', *E.J.*, 1892, p. 735; T. H. S. Escott, *England: its People, Polity and Pursuits*, London, 1879.

[2] See above, Chapter 2, section 10, pp. 65, 66.

seized the more intensely on the idea of Empire, for it seemed a means whereby they could postpone for a further period the diminution of their role in the British economy.

Nottingham was a notable example of an area prosperous in pre-industrial society, sinking to very low depths indeed because of the continuance of the traditional domestic system beyond its time.[1] Frame-knitted hosiery and other such wares, right down to the late forties, were produced by methods hardly changed since the seventeenth century. When demand fell off, the middlemen, the hosiers and bag-men, were able to pass the whole onus on to the workers. The latter, dispersed and vulnerable, could make no effective resistance to the wide range of petty peculations practised upon them. To a married couple it might well appear that the only way to meet the falling wage rate per person was to increase the size of the family. But this further reduced the reward to each worker. The entrepreneurs, in this pathetic situation, had no interest in bringing about change, for with the price of labour so low, there was no incentive whatever to invest in labour-saving machinery. Moreover the men of business were convinced that the technical problems of the industry were such that no dramatic change of method was possible. So long as innovation was precluded, and so long as no great recovery in demand could occur, there could be no other course of events but increasing degradation for the workers and comfortable profits for the entrepreneurs. But, as the danger point approached, both of these assumptions, so long and so firmly held, were disrupted. A general recovery in demand, both in Britain and abroad, began. In addition, the Chevalier Clausen, in Belgium, made available his 'roundabout' frame in 1845, which, with a series of accompanying inventions by others, put the whole technology of machine knitting onto a new basis. With the possibility of the application of power, factories appeared from 1850 onward, so that by the sixties the entire basis of the industry had changed from domestic to factory production. In the new situation effective workers' unions could come into being. The regional economy moved from its dilapidated base to a new level of prosperity.[2]

[1] W. Felkin, *History of the Machine Wrought Hosiery and Lace Manufactures*, London, 1867; S. D. Chapman, *The Life of William Felkin of Nottingham 1795–1874*, Nottingham Ph.D. thesis, I.T.A., 1959–60, no. 670; F. A. Wells, *The British Hosiery Trade, its History and Organization*, London, 1935; Edwin Hodder, *Life of Samuel Morley*, 2nd ed., London, 1887; R. A. Church, *The Social and Economic Development of Nottingham in the Nineteenth Century*, Nottingham Ph.D. thesis, I.T.A., 1960–61, no. 689.

[2] For another hosiery centre see P. Head, *Industrial Organization in Leicester, 1844–1914:*

Cotton and wool do not exhaust the story of nineteenth-century textile initiative. Two fibres of long standing continued to play their parts, silk and linen, and out of the latter came a new industry, the manufacture of jute. Silk had always been under threat of French and Eastern competition; it produced no great entrepreneurial figures. When, in 1860, with the French commercial treaty, protection was removed, English silk, especially in its traditional branches, was very hard hit. Coventry, insulated from the impact of change through its protected silk ribbon industry, was suddenly assailed by the blighting wind of competition. The effect was disastrous to the silk weavers and their ancient trade, but subjected Coventry to a sharp and beneficial discontinuity, making way for the development of a range of new metal-using light engineering products.[1]

Linen offered a better opportunity for the application of the new machines produced for the cotton industry. John Marshall of Leeds was the giant of linen, with a fortune of £400,000 by 1815.[2] He showed immense application and ingenuity, but his sons did not inherit his gifts or interest. From the forties the firm declined steadily so that by 1886, within a generation of its founder's death, the greatest enterprise of its kind expired, a classic example of loss of drive from one generation to another.

As linen could borrow from cotton, so jute could borrow from linen. Dundee in the early nineteenth century produced sailcloth and sacking, using linen from Prussia and Russia. But there was a mounting need for a packing material cheaper and stronger than cotton. The East India Company, among its many novelties, brought samples of jute from India. The men of Dundee exploited the new fibre with vigour, turning their town into 'Jutopolis'.[3] They went even further, for from the fifties onward they were busy building jute mills in India itself, raising up for Dundee its own great rival, in startling contrast to the behaviour of the cotton men.

[1] John Prest, *The Industrial Revolution in Coventry*, Oxford, 1961.

[2] W. G. Rimmer, *Marshalls of Leeds, Flax Spinners, 1788–1886*, Cambridge, 1960.

[3] See C. R. Fay, op. cit., chapter 6; Dennis Chapman, 'William Brown of Dundee, 1791–1864; Management in a Scottish Flax Mill', *Explorations in Entrepreneurial History*, vol. IV, no. 3, 1952.

a Study in Changing Technology, Innovation, and Conditions of Employment, Leicester Ph.D. thesis, I.T.A., 1960–61, no. 681a. For the later hosiery trade see Charlotte Erickson, *British Industrialists: Steel and Hosiery 1850–1950*, Cambridge, 1959; W. H. G. Armytage, 'A. J. Mundella and the Hosiery Industry: the Liberal Background to the Labour Movement', *Econ. H.R.*, 1948, *A. J. Mundella, 1825–1897*, London, 1951.

But the rôle of the textile men was becoming less important in the economy as a whole. Cotton and wool had enjoyed their greatest relative importance in the middle decades of the century. Though they continued to grow, and indeed in the early eighties still provided some 46 per cent of exports, it was to engineering and iron that the initiative was passing.

4. ENGINEERS, CONTRACTORS AND PROMOTERS

The textile men – the same is true of all industrial innovators – were dependent for the sinews of their industry upon the engineers: those who built the marvellous new machines, those who constructed the new mills, and standing behind them, the civil engineers who were unifying the country, and binding it to the world of trade.[1]

The post-Waterloo engineers did not constitute a profession in the sense that they had merely to offer their technical services to someone else.[2] In large measure they had to create their own opportunities, and bear the associated risks. The great machine makers from whom progress flowed were executants and risk bearers; indeed if unsuccessful in these two functions it was extremely difficult to make inventions, and almost impossible to profit by them.[3]

A great engineering enterprise called for a combination of genius and judgement, together with the ability to gain allies and to delegate to subordinates. James Nasmyth and William Fairbairn have left the fullest available accounts of the way in which this was done.[4] The former, locating his works adjacent to both the Liverpool and Manchester Railway and the Bridgewater Canal, intended them to serve primarily the rising Lancashire demand for machinery, but eventually the whole of Britain and beyond. Both he and Fairbairn prospered greatly. To their retirement in the fifties they were the arch-types of their generation of engineers, heading large concerns built by their own efforts, yielding high wages and good conditions of work, utterly impatient of the attempts of trade unionists to interfere with their operations.

The point of view of such men is understandable enough. Even more

[1] For general background see W. H. G. Armytage, *A Social History of Engineering*, London, 1961.

[2] For the history of the engineering professions see A. M. Carr-Saunders and P. A. Wilson, *The Professions*, Oxford, 1933.

[3] See above, Chapter 3, section 3, p. 86.

[4] *Autobiography*, ed. Smiles, London, 1883; A. E. Musson, 'James Nasmyth and the Early Growth of Mechanical Engineering', *Econ. H.R.*, 1957, pp. 121-7.

than the enlightened cotton masters, they thought in terms of a paternal relationship. In Nasmyth's case it was of the kind that had existed between himself and his beloved master, Henry Maudslay. But he perhaps forgot that not all engineering enterprises were under such a beneficent hand as his; the workers were often obliged to combine to assert their rights against their masters. Nasmyth, like Maudslay, was always constructing new tools in order to reduce dependence on 'mere manual strength or dexterity'. The result was that, in his own engineering works, ever higher standards were required from the élite of his workmen. But the quick execution of customers' orders required an ever-diminishing dependence upon the scarce elements of skill.

There is an air of amiability about the first great generation of engineering firms, down to the mid-century. Information and ideas passed freely between them (though not altogether without recrimination about priorities). The story of William Fairbairn is similar to that of Nasmyth, though his unhappy excursions into iron shipbuilding at Millwall in the thirties and forties showed how easily an engineer could over-reach himself. The demand for their products was great, with millowners, railway constructors, shipbuilders, all seeking their services. So the great engineering complexes of south Lancashire, of Leeds, of Clydeside, and of London came into being, with a potency for the growth of the economy that was enormously greater than was indicated by their size.[1]

By the mid-century the older easier conditions were being modified. Engineering firms, many with little real originality, multiplied to participate in good profits. There were 677 firms returned in the census of 1851. More than two-thirds had less than ten employees, but fourteen firms had upward of 350 men, reflecting the growth both in the number and in the size of firms that had come in the preceding decade; engineering took its place among the large-scale industries. But the machinery makers, though still retaining something of their aristocratic position, from the mid-century onward, had, in common with other industries, their periods of trial both in terms of slumps and of strikes.

Engineering was continually widening its technical bounds. New demands for the means for large-scale manipulation of nature were met. The Cornish Tangye brothers in the fifties specialized in hydraulic equipment of all kinds, scoring a great success in 1858 with their jacks

---

[1] See Charles Wilson and William Reader, op. cit., chapters I and II; J. Napier, *The Life of Robert Napier of West Shandon*, Edinburgh and London, 1904; A. E. Musson, 'An Early Engineering Firm: Peel, Williams and Co. of Manchester', *B.H.*, 1960.

which launched the stubborn Great Eastern, and so launched them-selves.[1] The firm of Burrell and Sons of Thetford pioneered the build-ing of traction engines for agricultural tasks in the fifties.[2] The makers of agricultural machinery like Ransome, Sims, and Jeffries, were important by the sixties; the machine tool makers had by this time achieved their typical modern position of being the last to feel the coming of crisis and among the last to feel recovery. Moreover by the third quarter of the century there had been great advances abroad: American machine makers had gained notable advantages in some lines.[3]

Engineering leadership from the thirties and forties was passing rapidly to firms directed toward the transport revolution represented by railways and iron steamships, both for peace and war uses.[4] The age of Rennie, Locke, Vignoles, Brassey, Stephenson, Brunel, and the other giants of construction, had begun. These were the men who, in conjunction with the new generation of civil engineering consultants, and the financiers, laid down the enormous new capital of Britain.[5]

The consulting engineers selected the routes for the new railways and designed the works; the contractors carried them out; the financiers and administrators concerned themselves with finding the resources. The contract system was not new. It had been developed in canal construction, and in the building of docks and harbours, by such firms as Joliffe and Banks. As early as 1816 the elder Rennie had laid it down that engineers should maintain this division of labour, sticking to their planning and supervising functions, 'free alike of contractors and contracts'.[6] George Stephenson rejected this distinction, for his new railway projects were so novel that he was obliged in large measure to do both jobs. But this could not go on; the old division of labour was bound to reassert itself. George's son Robert, in construct-ing the London and Birmingham Railway in the early thirties, used contractors on a large scale; Joseph Locke, another great engineer from the coalfields, did the same.[7]

[1] Rachel Waterhouse, *A Hundred Years of Engineering Craftsmanship*, Birmingham, 1957.

[2] Ronald H. Clark, *Chronicles of a Country Works*, London, 1952; *The Development of the English Traction Engine*, Norwich, 1960.

[3] D. L. Burn, 'The Genesis of American Engineering Competition', *E.H.*, 1931.

[4] Canal construction for barges was virtually ended by 1825; the Manchester Ship Canal of 1894 was a new concept. See Robert Payne, *The Canal Builders*, London, 1959.

[5] See D. B. Wardle, 'Sources for the History of Railways at the Public Record Office', *J. Tpt. H.*, 1955–56, p. 214. For a study of regional development see J. A. Patmore, 'The Railway Network of Merseyside', *T.I.B.G.*, 1961.

[6] *Autobiography*, p. 280.

[7] L. T. C. Rolt, *George and Robert Stephenson*, London, 1960; Joseph Devey, *Life of Joseph Locke, Civil Engineer*, London, 1862.

The engineers' task certainly tried their mettle. In this early phase of railway building the unknowns in terms of construction costs and engine efficiencies were probably greater than at any later time. It is too easily forgotten that, as things stood in the thirties, it was by no means obvious to promoters that the balance of advantage lay with the locomotive engine, as opposed to the stationary one, or even the horse.[1] Indeed the latter faithful animal very nearly powered the Newcastle and Carlisle line; it was only defeated by the high cost of sidings for servicing. Nor was the steam omnibus to be lightly dismissed as a rival. So fundamental a problem as the width of gauge could not be settled by simple reference to principle; George Stephenson insisted upon his 4 ft. 8½ in. spacing derived from the Killingworth Colliery in Northumberland; Brunel built his Great Western with a gauge of no less than 7 ft.[2] The first survey for the Edge Hill tunnel contained an error which would have caused the two bores to miss one another somewhere under Liverpool.

Unfortunately, the engineer had to perform as a public relations officer, spending much of his time in attending upon Parliamentary Committees. Indeed, this was sometimes more arduous than his true duties of construction, because of the hostility of rivals, and the serious concern by conscientious men about the new developments. Parliament, in order to ensure sound constructional practice, often interfered ineptly: many projects were rejected because of failure to conform to obsolete concepts.[3] The extent of railway legal and Parliamentary costs made it possible for Lord Grimthorpe to amass an immense fortune in handling railway affairs.[4] The essence of economy in construction on this unprecedented scale was to finalize the planning before beginning, a requirement that placed a further strain upon the engineer. It is not surprising that many died young: Robert Stephenson, the younger Brunel, and Locke all scarcely reached their middle fifties.

Many contractors paid the price of their own miscalculations. But there emerged a class of men the daring of whose enterprises it is difficult for a later generation to conceive. Thomas Brassey, encouraged by the elder Stephenson, and by Locke, became the greatest of the railway contractors.[5] He and his partners, Peto and Betts, and others, were not merely prepared to undertake the complete installation,

---

[1] Jack Simmons, 'For and Against the Locomotive', *J. Tpt. H.*, 1955/56.
[2] L. T. C. Rolt, *Isambard Kingdom Brunel*, London, 1957.
[3] Robert Stephenson, *Presidential Address, Institution of Civil Engineers*, 1856, p. 18.
[4] Peter Ferriday, *Lord Grimthorpe, 1816–1905*, London, 1957.
[5] Arthur Helps, *Life and Labours of Mr Brassey, 1805–1870*, London, 1872.

but also to relieve the engineer of the problems of finance, labour, and the provision of the heavy plant required. Brassey was instrumental in the construction of nearly 4,500 miles of railroad in three continents; he was the first British contractor to undertake construction abroad. He combined the traditional skills of the miner in tunnelling, and of the navvy in levelling and general handling of rock and earth, with the new engineering principles, now being rapidly developed. He curbed the notorious scamping and bribery of the sub-contractors, men, for the most part, of strong natural abilities, insight into the cost and method of executing work amounting to instinct, low tastes, violent habits, and grasping tenacity of purpose.[1] The railway builder felt the exhilaration, akin to that of the great soldier, who combines great bodies of men and materials to a vast but defined end, against all the hazards of nature and personality. Brassey with his enormous projects (he had thirteen heavy contracts under way in 1845), represented something new in entrepreneurial scale; to do so he became a master of delegation. Even so, in the financial crisis of 1866, calamity was only narrowly avoided. But too often the contractor accepted no responsibility for the navvies and their unskilled helpers of the railway gangs, allowing the most appalling condition to arise.[2]

The railways provoked new large undertakings to service them and to connect them with the world beyond the British Isles. The stimulus to dock and harbour construction, already felt, was greatly strengthened, especially on the Mersey, under the sponsorship of the Docks and Harbour Board, eventually producing by the eighties some seven miles of monumental granite sea-wall. The great railway terminals came to depend upon the engineering and metallurgical works they gained. Crewe, an entire community, sprang from the railway; the old towns of Carlisle and Rugby underwent change and expansion.[3]

Nor did the railway merely connect centres of population: it operated powerfully within them. In January 1863 the promoters of the Metropolitan Railway Company introduced London and the world to subterranean travel. Omnibuses plied the streets from the late twenties, largely following French initiative; the horsedrawn tram came from America in the late fifties and early sixties, provoking the

[1] Obituary of Brassey, *Proceedings, Institution of Civil Engineers, Session 1871–72*, part I, 1872, p. 248; R. K. Middlemas, *The Master Builders*, London, 1963.

[2] See R. A. Lewis, 'Edwin Chadwick and the Railway Labourers', *Econ. H.R.*, 1950; J. A. Patmore, 'A Navvy Gang of 1851', *J. Tpt. H.*, 1962.

[3] W. H. Chaloner, *The Social and Economic Development of Crewe, 1780–1923*, Manchester, 1950.

Tramways Act of 1870.[1] These enterprises had profound effects on the shape and extent of English cities.

The railway itself was a most potent element in creating new railways; the English astonished Frenchmen by building a temporary line first, parallel to the permanent way, to make construction easier. But for the lifting of the fabulous quantities of earth, human muscles served, as in the days of the Pharaohs. In the late fifties steam excavators were used in the construction of the Grand Trunk Railway in Canada because of labour shortage, but elsewhere the navvy was the great remover. By an extraordinary convergence of initiative an average army of over 100,000 of the most physically powerful men in Britain, each reputed to be able to lift some 20 tons of earth to a height of six feet in a day, worked for twenty years to produce some 10,000 miles of railway. Their work was not continuous either in time or space, for there were great bursts of building followed by lulls, and all the while the place of their labours was changing as the railhead crawled across Britain or a new start was made.

One of the great keys to the enormous success of the gentle Brassey lay in the way these men accepted his leadership. In large measure this rested upon fairness, for Brassey consulted and often deferred to the opinions of his gangers. He worked through sub-contractors, providing them with all necessary materials and equipment, but leaving the hiring of manual labour to them, operating in effect the butty system taken from the coal mines, but keeping watch against abuses. In a sense the railways of Britain were carried through on the same basis as the development of the West Indian Islands a generation or more earlier — the application of labour gangs. It was an astonishing feat, consisting in the supply, control, and guidance of men, aided by ancient tools and the modern road they themselves were creating.

The earliest lines, the Stockton and Darlington of 1825 and the Liverpool and Manchester of 1830, were conceived as entities in themselves — the one to remedy a traditional defect of the sailing ship by making it possible to bypass the tedious windings of the river Tees up to Stockton, the other to offer a profitable alternative to costly canal transport between the two great Lancashire towns.[2] Their promoters, local men of substance and initiative, made their calculations as careful and realistic as possible.[3]

[1] Clapham, vol. II, pp. 202–4.

[2] Sir Alfred E. Pease, Bart, ed., *The Diaries of Edward Pease the Father of English Railways*, London, 1907, p. 83.

[3] C. F. Dendy Marshall, *A Centenary History of the Liverpool & Manchester Railway*,

Estimating potential revenue, though by no means easy, was simpler than it later became. Even so, in the great majority of cases, railway promoters consistently underestimated both the capital cost and future revenue. This meant that though the project might eventually be profitable, the degree of commitment was much higher than was originally expected. In the next phase of railway creation, that of the construction of trunk lines, linkages between lines became possible, and even alternative routes were offered. This meant that charges and the volume of carriage became even more difficult to assess. As longer lines, with correspondingly greater risks were projected, an ever more careful calculation was called for.

The response to the railway challenge was not of this rational sort at all; indeed it was a lapse into irrational optimism, culminating in the railway manias of 1836 and 1845. Railway sponsorships passed from the cautious hands of provincial men of business to a new generation of projectors, for whom plausibility was more important than proof. Whereas many of the factory masters had allowed the capital question to dominate their minds, many of the new buyers of shares in the joint-stock railway companies showed a frivolous disregard for the problem. During 1845 Parliament authorized the construction of new railways that would have doubled the then existing lines of some 2,200 miles. For the careful projector this extravagant planning of new lines made the problem of calculation much more difficult. It rendered whole areas of the system vulnerable to the high fixed costs which had been so rashly incurred. Increased traffic was the only escape from bankruptcy. Hence began the long struggle for control of those lines which might provide profitable new connections. With it came the struggle of pre-emption: the building of lines merely to preclude a rival company.

Just as the railways produced a new generation of constructive tycoons, so they produced ownership and management titans on a new scale. George Hudson created the greatest and most resounding reputation, and suffered the greatest fall.[1] This draper of York was relatively early on the scene, using a fortunate legacy, and a chance meeting with George Stephenson in 1835, to launch himself upon the dangerous game of railway promotion and amalgamation. Hudson was not a railway man in the sense of having building or operating

[1] R. S. Lambert, op. cit.

London, 1930; Brother Lewis Donaghy, 'Operational History of the Liverpool and Manchester Railway, 1831–1845', *Dissertation Abstracts*, vol. XXI, 1961, no. 11, item 3758.

experience; he was a manipulator. He was the first company promoter to operate on the scale now made possible by the railways.

The men he handled comprised the Boards of the infant undertakings. Democratic procedures still ruled to a considerable extent in such bodies; but their members, not surprisingly, had little notion that, for democracy to be effective in business, there must be adequate data, and this in turn rested upon the development of adequate accountancy techniques. It was this ignorance, together with the inherent vulnerability of the railways, that made Hudson's transactions possible. The great bankers had for generations exercised financial control over many enterprises, but it was Hudson who ushered in the new era of big business, characterized by the gaining of control of real resources on an enormous scale through the manipulation of companies.

The collapse of Hudson was by no means the end of great names in the railway world. Indeed, the age of the colossus was succeeded by the age of the matching giants, for each major company produced its champion. Usually men of humble origins, they found themselves in command of systems, but confronted by rivals as great and as ruthless as themselves.[1] A state of relentless war often existed between them in the fifties and early sixties, even reaching the absurd lengths of hand-to-hand battles between train staffs, and the competitive blockading of termini, together with endemic rate warfare. Simple ideas of gentlemanly behaviour within the rules of a free market were hopelessly inadequate as a guide to the understanding of the new age of heavy fixed capital.[2] Bitter complaints continued in the later sixties and seventies about the ineptitude of railway management.[3] Nor were the railway promoters and operators blameless in subsequent decades.[4] The collapse of the Tay Bridge in 1879, carrying with it an entire passenger train, was the outcome of the rivalry between the North British and the Caledonian Railway; it ended with the North British paying £5 for each human body hooked out of the Tay.[5]

But for all the zest for battle shown by many of these men, it was clear that if a formula for peace was not found mutual destruction

---

[1] See C. Hamilton Ellis, *British Railway History: an Outline from the Accession of William IV to the Nationalization of the Railways, vol. I, 1830–1876*, London, 1954, parts II and III.

[2] Phillip S. Bagwell, 'The Rivalry and Working Union of the South Eastern and London, Chatham and Dover Railway', *J. Tpt. H.*, 1955/56.

[3] A Civil Engineer, [Francis R. Conder] *Personal Recollections of English Engineers, and of the Introduction of the Railway System into the United Kingdom*, London, 1868. 'The Board of Trade and the Railways', *Econ.*, 28 February 1874, p. 257.

[4] C. Hamilton Ellis, *British Railway History, 1877–1947*, vol. II, London, 1959.

[5] John Prebble, *The High Girders*, London, 1956.

must eventually come. Its nature was apparent from the form of the struggle itself: tactics had suggested the absorption of smaller lines; absorption led on to amalgamation.[1] Hudson had shown the way with the creation of the Midland in 1843-44. By 1854 the North Eastern Company had been formed, which by the mid-sixties had a monopoly of its area. A similar arrangement for the south-west first failed as early as 1848. The early sixties was a great period of amalgamation. By the crisis of 1866 East Anglia was unified, the Great Western had swallowed some 400 additional miles, and the movement had gone a long way in Scotland. Public misgiving was reflected in the Report of the Royal Commission on the Railways in 1867, and in setting up of a Select Committee of both Houses of Parliament in 1872. Thereafter, apart from minor changes, the situation remained unchanged until the nineties. The railway now had found a rough and ready *modus vivendi*; after 1867 they learned to afford one another's facilities rather than to wage irreconcilable war.[2] From 1850 the Railway Clearing House had operated under Act of Parliament to arrange for through-carriage using different lines.

By 1870 the great days of railway construction in Britain were nearly over — though a good deal was done in the next decade or so in building branch lines, links, and local connections. The railways of Britain, now covering most of the country, produced their last great burst of engineering initiative, first with the Tay Bridge, stretching its rickety length over two miles, but more soundly with the Severn Tunnel and the splendid Forth Bridge. Sir Edward Watkin, the Railway King of his day, consumed since boyhood with the ambition to run a train from his native Manchester to Paris, caused work to be begun on a Channel tunnel in 1881 only to have the British government stop it in the following year.[3]

With such a rate of progess in engineering it is not surprising that the industry should, from the fifties onward, become increasingly involved with production for war. The navy was being remade by armour plating and iron construction, with ordnance and control mechanisms becoming increasingly complicated. The infantry man and gunner also were affected. Whitfield in 1854 was invited by the government to produce rifle-making machinery. His developments extended to

---

[1] C. I. Savage, *An Economic History of Transport*, London, 1959, p. 65.

[2] For railway legislation see below, Chapter 9, section 7, p. 361.

[3] Humphrey Slater and Correlli Barnett, *The Channel Tunnel*, London, 1958; P. A. Keen, 'The Channel Tunnel Project', *J. Tpt. H.*, 1958; A. W. Currie, 'Sir Edward Watkin, a Canadian View', *J. Tpt. H.*, 1957-58. See above, Chapter 3, section 1.

heavy guns, using hydraulically compressed steel. Inspired by the Crimean War, the local master gunsmiths of Birmingham, so long independent, combined to form the Birmingham Small Arms Company. In consequence of this trend, signs of monopoly in ship construction were beginning to appear, for only the largest firms could produce the new compound armour of the seventies and eighties: two firms, Cammel's and Browns, under cartel arrangement, divided the contracts. The day of massive armaments in iron and steel was about to dawn, coinciding with colonial expansion in the eighties.

5. MAKERS AND OWNERS OF SHIPS

Though the size of ships was continuously increasing, scale was not the principal challenge confronting the shipbuilders.[1] New construction materials — iron and later steel, and new modes of propulsion — the steam-engine, with eventually the triple compound engine, and the screw to replace the paddle, were the elements of the new age.[2] The railway contractors could work to the plans of others; the shipbuilders, with the conservatism nurtured by generations of traditional construction in an older idiom, were called upon both to plan and execute the new kind of ship.

British ship design and construction had lagged behind the French during the Napoleonic Wars, relying still, very largely, on the empirical methods of the days of Pepys. Each ship was unique — the frames were set at roughly estimated distances, and the rest done at the pleasure of the builder. Though steam and iron patently called for a more thoughtful treatment, the construction of hulls continued on the old basis of eye and judgement. Only the Royal Dockyards sought, in spite of lack of political support, to evolve a scientific system and to train apprentices.[3] The products of the school of naval architecture founded in 1811 at Plymouth were the men who met the challenge of the naval scares inspired by fears of the French in the late fifties and sixties. Even the invaluable contributions of the general engineers from outside — Maudslay, the Rennies, the Napiers, and others, did not shake the complacency of the private shipbuilders for thirty years after 1840.

[1] For general bibliography, Rupert C. Jarvis, 'Sources for the History of Ships and Shipping', *J. Tpt. H.*, 1958.

[2] Sir William Fairbairn, *Treatise on Iron Shipbuilding, Its History and Progress*, London, 1865; R. H. Thornton, *British Shipping*, Cambridge, 1939; W. S. Lindsay, *A History of Merchant Shipping*, London, 4 vols., 1874–76, vol. III.

[3] S. Pollard, 'Laissez-faire and Shipbuilding', *Econ. H.R.*, 1952.

The Crimean War in the fifties signalled the arrival of the iron ship, screw powered for commercial carriage. The battle of Hampton Roads in the American Civil War, in 1862, dramatically demonstrated the superiority of the iron-clad for war; the end of the wooden walls was in sight.[1] In 1860 the Institution of Naval Architects was formed, largely by Admiralty men. Though as late as 1879 its president had not outgrown his earlier idea that naval officers were fully competent to design warships, it did much to improve design.[2] Slowly the shipbuilders accepted conversion. Science, calculation, and draughtsmanship were finding their way into the yards, so long ruled by the empirics of wood and wind, though progress was still slow down to 1873.

The great depression of the seventies was a rude shock to complacency; thereafter ship construction was much more a matter of taking thought. The experiments of Beaufoy and Bentham on hull shapes had been resumed by William Froude in the later forties; by the seventies Admiralty money was available for a testing tank. The private builders still lacked initiative, for it was the navy that first brought steel into general use. Yet the private shipbuilders were never tired of slighting the naval dockyards. This was especially so in the sixties, when new joint-stock companies had created excess building capacity, and sought to maintain orders by propaganda against the navy's own establishments.

National success in the new age of shipbuilding depended upon the interplay of many factors, especially the productivity of the iron and coal masters who made available cheap materials to British shipbuilders, and the marine engineers on the Thames, the Clyde, and the Mersey, who provided the marvellous engines.[3] The shipbuilders' attitude to the sea was not unlike that of the peasants' attitude to the land; they had to be pushed ahead. The engineers and the men of coal and iron provided the impetus. But the joint product of their efforts with that of the builders was much superior to the ships of other nations. The world bought its ships in Britain; British shipbuilding from the sixties onward enjoyed an extraordinary supremacy, in startling contrast to its humiliating inferiority to the Americans in the earlier part of

[1] J. Scott Russell, *The Fleet of the Future; or England without a Fleet*, London, 2nd ed., 1862.

[2] K. C. Barnaby, *The Institution of Naval Architects, 1860–1960; an Historical Survey of the Institution's Transactions and Activities over 100 Years*, London, 1960.

[3] See D. Napier, *David Napier, Engineer, 1790–1869, an Autobiographical Sketch*, Glasgow, 1912.

the century. They sailed faster than before, carrying bigger pay loads.[1] Early in the century British shippers had developed the habit of chartering rather than owning. Many West-India men had been owners, but the new American trade was largely carried in American bottoms. This hiring of ships had contributed to the decline of the British merchant marine, and the American superiority in cost and construction accentuated the trend. In the thirties at Liverpool the American packets got ten passengers for one that went by a British vessel. But steam brought a new situation. Led in the Spanish and Portuguese trade in the thirties by Wilcox and Anderson, who demonstrated the inefficiency and corruption of the old Admiralty packets to Falmouth by founding the Peninsular and Orient Line, a new generation of shipowners began to arrive, attracted from the thirties onward by the grant of fairly generous subsidies for the carriage of mails. This was a vital step, for long-distance carriage by steamship was still far from economic, as the failure of those not enjoying mail contracts abundantly proved. In this age of *laissez-faire* it was state intervention that provided the first great impetus to British maritime leadership.[2] It was no new principle, for governments had always been responsible for the mails, but it represented a great extension of older practice.

A battery of great names began to appear. The virtuoso of construction, I. K. Brunel, conceived the Great Western Steamship as an Atlantic extension of the Great Western Railway; it was followed by the legendary Great Eastern.[3] The Great Western inspired Nasmyth's steam hammer, for a new tool was necessary to forge the great shafts. Samuel Cunard's success in 1839 in gaining the Atlantic mail contract defeated and enraged the Bristol sponsors of the Great Western, and made Liverpool the great Atlantic passenger terminus.[4] To it fell much of the emigration trade, so busy in the forties and later. The Pacific Steam Navigation Co. and the Royal Mail Steam Packet Co. were soon in operation. By 1860 £1 million a year was being paid in mail contracts, keeping at sea some fifty-three regular ocean liners.

Many of the new steamships were lost, grounding as they were

---

[1] For shipbuilding firms see: Sir Allan Grant, *Steel and Ships: the History of John Browns*, London, 1950; Anon., *Alexander Stephen and Sons Ltd., a Shipbuilding History, 1750–1932*, Glasgow, 1932.

[2] See below, Chapter 9, section 7, p. 361.

[3] James Dugan, *The Great Iron Ship*, London, 1953.

[4] N. R. P. Bonsor, *North Atlantic Seaway*, Prescott, 1955; C. R. Vernon Gibbs, *Passenger Liners of the Western Oceans*, London, 1952; A. Basil Lubbock, *The Western Ocean Packets*, Glasgow, 1925.

entering or leaving port, sinking on the high seas amid dramatic scenes recounted by survivors, or simply disappearing without trace. Speed was a great consideration especially on the North Atlantic, bringing the premature end of many a ship. Thus did the passengers bear an important part of the risk of development. Only Cunard placed safety first, not losing a life other than in wartime, but losing custom to its more exciting rivals, especially in the later sixties and early seventies.

But the steamships were merely the élite of the sea.[1] Wood and sail still represented the bulk of shipping.[2] British sailing tonnage reached its peak of just under five million in 1865; steam comprised well under a million, but was much more effective. The American response to steam was a great development of more efficient sailing vessels. Liverpool shipowners, especially, bought many of such ships, to launch the famous Black Ball line and others. While the new steam liners were ensuring themselves with mail contracts, it was the British sailing ship that bore the effect of the repeal of the Navigation Acts in 1849 leaving carriage to and from Britain open to all.

As late as 1860 the shipowners were prepared to blame their difficulties upon the repeal of the Navigation Laws.[3] They had become inefficient and irresponsive, failing to make any real extension to tonnage since 1815. Virtually a new industry came into being in face of the new challenge in a surprisingly short space of time. Every aspect of shipping was affected. The training and behaviour of British shipmasters, formerly notoriously bad, was placed on a new footing of professional dignity and skill. The shipbuilders of the Clyde showed an especially vigorous response to the new conditions.[4] In 1853 they had some 60,000 tons on the stocks, mostly using iron and steam. Many of the employers had themselves been workmen within the previous thirty years.

The discovery of gold in Australia in the early fifties gave an enormous new importance to the Antipodes, and to designing for speed. London and Liverpool and Glasgow shipowners quickly entered the

---

[1] J. R. T. Hughes and Stanley Reiter, 'The First 1945 British Steamships', *Jr. American Statistical Association*, 1958.

[2] G. S. Graham, 'The Ascendancy of the Sailing Ship, 1850-85', *Econ. H.R.*, 1956.

[3] *Econ.*, 1860, p. 982, quoting extracts from the Report of the Select Committee on Merchant Shipping.

[4] W. S. Cormack, *An Economic History of Shipbuilding and Marine Engineering*, Glasgow Ph.D. thesis, 1931; also his contribution to C. R. Fay, *Round about Industrial Britain*, chapter 7.

field. They bought American ships — an additional goad to British builders. Sail had still a great advantage in such long journeys, striking the public imagination with annual races, carrying the season's wool crop to London for forwarding to the West Riding, and later engaging in the tea races from Foochow. But steam was catching up, greatly aided by the enormous fuel economies of the new compound engines of Elder and Randolph, fitted in 1856 in two ships of the Pacific Steam Navigation Campany. In nine years after 1863 fuel consumption of good engines was halved.

Protagonists of sail struggled to reduce operational costs by making sail more manageable through the shortening of masts and the reduction of the size of sails. Compromises were tried — composites of iron frame and wooden planking, and sail aided by auxiliary engines. But compromise was as futile as the fierce clinging to sail. In spite of the conservatism of shipowners and government officials, sail and wood were doomed to yield to steam and iron. Even so, conservatism could impose paddle wheels on British steamers down to the sixties, in spite of the obvious superiority of the invention by the Swede, Captain Ericsson, of the screw propellor.[1] Ship management became a highly developed branch of trade, involving the disappearance of most of the welter of petty firms with which the seaports were littered.[2] Even the staggering prestige of the American ships, dazzling the British adolescent mind in the mid-century, was fully matched in the sixties.

British shipbuilding and the British merchant marine had achieved an outstanding position by the seventies. Under competitive stimulus, and aided by the obstacles confronting Britain's rivals — the Civil War in America, the political fragmentation of Germany, the inadequacies of the Continental ports — success had been achieved. Britain had as much commercial sea-going shipping as all the other nations of the world put together; between 1852 and 1864 her tonnage had increased by no less than 50 per cent. With the increase of the merchant fleet came a new code of control, beginning in 1850, defining safety rules and working conditions. Iron ships constituted some five-sixths of the fleet by the seventies; indeed 1860–80 was their great apogee, prior to their eclipse by steel. The new modes of production meant the virtual end of the small firm; by the eighties shipbuilding was in the hands of large enterprises.

---

[1] William Conant Church, *The Life of John Ericsson*, London, 1890.
[2] David R. Macgregor, *The China Bird: the History of Captain Killick and one hundred Years of Sail and Steam*, London, 1961.

But London, once dominant, was unable to hold its lead after 1850. Distance from coal and iron, together with strong unions and a rigid wage system, weakened the Thames industry until in the crash of 1866 it received its virtual death-blow. The northern builders, on the Clyde, the Mersey, the Tyne, and Wear, were supreme.[1] The sponsors of British shipbuilding ascendancy were the new ship-owners who had emerged under the post-Navigation Act conditions – Holts, Bibbys, Inmans, Booths, Leylands, Thompsons, Brocklebanks, Donaldsons, Curries, and others, largely based on the western ports, trading in all the seas of the world.[2] Their fortunes did not rest upon speculations and quick gains, for the plungers almost invariably failed. The formula for success lay in the long continuous struggle for efficiency and low costs, financed by the ploughing back of profits, and involving modest personal consumption, at least at first. For them the calculation upon which all else rested was the future course of freight rates, and the consequentially necessary adjustments in fleet size, including the placing of orders for new building.[3] Their families formed a new kind of commercial aristocracy by the seventies, abounding in confidence, eclipsing the families that had rested upon trade only, and furnishing the outports with a new generation of leaders in their rivalry with London. Each great branch of trade had its peculiar problems to which these men addressed themselves with close applica-tion: the North Atlantic, the South American, the Indian, and Far Eastern and Antipodean.[4] Moreover, under their management, the conditions of work in the merchant service steadily improved.

In strong contrast to their predecessors of Navigation Law times, they had embraced the need for vigorous competitive rivalry. By the mid-eighties, however, the first signs of a return to the principle of regulation began to appear, with tentative experiments with shipping rings. The shipowners, even more than other great capitalists, had found the problem of erratic trade very difficult.

[1] S. Pollard, 'The Decline of Shipbuilding on the Thames', *Econ. H.R.*, 1950.

[2] F. E. Hyde, written with the assistance of J. R. Harris, *Blue Funnel: A History of Alfred Holt and Company, 1865 to 1914*, Liverpool, 1956; A. H. John, *A Liverpool Merchant House*, London, 1959; G. Blake, *The Ben Line 1825–1955*, London, 1956; Alastair M. Dunnett, *The Donaldson Line: a Century of Shipping, 1854–1954*, Glasgow, 1960.

[3] Douglass North, 'Ocean Freight Rates and Economic Development, 1750–1913', *J. Econ. H.*, 1958.

[4] M. A. Jones, *The Role of the United Kingdom in the Transatlantic Emigrant Trade, 1815–1875*, Oxford D.Phil. thesis, I.T.A., 1955–56, no. 741; Kwang-Ching Liu. *Anglo-American Steamship Rivalry in China, 1862–1874*, Cambridge, Mass., 1962.

6. THE LESSER USERS OF METALS

Vast quantities of metal were used in the great construction works of the century. But in Birmingham and Sheffield there were complexes of entrepreneurs famous for their manipulation of metals on a smaller scale into tubes, nails, and wires, for their skills in plating and jewellery making, and for their 'toys' of brass and iron – tools, weapons, and household goods of all kinds.[1] Sheffield, with its leadership in steel, was increasingly celebrated for tools and utensils requiring special tempers and cutting edges.[2] In both cities the metal users were typically men of modest means, with only a few employees. The distinction between master and man was much less developed than elsewhere, with something like a pre-industrial relationship long surviving in strong contrast to the Manchester gulf between masters and men.[3] Down perhaps to the mid-century these small men gave the life of these towns its typical flavour. Radicalism in Birmingham and Sheffield found much support among them.

It was these lesser users of metal who brought the necessary contact with the domestic metal consumer. Their thoughts dwelled not upon vast constructional undertakings or the increase in the flow of materials, but upon the kind of ends that might be served, especially in the home – looking-glasses, fire guards, brass bedsteads, pots and pans, water taps, buttons, lamps, gas brackets, needles, screws, razors, scissors, pocket knives, cutlery, and an infinity of other things.

The makers of tinplate were in a peculiar situation: they stood between the iron masters and the users of thin sheets.[4] By 1800 Britain had a tinplate industry that was small, but which was nevertheless the world's largest source of material for tin trunks, baths, dairy equipment, lighting fixtures, household ornaments, and all manner of domestic and industrial goods in which wood had yielded to metal plate. The Crimean War brought a great increase in tinned food for the sustenance of the nation's defenders; tinplate had profound effects

[1] See W. H. B. Court, *The Rise of the Midland Industries, 1600–1838*, Oxford, 1938; G. C. Allen, *The Industrial Development of Birmingham and the Black Country, 1860–1927*, London, 1929; Conrad Gill and Asa Briggs, *History of Birmingham*, Oxford, 1952.

[2] Sidney Pollard, *Three Centuries of Sheffield Steel*, Sheffield, 1954; M. W. Flinn and Alan Birch, 'The English Steel Industry Before 1856; with Special Reference to the Development of the Yorkshire Steel Industry', *Y.B.*, 1954; G. I. H. Lloyd, *The Cutlery Trades: an Historical Essay in the Economics of Small Scale Production*, London, 1913; J. B. Himsworth, *The Story of Cutlery*, London, 1954.

[3] Alan Fox, 'Industrial Relations in Nineteenth Century Birmingham', *O.E.P.*, 1955.

[4] W. E. Minchinton, *The British Tinplate Industry*, Oxford, 1957.

upon the shippers and sellers of food for civilian use, for by the eighties, there was available, in addition to refrigeration, the means for cheap and effective canning. The world petroleum industry added to the demand for tinplate: after 1859 it required a new container to replace the traditional liquid carrier, the wooden barrel. Britain enjoyed almost a world monopoly in exports of tinplate down to the nineties. So was added to the regional economies of the midlands and South Wales a valuable adjunct to iron and steel.

Increasingly after the mid-century both Birmingham and Sheffield were producing or attracting men interested in developing larger enterprises. As their initiatives succeeded, small-scale manipulation of metal yielded pride of place to large light engineering firms, employing not craftsmen but the new army of the semi-skilled.[1] From America came self-acting machinery to make the handcraftsman obsolescent for many tasks. In 1865, at long last, the revolution in the production of firearms, stimulated by Colt's famous revolver, mass produced to meet the needs of the Californian gold diggers in '49, reached Britain.[2] The Birmingham Small Arms Company in which the Birmingham handicraftsmen came together to form a factory, began machine production, on the principle of interchangeable parts.[3] As the heavier branches of engineering also assumed a more important role in these areas, toys and cutlery found themselves joined by heavy products in both metalware cities.[4] In the seventies Sheffield had a practical monopoly in the production of armour plates.[5] With the coming of mass production and heavy units the quality of local life underwent change, with the older intimacy of master and man yielding to the impersonal relationships that had come so much earlier in south Lancashire. But the semi-skilled man was in a stronger position, in general, than the cotton operative.

The builders of railway rolling stock were the first men of business to turn out highly complex products meeting two conditions – production in long runs, and bringing together a wide range of components, the products of other industries.[6] For this the midlands were a great centre, as were the new railway towns of Crewe, Swindon, and Earls-

---

[1] J. T. Bunce, *Josiah Mason, a Biography*, 2nd ed., London, 1890.
[2] For the state of affairs in the weapon manufacture in the fifties and sixties, see Richard Cobden, *Speeches*, London, 1878, pp. 294–309.
[3] J. G. Goodman, 'Statistics of the Small Arms Manufacture at Birmingham', *R.B.A.*, 1865, Section F, p. 150.
[4] R. E. Waterhouse, *A Hundred Years of Engineering Craftsmanship*, Birmingham, 1957.
[5] J. D. Scott, *Vickers, a History*, London, 1962, chapter 5.
[6] C. Hamilton Ellis, *Nineteenth Century Railway Carriages*, London, 1949.

town. There the demands of railways in Britain had been met since the fifties; there by the seventies were placed large export orders for India, and in the eighties for South America. Engines and carriages required the mobilizing of rubber, glass, leather, fabrics, fixtures, locks, lighting devices, and a complex set of constructional parts in wood and metal.[1] All these had to be assembled to produce land vehicles of unprecedented complexity.

But it is doubtful whether we can regard railway rolling stock as the true beginning of the mass-produced vehicle, for it was not from it that other modes of carriage came. The first bicycle of modern pattern was the Rover Safety Cycle of 1885.[2] Hans Renold's bush roller chain of 1880, and J. B. Dunlop's development of the pneumatic tyre by 1888, caused the bicycle to move toward true mass production and prepared the way in industry for the automobile.[3] Goldsworthy Gurney in the forties had built a steam car but no effective rivalry of the railway was possible until the internal combustion engine.[4]

THE METAL MAKERS

The engineers, builders, and machine and tool makers were in their turn dependent on the producers of metal. It was the immense development of the iron industry that provided the new material frame for society.[5] M. Chevalier in the thirties was astonished how far Britain had gone in replacing wood with iron: for machinery, props, pillars, roofs, lamp posts, reservoirs for gas and water.[6]

There had been great progress in the last quarter of the eighteenth century in iron production, and the long war against France, with its demand for weapons, had further stimulated development.[7] There were four leading iron complexes in 1815. The iron masters of Shropshire, south Staffordshire and Derbyshire, led by the Darbys and Wilkinsons, had seized the initiative from the charcoal users of the Forest of

---

[1] P. L. Payne, *Rubber and Railways in the Nineteenth Century*, Liverpool, 1960.

[2] A. Davis, *The Velocipede, its History and How to Use It*, London, 1869; Viscount Bury and G. Lacy Hillier, *Cycling*, London, 1887, chapter II.

[3] Basil H. Tripp, *Renold Chains, 1879–1955*, London, 1956.

[4] Anthony Bird, *The Motor Car, 1765–1914*, London, 1960.

[5] For output from 1840 see *Mineral Statistics of the U.K. of Great Britain and Ireland;* for the production of pig iron in the U.K. 1740–1934 see *Statistics of the Iron and Steel industries*, British Iron and Steel Federation, 1935, pp. 4–5; for steel p. 10.

[6] Elie Halévy, *History of the English People in the Nineteenth Century*, vol. III, *The Triumph of Reform, 1830–1841*, London, 2nd rev. ed., 1950.

[7] T. S. Ashton, *Iron and Steel in the Industrial Revolution*, Manchester, 1924.

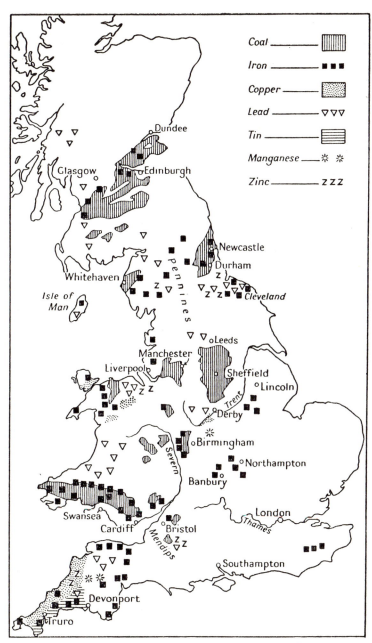

Coal ———— ▦
Iron ———— ■■■
Copper ———— ▨
Lead ———— ▽▽▽
Tin ———— ▤
Manganese ——— ❋ ❋
Zinc ———— ᴢᴢᴢ

Dundee

Glasgow
Edinburgh

Newcastle
Durham
Whitehaven
Cleveland

Isle of
Man

oLeeds

Manchester
Sheffield
Liverpool
oLincoln
ᴢ ᴢ
Derby

Birmingham
Northampton
Banbury

London
Thames
Swansea
Cardiff
Bristol
Mendips
Southampton

Devonport

Truro

THE MINERAL RESOURCES OF BRITAIN, 1851

After R. Hunt

151

Dean and the Weald of Kent.[1] In South Wales, the intense rivalry of the Guests of Dowlais and the Crawshays of Cyfarfa produced an astonishing output.[2] In the north-east Ambrose Crowley had been a pioneer, especially in large-scale imports of Swedish ores.[3] In Scotland Carron Company was a leading enterprise from 1759.[4]

The nineteenth century brought a series of new discoveries, causing the regional initiative in the production of iron to move in a most dramatic fashion.[5] First came a challenge from Scotland, based upon the development by Neilson of the hot-blast in 1828, making it possible to work the enormous black-band ironstone deposits of Lanarkshire and Ayrshire, and to do so with unprecedented economy using Scottish 'splint' coal without the necessity of first converting it to coke. Fuel costs were suddenly halved. Labour costs, too, were low, with an influx of Irishmen and west Highlanders to keep wages down. New enterprises sprang up, sponsored by men new to the game and therefore uninhibited by older ideas. By the mid-forties the Glasgow Pig Iron Market had attained a leading position, and Scots iron dominated British exports.[6] Prejudice also had helped the Scots, for other British producers were slow to take up the hot-blast.

But men in other regions, aware of the fortunes made in iron, were eagerly seeking new possibilities as the demand for iron increased. South Wales had been unable to bring its anthracite to bear in the competition; after 1837, following George Crane, anthracite was successfully used with the hot-blast, and the Welsh challenge renewed.[7] By 1851 British producers turned out $2\frac{1}{2}$ million tons of iron, about one-half of the world total, with Scotland, South Wales and south Staffordshire responsible for about 80 per cent of this. By this time the railways had destroyed the local protection that high transport costs had provided for many small producers.

Other men sought advantage in terms of the discovery of new ores. H. W. Schneider and his partners had been searching for some ten

---

[1] Arthur Raistrick, *Dynasty of Iron Founders: the Darbys and Coalbrookdale*, London, 1953.

[2] Madeleine Elsas, ed., *Iron in the Making: Dowlais Iron Company Letters, 1782-1860*, Cardiff, 1960; J. P. Addis, *The Crawshay Dynasty: A Study in Industrial Organization and Development, 1765-1867*, Cardiff, 1957.

[3] M. W. Flinn, *Men of Iron: the Crowleys in the Early Iron Industry*, Edinburgh, 1962.

[4] R. H. Campbell, *Carron Company*, Edinburgh, 1961.

[5] J. C. Carr and W. Taplin, *History of the British Steel Industry*, Oxford, 1962.

[6] R. H. Campbell, 'Developments in the Scottish Pig Iron Trade, 1844-1848', *J. Econ. H.*, 1955, pp. 209-26; 'Fluctuations in Stocks; a Nineteenth-Century Case Study', *O.E.P.*, 1957, pp. 41-55.

[7] A. H. John, *The Industrial Development of South Wales, 1750-1850*, Cardiff, 1950.

years in the Furness area of northern Lancashire for new deposits of haematite ore, when in 1850, he came upon the largest such deposit yet known.[1] Six years later, not far away, in west Cumberland, the greatest haematite deposit in British history was found.[2] So was made possible a hectic development of iron production in the north-west of England. North Lancashire and Cumberland together, with what were virtually mountains of high grade phosphorus-free iron, became a boom area, with most vigorous and ambitious plans for expansion.

Next, it was the turn of the north-east coast, in the Cleveland region. Ironstone there was newly exploited in 1850, pioneered by Bolckow and Vaughan at their Eston mines.[3] By 1862 Cleveland was the greatest of the iron-ore areas, and was producing the cheapest pig-iron in the world. But the ores of the Cleveland Hills were unsuited to wrought iron or steel making because of their phosphoric content; indeed it was cheaper to bring high grade iron from Spain to Middlesbrough. Finally came the development of the ore fields of Northamptonshire and Lincolnshire, responsible by 1875 for some 7 per cent of U.K. output.[4]

Thus both nature and technology had blessed Britain as an iron producer for most of the nineteenth century. Never had there been such opportunities for British iron masters, the objects of the envious admiration of the world. Following their initiatives whole new complexes of industries had been induced by the cheapness of their product. But these advantages could not be permanent. Shropshire and south Staffordshire, Scotland and the north-west of Britain each in turn found their deposits of ores approaching exhaustion and their iron industry diminishing.

Coal mining responded, especially as, down to the sixties most home-produced ore came from the coal measures. Engineering also felt the stimulus where, as in the west of Scotland, there were favourable conditions, especially a skilled artisan class raised up by the textile industry and the pioneers of steam for ships. But this might not occur where, as in South Wales, no such conditions applied. There the dependence upon the iron and coal complex mounted steadily. In Shropshire and south Staffordshire, though engineering initiative was

[1] J. D. Marshall, *Furness and the Industrial Revolution*, Barrow-in-Furness, 1958, p. 202.

[2] O. Wood, *The Development of the Coal, Iron and Shipbuilding Industries of West Cumberland 1750–1914*, London Ph.D. thesis, I.T.A., 1951–52, no. 379.

[3] J. S. Jeans, *Pioneers of the Cleveland Iron Trade*, Middlesbrough-on-Tees, 1875, chapters II and III.

[4] Carr and Taplin, op. cit., p. 89.

certainly not lacking in the eighteenth century, the sending of the iron elsewhere, especially to the metal workers of Birmingham and the Black Country, meant that effective regional diversity did not take place. In north Lancashire and Cumberland, however, the iron masters made great efforts to build new enterprises upon the thriving iron foundation; indeed they were over-optimistic in such plans.

The fate of the regional groups of enterprises varied greatly as conditions deteriorated against each in turn. By 1885 no less than three-quarters of British iron ore came from non-coal-measure seams — from the north-east and north-west coasts. The west midland area declined to the status of a marginal producer, enjoying no great development from the mid-century onward. Darby's works, like his iron bridge, are still part of the landscape, with no new industries contending for the site; but though there was complaint and unrest, growth under nineteenth-century conditions had been modest, so that decline could occur without frightening disturbance. In South Wales the immense increase in the demand for coal sustained a vigorous continuation of growth but further narrowed the economic base of the region. The north-west, especially Barrow, was poorly sited for fuel, so that its ore tended to move coastwise to other places, thus keeping the area on an extractive basis and disappointing the hopes of its industrialists.

British iron masters in general in the sixties do not appear to have shown any very high order of initiative.[1] Those who, as in Scotland, had innovated with such vigour, were now ageing.[2] They had no real data about themselves and their industry, they were hostile or indifferent to iron technology for some forty years down to 1870, ridiculing chemical analysis in the sixties. Many were unresponsive to the need for scientific and technical education, and appeared to no advantage alongside their competitors at the Paris Exhibition of 1867. The Iron and Steel Institute was founded in 1869, partly as an attack on this obtuseness.

There appears to have been a strong disposition to confine attention to problems of selling, rather than to think about new products and new processes. The iron masters were subject to very great vicissitudes of trade, so that, in improving times the urgent task was to get the works into full operation, while in bad times, with excess capacity, there was neither the incentive nor the means for improvement in

---

[1] D. L. Burn, *The Economic History of Steelmaking 1867–1939*, Cambridge, 1940, p. 65; T. H. Burnam and G. O. Hoskins, *Iron and Steel in Britain 1870–1930*, London, 1943.

[2] R. H. Campbell, *The Growth and Fluctuations of the Scottish Pig Iron Trade 1828–1873*, Aberdeen Ph.D. thesis, I.T.A., 1955–56, no. 720.

plant. Throughout the history of the industry there was usually considerable reserve capacity overhanging the market. It was part of the game that iron masters must accept losses in depression, to be made up in good times. This meant that, as prices began to recover, the producer sought to hold back output until a remunerative price was obtained. The man who, in his haste to take orders, accepted a price prematurely, was the enemy of all. Moreover, to make the best of a recovery, the iron master had to enter upon the expansion with as few old orders on his books as possible, for new orders at higher prices were his aim. This meant a very delicate problem in keeping his enterprise together during bad times. It might be eased by making for stock, but this was a costly and risky tactic. Well into the sixties the British predominance in iron manufacture was such that if British producers followed a sensible price policy all could gain; an association existed to determine 'what was to be deemed the right price'. But with the rise of foreign rivals this was no longer so, for the newcomers might seize for themselves the gains made possible by self-discipline among British producers.

The Scots iron merchants developed a novel solution to the problem of stocks. They formed the Scotch Pig-Iron Trade Association (the Glasgow Iron Ring); merchants, iron masters and manufacturers deposited their iron with Messrs Connal & Co., receiving warrants as receipts. These were transferable and hence saleable. Though the system could not guarantee stable prices, it meant that iron could always be realized for money. By the seventies there were very large transactions daily, with the price of iron warrants a barograph of trade, for Scots iron still dominated the export market.

In 1856 Bessemer showed how steel could be made in a converter, thus obviating the earlier costly processes, opening the way to mass production, firstly undertaken by John Brown in the manufacture of ships' plates and steel springs, and placing an additional premium upon the haematite ores of the north-west. After 1868 the open-hearth process developed by Siemens was also available. But the boom of the early seventies was a matter of iron rather than steel. Indeed malleable iron was a kind of bridge between iron and steel. In the relapse, between 1873 and 1879, steel prices fell to something like one-third of their former level.[1] This had the effect of bringing iron and steel prices closer together. By 1876 the iron rail had yielded to steel; steel shipbuilding began in 1878. Serious though the difficulties were, those

[1] For conditions in the seventies see I. Lowthian Bell, *The Iron Trade of the United Kingdom*, London, 1886.

who turned to the production of steel did better than the iron men in the depression years, and as construction momentum was regained, they rapidly surpassed them.[1]

From the seventies onward two circumstances dominated all else: the coming of large-scale steel production, and the ability to smelt phosphoric ores, so much more plentiful than haematite, following the invention of Gilchrist and Thomas in 1879.[2] It was the younger districts of the north-east and north-west that showed the greatest vigour. Bolckow and Vaughan led the Cleveland move into steel in 1871. The boom in steel first caused those with the greatest initiative to import foreign high grade iron on a rapidly increasing scale and to carry ore from the north-east coastwise; this meant that much steel production moved to tide-water.[3] The Bessemer converter and the Siemens open-hearth were increasingly brought into operation.[4]

But though steel could bring a new vigour to British metallurgy, nature, so generous in the past, now favoured others.[5] The Gilchrist-Thomas process, the answer to such a longstanding problem, brought not only a new expansion in Britain, but also one more great geographical shift, this time to the Cleveland Hills. But it brought even greater growth to Germany and the United States. Whereas Britain still produced about one-half of the world's pig iron in 1870, this fraction was down to less than one-third by the early nineties.

In pre-industrial times Britain had enjoyed a fairly wide range of non-ferrous metals; these, though continuing to make some contribution to the economy in the nineteenth century, could not hold their position. Though the demand for tin increased vastly as the tinplate industry expanded, the old Cornish sources were far eclipsed by imports.[6] So too with lead and with copper. The construction of the electric telegraph and other uses created a very great demand for copper, but home resources could not respond. South Wales had a considerable non-

---

[1] W. T. Jeans, *The Creators of the Age of Steel*, London, 1884. For histories of steel enterprises see: A. C. Marshall and Herbert Newboult, *The History of Firth's 1842–1918*, Sheffield, 1924, R. F. Butler, *The History of Kirkstall Forge through Seven Centuries, 1200–1954, A.D.; the story of England's Oldest Ironworks*, 2nd ed., York, 1955.

[2] For Scotland see I. F. Gibson, 'The Establishment of the Scottish Steel Industry', *S.J.P.E.*, 1958. Also his *Economic History of the Scottish Iron and Steel Industry*, London Ph.D. thesis, I.T.A., 1954–55, no. 723.

[3] M. W. Flinn, 'British Steel and Spanish Ore, 1871–1914', *Econ. H.R.*, 1955.

[4] See W. A. Sinclair, 'The Growth of the British Steel Industry in the late Nineteenth Century', *S.J.P.E.*, 1959.

[5] For the new steel men see Charlotte Erickson, op. cit.

[6] John Rowe, *Cornwall in the Age of the Industrial Revolution*, Liverpool, 1955; A. K. H. Jenkin, *The Cornish Miner*, London, 1927.

ferrous metal industry, centred on Swansea, largely based on imported materials.[1] In the main Britain came to depend for such metals upon other countries. But though they figured increasingly among imports, they involved exports also in the form of British mining initiative and enterprise, especially to South America.

8. THE COAL MASTERS

By 1815 the coal masters were operating powerfully on the economy and society of Britain.[2] But growth had only begun. Output, about thirteen million tons at Waterloo, was over one hundred million by 1866, almost 60 per cent of the world total. Of this it is likely that about half was consumed by the various branches of the iron trades. By 1885 Britain produced no less than one hundred and sixty million tons. The entire economy rested upon this source of heat and power, a fact that struck the country with frightening clarity in the late sixties and early seventies.[3]

At the time of Waterloo it was only in the exporting coalfields of Cumberland and Northumberland that mining enterprise had developed intensive mining communities.[4] In Cumberland the Lowther family was working at surprising depths and below the sea, dominating the Irish market.[5] Here in isolation, under the paternal eye of Lowthers and Curwens, the colliers were a distinct community. But only in the Tyne area — the greatest of the coalfields, supplying the largest of the national markets, London, under monopolistic conditions — were deep sinkings and a very large output typical. In Lancashire the pits were usually small, operated by many petty owners, bringing their coal by the old-fashioned carts or the new canals into Manchester and Liverpool, supplying the Cheshire salt-boilers, and acquiring salt interests. In Shropshire too the scale of operations was modest, with the main market and the main initiative coming from the iron masters. In south Staffordshire, stimulated by coke smelting and puddling,

---

[1] R. O. Roberts, 'The Development and Decline of the Copper and Other Non-Ferrous Metal Industries in South Wales', *Transactions of the Honourable Society of Cymmrodorion*, 1956.

[2] For earlier times see T. S. Ashton and J. Sykes, *The Coal Industry of the Eighteenth Century*, Manchester, 1929.

[3] W. S. Jevons, *The Coal Question, an Inquiry concerning the Progress of the Nation, and the Probable Exhaustion of our Coal Mines*, London and Cambridge, 1865.

[4] For general history see R. L. Galloway, *Annals of Coal Mining*, First Series, London, 1898; Second Series, 1904.

[5] A. E. Smailes, *North England*, Edinburgh, 1960.

many tiny pits huddled together supplying the host of petty lessees working iron on a small scale. This was made possible by the extraordinary 'Ten Yard' seam, so rich and accessible that no great capital was required to launch a profitable enterprise.[1] A Black Country colliery might cost two or three thousand pounds; in the north-east very much more capital might be needed.[2] Small pits were typical in Yorkshire too, but were more widely dispersed.[3] The war had greatly stimulated enterprise in South Wales, causing canals to be built and the main ports to be developed – Llanelly, Swansea, Neath, Cardiff, and Newport.[4] But the expansion was narrowly confined to the bituminous and steam coal along the southern rim of the coalfield, and to Monmouthshire, where the developers enjoyed a monopoly of the Severn market. In Scotland the ancient fields of the Forth estuary were still in production, together with the Ayrshire field, but the central area remained pastoral.[5]

The costs of working at the depths then attainable on the Tyne were becoming prohibitive. A series of fearful explosions occurred in 1815. In London people were momentarily shocked that within one month nearly three hundred widows and orphans were thrown upon charity and parochial assistance, but they were more lastingly impressed by the unheard-of price of coal. The men of science and engineering were brought in: the Davy lamp and Stephenson's Geordie came to the rescue, allowing the pits to be driven ever deeper. But the Tyne could not hold its lead; the initiative was passing to the Wear.

The owners' position in the north-east in the twenties was not easy. The coal-field had always been susceptible to 'cree' – the buckling up of the seams and workings – this became the more serious with greater depth, calling for improved engineering skill and great discipline. It was impossible to effect insurance against fire, water, or any other calamity. A serious accident meant that the pit might be stopped for twelve months; in addition the restoration of the pit might cost almost as much as a new sinking. Pumping was a great expense which had to be met, working or not. In one colliery the weight of water raised was

[1] A. J. Taylor, 'The Sub-Contract System in the British Coal Industry', in L. S. Pressnell, ed., *Studies in the Industrial Revolution presented to T. S. Ashton*, London, 1960.

[2] Taylor, op. cit., p. 221.

[3] Gordon Rimmer, 'Middleton Colliery, near Leeds, 1770–1830', *Y.B.*, 1955. For Derbyshire see J. W. Williams, *The Derbyshire Miners: A Study in Industrial and Social History*, London, 1962.

[4] E. L. Chappell, *The History of the Port of Cardiff*, Cardiff, 1939.

[5] A. S. Cunningham, *Mining in the Kingdom of Fife*, Edinburgh, 2nd ed., 1913.

eighteen times that of coal. In deep mines, made accessible by the new lamps, the pillars of coal supporting the roof were sometimes as much as three-quarters of the coal. To make working economical these had to be removed according to some plan, which greatly increased the danger of falls and impaired the ventilation.

The London market required very delicate regulation. Though London and its environs consumed something like one-quarter of the national output, the collieries of the north, in bad times, had the power of sending as much coal in one month as would last for two. Further, the greater the potential excess, the greater the strain placed upon the owners' regulating agreement: it was always in danger of collapse, and did break down from time to time under the drive of those who either preferred to chance their arm in open competition, or those who, by secret concessions, sought to increase their share of the market.

So great were the risks that the owners of the coal were often deterred from exploiting it, but chose to negotiate leases in return for royalties. On Tyneside the typical form was the company of adventurers holding shares, but with unlimited liability and no privileges of incorporation, dividing the enterprise into fractions as small as sixty-fourths. A colliery, with sinkings and machinery, could cost from £10,000 to £150,000. John Buddle estimated the capital sunk in the Tyne collieries in 1829 as some £1½ million.[1] He denied excessive profits, maintaining that the return was on average less than 10 per cent annually. Bankruptcy was sometimes brisk, which meant that new adventurers took over at knock-down prices. Not all who embarked as mining capitalists emerged with fortunes, though among the successful there were some of great wealth.

But in the later twenties and early thirties, in spite of all their difficulties, the north-eastern coal men, together with outsiders, began to make preparations for the greatest advance in the history of the industry. They were at last prepared to face the challenge of opening up the enormous untouched resources of south and south-east Durham.[2]

It had long been known that beneath the magnesian limestone of south Durham lay coal, but the depths were too great for successful working; moreover the engineers of Tyne and Wear held that coal beneath limestone was inferior. Various attempts had been made in

[1] *Report on the Coal Trade of the U.K.*, 1830, P.P., vol. VIII, p. 34.
[2] A. E. Smailes, 'Development of the Northumberland and Durham Coalfield'. *Scottish Geographical Magazine*, 1935.

1810, 1813, 1816, to pierce the limestone, but without success. But Monkwearmouth, costing something like £100,000, began to raise coal in 1831. Thereafter exploitation accelerated – Colonel Braddyll and his partners in the South Hetton Coal Company opened the Hetton Seam in 1833. The opening of the Stockton and Darlington Railway in 1825 had given access to the maritime trade. Seaham Harbour, begun in 1828, shipped coal from 1831 onward. In the latter year Middlesbrough Dock was opened; four years later Hartlepool Dock and Harbour were in operation. Huge, highly capitalized collieries spread over the area. There was a rush to apply the joint-stock principle to coal getting. In the boom of 1836 the expansion of the lower Wear and of south Durham contributed substantially to the general speculative crisis: something like £1½ million was lost with the failure of the Durham County Coal Company and the Northern Coal Mining Company.

The new depths confronted the mining engineers with a new challenge. In many disputes about the level at which workings could be economically successful the opinion had been strongly held around 1830 that 1,200 feet was the limit. But Monkwearmouth had attained almost 1,600 feet by 1831. Winding engines had to be increased in power and reliability, and new ropes, with iron and steel replacing fibres, perfected. The invention that contributed most to the new potential had to do with the mode of raising the coal itself. The universal method was to use great wickerwork baskets or corves hanging free on hempen ropes. T. Y. Hall, who worked his way up from pit boy to become apprentice to the celebrated Buddle, perfected the modern system of a double-decked metal cage running in guides.[1] On the new basis double the quantity of coal could be drawn.

The influences that inspired the northern coalmen to wider initiative had their effects elsewhere. The growth of population and the extension of both domestic and industrial demand provided the basis for steady expansion in most coalfields, especially Lancashire and Yorkshire. The iron trade had made a spectacular burst of progress, pulling coal forward with it, a factor to be of particular importance in Scotland. At last Parliament was prepared, in 1831, to remove the anomalous duties on coals consumed in the United Kingdom, and to lower those on coal exports. The manufacture of coal gas added to the demand.[2] At the

---

[1] Clapham, op. cit., vol. I, p. 436.

[2] D. Chandler and A. D. Lacey, *The Rise of the Gas Industry in Britain*, London, 1949; Everard Stirling, *History of the Gas, Light and Coke Company*, London, 1949.

retailing end the coal merchants were showing new initiatives, especially in London.[1]

It was the railways, springing from the collieries of the north-east, that brought the great revolution. They were the greatest demanders of iron, and with an increase in iron output, up too went that of coal. Domestic consumption also rose as the coal merchants in conjunction with the railway companies opened coal 'wharves' up and down the country. With expansion came a new kind of competition. Until the late thirties regional markets had been fairly safely insulated, with canals the only effective means of movement. Thereafter rivalry became much more fierce, with one area aggravating the problems of another as the contest for expansion was joined. The railways were free of lighterage charges, and did less damage through breakage than did carriage by sea. Down to 1851 the railways had almost no part in supplying London with its 3·5 million tons of coal; by 1880 no less than 6 million of the 9 million tons delivered went by rail. Differences in natural conditions had always meant differences in cost, but these had been masked by natural protection. Now, as the struggle sharpened, a new premium was placed upon the ability to try new methods. Though some areas, like Lancashire, second in output, continued to depend mainly on local demand, even this could only continue if costs were competitive.

In Scotland coal output showed the same fabulous increase as did iron.[2] The semi-anthracite steam-coal trade of South Wales had effectively come to birth with the winning of the Waenwyllt Colliery in the Merthyr Valley in 1828–29; collieries rapidly spread in the Aberdale and Merthyr districts, sending their output by the Taff Vale Railway, built in 1840–41, and the West Bute Dock at Cardiff, opened about the same time, to power engines for use on land and sea.[3] Simultaneously the ironstones of north Stafford began to be worked. In the established centres of iron production also, the new stimulus was felt, for south Staffordshire expanded its output substantially, though not to the same relative degree as Scotland and Wales.

As in the Northumberland and Durham fields, there was a general

[1] Elspet Fraser-Stephen, *Two Centuries in the Coal Trade: the Story of Charringtons*, London, 1952.

[2] A. J. Youngson Brown, *The Scots Coal Industry 1854–1886*, Aberdeen D. Litt. thesis, I.T.A., 1952–53, no. 717.

[3] See J. H. Morris and L. J. Williams, *The South Wales Coal Industry 1841–1875*, Cardiff, 1958; E. D. Lewis, *The Rhondda Valley*, London, 1959; L. B. Collier, *History and Development of the South Wales Coal Industry*, Univ. of London Ph.D. thesis, 1940.

drive to greater depth and larger output per colliery where possible. In Lancashire new sinkings, very much deeper than anything known earlier, began in the late thirties, with the older, shallow workings being eclipsed by enterprises on a new scale. In south Staffordshire new sinkings were driven down almost a thousand feet, through the red sandstone, far below the Ten Yard Seam. So too in north Staffordshire, where the Apedale Pit near Newcastle-under-Lyme reached the greatest depth known in the thirties – some 2,100 feet.

At the beginning of the eighties the several coalfields ranked in output more or less as they had at the beginning of the sixties. South Wales was an exception moving up from fifth place to rival Lancashire as second largest producer. In the intervening years Scotland had risen briefly to second place but had later subsided to fourth. The 'Great Northern Coalfield' was still far ahead of the rest, with almost a quarter of total output.

The colliery owners were a diverse lot, with diverse problems. They ranged from the great landowners of noble lineage, presiding over the northern coalfield, with their patriarchal attitude to the pitmen, to the petty iron masters of south Staffordshire, short of capital, working tiny pits, isolated from their colliers by the ubiquitous butties.[1] They ranged from those with large, well-established concerns with a comfortable future supply of coal, to those harassed by insufficient capital, narrowing seams, and rising costs. But all of them, like the millowners, were providing employment on an increasing scale, and pointed to this achievement with some justification. The increasing supply of coal had to be forthcoming to sustain the economy; whatever the working conditions of the colliers might be, to impair the profit incentives of the coal producers would be a dangerous business.

Yet there were misgivings. Fears of coal exhaustion had been expressed as early as the late forties, and all colliery owners knew the problem in the particular form of increasingly difficult access. Steam-engines, lamps, cages, and handling gear did not affect the mode of winning coal at the face. This continued to depend on the old way of doing things, more intensively applied: blows deftly struck by hand against

---

[1] See David Spring, 'The English Landed Estate in the Age of Coal and Iron: 1830–1880', *J. Econ. H.* 1951; 'The Earls of Durham and the Great Northern Coal Field', *Canadian Historical Review*, 1952; Edith Vane-Tempest-Stewart, Marchioness of Londonderry, *Frances Anne: The Life of Frances Anne, Marchioness of Londonderry, and her husband Charles, Third Marquess of Londonder y*, London, 1958; F. M. L. Thompson, *The Economic and Social Background of the English Landed Interest, 1840–1870; with particular reference to the Dukes of Northumberland*, Oxford D.Phil thesis, I.T.A., 1955–56, no. 377.

the coal face, and underground transport very largely by human muscle. This remained true, even though the increasing use of gunpowder reduced the amount of human effort on the part of the hewers (though at the cost of greater risk) from the mid-century onward.

The new large undertakings were involved in heavy fixed commitments; they were conceived as long-term ventures, and their sponsors were anxious to embody new techniques of coal handling in order that their high fixed charges might be spread over a maximum output. The most effective source of economy was in underground transport, and large sums were spent on handling equipment. With the smaller mines, these economies could not be gained. With shallow pits and narrow seams, human effort, with a minimum of specific equipment, was the cheapest way of tackling the job. The more narrow and difficult the seams, the greater the need for flexibility. Like the small cotton factory owners the operators of small collieries found technology and capital working against them, leaving the exploitation of labour as their only resource.[1] But even the smaller owners, faced with legislation, found that something could be done by taking thought about the conditions of work.

Restriction of output was not uncongenial, especially to the northern owners who had practised it so long.[2] But it was inconsistent with an increase in capacity and output. By 1843 the 'quota' in the north-east, with a basis of 1,000, had sunk to 414; that is to say, a colliery capable of producing 50,000 tons was allowed a sale of only 20,700. But free competition was not to the liking either of the small coal masters or the large. The former with their high costs were cut from the market; the latter soon began to wonder whether they were exploiting their pits too rapidly, to their own long-term detriment. Moreover, there was increased unemployment as production concentrated on the more efficient pits. The upshot was proposals for a return to control. But it was too late.

When regional markets had been relatively secure, and the trend of output steady, the masters had tended, in the main, to accept reasonable labour costs as a datum. But the rise of new conditions – competition in London, with the railways rapidly widening and perfecting coal markets, and the influx of new capital into the coalfields – all

[1] P. E. H. Hair, *The Social History of British Coalminers, 1800–1845*, Oxford D.Phil. thesis, I.T.A., 1955–56, no. 644.
[2] See P. M. Sweezy, *Monopoly and Competition in the English Coal Trade, 1550–1850*, Cambridge, Mass., 1938.

disturbed the old relative placidity. To an increasing degree wages became subject to competitive criteria, for the colliery owners were no longer content to confine the search for economies to improvements in engineering skill, but also put vigorous downward pressure upon wages in bad times.

Though the demand for coal and iron tended to move together, they were often inversely related in terms of profitability. If coal costs rose, the iron master was bound to feel the pinch. During the 'Coal Famine' that accompanied the boom of 1870–73 the vulnerability of iron producers to coal prices was frighteningly demonstrated. The obvious course for iron masters was to acquire coal enterprises of their own. In Scotland, for example, by 1872 every large iron concern was in fact an integrated enterprise; the same trend was apparent in most areas. This meant that the regional economies in the heavy areas were under the dominant ownership and control of a quite small number of men: the Pease family in the north-east, the Bairds and Dixons in Scotland, the Guests and Crawshays in South Wales.[1]

As the pace of coal production mounted, misgivings about the cumulative degradation of life and work became stronger. Englishmen studying their own society found much ground for criticism. G. R. Porter, seeking to compare the English coal industry with that of France, found he could not do so because English data was lacking — provoking him to condemn resistance by private interests to legislative authority in the gathering of important information.[2] The Victorian urge for order and economy was outraged by the signs of carelessness in many coalfields.[3] Colliery workers, as in Lancashire, complained that they had no alternative but to strike if they wished to make contact with the owners.[4]

Of the immense power of the iron and coal masters over their workers there can be little doubt.[5] By the third quarter of the century, owners of very extensive enterprises had often lost touch with the mass of their employees. They knew that a minimum degree of ruthlessness was essential if the enterprise was to hold its own over the successive

---

[1] A. McGeorge, *The Bairds of Gartsherrie*, Glasgow, 1875; Earl of Bessborough (ed), *The Diaries of Lady Charlotte Guest . . . 1833–1852*, London, 1950; also Charles Wilkins. *The History of Merthyr Tydfil*, Merthyr Tydfil, 1867; J. P. Addis, op. cit.

[2] *J.S.S.*, 1838, p. 327.

[3] Robert Hunt, *J.S.S.*, 1856, p. 323.

[4] *Select Committee on Mines*, 1866, pp. 28, 44.

[5] Frederic Harrison, 'The Iron-Masters' Trade Union', *F.R.*, 1865; William Robinson Hopper, 'An Iron Master's View of Strikes', *F.R.*, 1865.

phases of the cycle. There seemed no other sensible rule but to resist claims for wage increases in good times and to enforce decreases in bad. Many of them quite literally could not afford to become involved in the human effects of policy, for to do so would have meant the passing, first of wealth and power, and then of the viability of the enterprise. The result was a slowly mounting tension, with the eventual crisis staved off by one means or another including the higher wages of boom and the growth of coal exports.[1]

9. THE BUILDERS

While industry and population were expanding at the astonishing rates apparent in the nineteenth century, it was necessary that a group of men should respond to the need to provide shelter for both. The building trades were active in all areas of expansion; it is often possible to correlate regional bursts of industrial growth with new housing. Moreover the output of the builders represented a very high proportion of the new real capital that was being created. With the exception of agriculture, the building trades were the largest employers of males during the first half of the nineteenth century.

Great builders of warehouses and mills like Rennie and Fairbairn, and a few major master builders like Thomas Cubitt (who enjoyed the warm admiration of the Queen) made their impression on the public mind, organizing and deploying vast resources.[2] The speculative builder of middle-class suburbs has received some attention, operating on a considerable scale on open sites, from the seventies onwards.[3] But the numerical majority of builders, providing for the workers' needs in areas of congestion, operated upon so small a scale as to fail to impress either their own contemporaries or later students of the period.

A large number of small firms has been typical of the industry to the present day; the great changes of scale, apparent in other industries, hardly affected the building of houses. So, too, with construction method: no dramatic innovations took place. Building for the workers

---

[1] For factors affecting productivity see A. J. Taylor, 'Labour Productivity and Technological Innovation in the British Coal Industry, 1850-1914', *Econ. H.R.*, 1961.

[2] J. M. Richards, *The Functional Tradition in Early Industrial Buildings*, London, 1959; E. W. Cooney, 'The Origins of the Victorian Master Builders', *Econ. H.R.*, 1955.

[3] H. J. Dyos, *Victorian Suburb: A Study of the Growth of Camberwell*, Leicester, 1961, pp. 122-37.

represented the leading case, during this phase of great changes, of an industry extending its output greatly, but doing so mainly on the basis of multiplying older organization and techniques.

There was plenty to be done in the construction of mansions for the Victorian trading and industrial grandees, both round the great London squares and in the countryside.[1] Big men like Cubitt, who began as a journeyman carpenter, responded to this need.[2] Houses for humbler people were usually built by men of their own number who had saved a little money and risen from the working class – artisans, bricklayers, carpenters, masons – the much abused jerry-builders. Having little or no capital, they bartered their labour for materials, combining to build houses on the principle of exchange of work. This was true to greater or less extent in London, Manchester, Birmingham, and Liverpool (where the Welsh builders, often Caernarvonshire quarrymen, were active). They were often fearfully inefficient, without building standards and often without conscience, so that better houses might have been built at less cost, had the organization been sounder. The commercial casualties among them were high. The upper income groups were able to monopolize building skill and capital, to the detriment of those who, though paying a high price for housing relative to their incomes, could not compete for efficient building organization.

There were other reasons why housing for the masses did not lend itself to the economies of scale. There were many difficulties of a legal kind stemming from the awkward state of the law governing land acquisition. The spacial aspect involved a special set of difficulties, for if housing was to gain from the new skills in organization it was necessary to execute very large areas as units; this was seldom possible. Technically, house construction produced no great innovations, though the production of certain building materials did.[3]

By the sixties the opinion was fairly general that housing for the labourers was an uneconomic venture for capital. As a result the rate of interest charged on building loans was notoriously high. Various

[1] John Summerson, *Georgian London*, London, 1945, chapter 14.

[2] See Cubitt's obituary, *Minutes of Proceedings of the Institution of Civil Engineers*, Session 1856–57, London, 1857, pp. 158–62.

[3] Marion Bowley, *Innovation in Building Materials*, London, 1961, part II, for accounts of bricks, cement and glass. For slates see A. H. Dodd, *The Industrial Revolution in North Wales*, 2nd ed., Cardiff, 1951. For glass, see T. C. Barker, *Pilkington Brothers and the Glass Industry*, London, 1960. In general, Norman Davey, *A History of Building Materials*, London, 1961.

societies were formed in the attempt to meet the problem – buying land for estates, and generally seeking to sponsor development projects. Much emphasis was placed on flatting or the 'cellular' plan. In spite of the relative success of Peabody's Trust, all this effort came to little.[1] Housing remained beyond the reach of commercial initiative.

To all this must be added the power of the building operatives, in spite of their geographical dispersal, to keep the supply of labour down by restrictions of recruitment, thus keeping wages up. Indeed, before the mid-century builders and employees had a developed system of 'working rules'. It seems likely that the very factors which kept the industry on a small-scale basis further added to the price of houses by minimizing distinctions between master and man, and thus making general collusion about wages and costs easier.

Yet it must not be forgotten that the emergence of the large builders, though they were relatively few in number, was a significant development. The master builder, wielding all the elements of a major building construction project, was a new figure. Alexander Copland executed contracts for the construction of army barracks to the extent of £1,300,000 between 1796 and 1806. Cubitt, starting as a master carpenter, developed his firm and his methods in the decade or so after Waterloo, and had a number of imitators. Such men set up their permanent workshops, wagons, and storage areas. They brought together representatives of all the trades involved in building, and sought to provide them with continuous employment. They had their own draughtsmen, a boon to a busy but ambitious architect. Of special importance, they collected and retained their own set of trusty foremen. In short, they organized continuous large-scale building firms. These could draw upon a growing range of specialized ancillary services.[2]

Builders were speculators in their own products on a considerable scale, acquiring land and building for sale at their own risk. Architectural unity was often preserved in the new squares by the design and control of architects and surveyors, as with the Westminster estates. But, there was also a good deal of simple standardization – one householder of the sixties expressed his detestation of retiring to bed, conscious that 'perhaps eight hundred masters of households were slumbering in eight hundred bedchambers exactly the same'. Church

---

[1] See below, Chapter 7, section 6, pp. 242, 243.
[2] J. Nisbet, 'Quantity Surveying in London during the Nineteenth Century', *Journal of the Royal Inst. of Chartered Surveyors*, 1951–52.

construction was subsidized by the government, docks and public building added to demand, helping the larger builders to obtain continuity of work. As London grew, there was much discussion about the architectural and amenity standards of the imperial city.[1]

It seems probable that during the first half of the nineteenth century, both official and private sponsors of building in London and the larger cities, affected by competitive doctrines, adopted the practice of seeking competitive tenders for the entire job, a development which favoured the large firm. In the countryside on the other hand a landlord about to build might himself seek out the masters of the various trades. By the seventies there were some 250 building firms in London operating on a fairly large scale, and to a greater or lesser degree using Cubitt's methods. For though house construction continued on a petty basis throughout the century, building contracting, though risky, did provide a ladder to bigger things, of which a good many men of humble origin took advantage.

Cubitt was the great prototype of the Victorian builder, in terms of the accepted architecture of the day – the revived classical style of Nash, and his successors, and the Gothic of Pugin, which very quickly became eclectic. But another style – the functional, was making its appearance, and with it, the great engineering firms, skilled in the use of iron, were moving into building. In the first decade of the century men like the Liverpool iron master, John Cragg, had experimented with cast-iron elements for building construction. But Cragg had been a devotee of the elder Pugin, and had cast Gothic arches and crocketed pinnacles, demonstrating impressive virtuosity, but depressing lack of insight into his material. It was the building of railway stations that brought the girder into prominence. The railway companies sought to demonstrate their solidity by great exercises in classical architecture in the public halls of their major termini; Euston impressed its patrons with a gateway derived from the Athenean Agora. But this would not do for the great vaults necessary for covering the platforms. Engineering firms like that of Fox and Henderson learned a great deal about the covering of such great areas with glass. The arsenal at Woolwich had escaped from the tyranny of bricks and beams by throwing light roofing over the roadway spaces. The genius of Paxton's design for the Crystal Palace, based upon the greenhouses of Chatsworth, was matched by the organizing skill of Sir Charles Fox and his firm, the great exponents of the new method; from design to completion the

[1] Henry Russell Hitchcock, *Early Victorian Architecture in Britain*, London, 1954.

building of a home for the Great Exhibition took a mere seven months.[1] But the engineering firms were not yet a serious threat to men like Cubitt, for the architects did little to apply the new constructional concepts in the expanding cities, much less to integrate the engineer with the architectural profession.[2]

By the sixties a third category of builder was well established, who combined some of the economies of scale with building programmes of a socially questionable kind. These were the men who left the centres of the cities to the giants and the pygmies, moving right to the outskirts and building new villages. The houses were mortgaged while under construction, at high rates of interest, to the new suburbanites, who then undertook the daily train journey to work, often dropped far short of their destination by railways not yet able to gain access to the centre of the city.

10. THE PRODUCERS OF CHEMICALS

Throughout the nineteenth century the chemical manufacturers were concerned with bringing together vast quantities of bulky materials — limestone, coal, salt, kelp, sulphur, and pyrites, and in processing them in various combinations to produce 'heavy' chemicals — alkalis and acids of various kinds.[3] At first because the manufacturer aimed at a single product from all this bulk there was immense waste, accompanied by dire pollution of air, water, and earth. The twin problems of bulk transport and unprecedented filth had two great effects upon the industry. Its location altered as manufacturers chose sites convenient by sea, canal, and rail, and where wastes could be disposed of and fumes and effluvia distributed where complaint would be least effective. The manufacturers found that their mounting heaps of waste could be re-worked for by-products, thus reducing the social offence and also making it possible to spread transport costs over a wider range of output.

The first of the great enterprises was the creation of Charles Tennant,

[1] Violet Markham, *Paxton and the Bachelor Duke*, London, 1935; George F. Chadwick, *The Works of Sir Joseph Paxton, 1803–1865*, London, 1962; Christopher Hobhouse, *1851 and the Crystal Palace*, London, 1951; C. R. Fay, *Palace of Industry 1851*, Cambridge, 1951.

[2] Frank Jenkins, *Architect and Patron*, Oxford, 1961.

[3] L. F. Haber, *The Chemical Industry in the Nineteenth Century*, Oxford, 1958; A. and N. Clow, *The Chemical Revolution; a Contribution to Social Technology*, London, 1952; T. C. Barker, R. Dickinson and D. W. F. Hardie, 'The Origins of the Synthetic Alkali Industry in Britain', *Econa.*, 1956.

at St Rollox in Glasgow.[1] By the thirties this chemical works, less than a mile from the ancient cathedral of St Mungo, the nucleus of the city, was the largest in the world. Tennant, aware of the costly nature of existing bleaching processes as they were practised by the growing textile industry in Glasgow, had applied a discovery of Berthollet, communicated to him by Watt, to produce a chemical bleach by the use of lime. This was the basis of the immense growth of the works, soon of European celebrity. Other products, especially soap production, were soon added.

Challenges quickly came from other areas. By the mid-century Tyneside had taken the lead, following the initiative of William Losh, the first producer in Britain to adopt the new Leblanc process. The repeal of the Salt Duty in 1823, effective in 1825, made this initiative even more rewarding. The firm of Losh, Wilson and Bell and its successors vigorously exploited the markets available to them by sea – especially to the glass makers and soap boilers of London and the Continent; sea communication also helped greatly with the import of materials. The Leblanc process required large quantities of sulphuric acid, the second in importance of the heavy chemicals.

But Tyneside was not to maintain this ascendancy. The manufacturers of south Lancashire too were making their bid in chemicals, in the St Helens area, where in the midst of coal mines, copper smelters, and a foundry, the basis of leadership in the glass industry was being laid.[2] But there were serious local limitations. Of the heavy materials required, only coal was available. Moreover fierce litigation was waged by landlords against the chemical manufacturers as the damage to the countryside increased. A much more economical site was found at Widnes and Runcorn where all the leading requirements were combined, including convenient transport and access to the great Cheshire salt field.[3] Upon this area converged leaders of the industry after false starts elsewhere, including Gossage, perhaps the greatest inventor-entrepreneur in chemicals, and James Muspratt.[4] The latter, though not the first on the site, soon became the greatest. As a producer of Leblanc soda for the soap trade he was without rival. But the social effect of

---

[1] E. W. Tennant, 'The Early History of the St Rollox Chemical Works', *Chemistry and Industry*, 1947, pp. 667–73.

[2] T. C. Barker and J. R. A. Harris, *A Merseyside Town in the Industrial Revolution: St. Helens, 1750–1900*, Liverpool, 1954.

[3] D. W. F. Hardie, *A History of the Chemical Industry in Widnes*, Liverpool, 1950.

[4] J. Fenwick Allen, *Some Founders of the Chemical Industry*, Manchester, 1906; D. W. F. Hardie, 'The Muspratts and the British Chemical Industry', *Endeavour*, 1955, p. 29.

the success of chemicals on the Mersey was appalling, with mean houses and a landscape over which life had been destroyed. In the midlands there was chemical initiative too, especially in the production of glass and phosphorus.[1] British heavy chemicals after 1850 entered upon a period of immense success, exporting vast quantities of alkali. The degradation of the chemical areas under their fumes, effluents and tips horrified observers, but the increasing sense of the importance of chemicals to modern industry inhibited criticism. Even so, the social damage was becoming so obvious in Glasgow, Tyneside, St Helens, and Widnes that the Alkali Act, the first provision for the control of chemical nuisances, with the aid of an inspectorate, came into operation in 1864. A Royal Commission sat in 1878, and discovered a maturing disposition among the manufacturers to cooperate; a new Alkali Act of 1881 was the outcome.

By the seventies the Leblanc soda process, as practised in all the leading chemical areas, was the backbone of the chemical industry. New by-products were found, in part in the attempt to deal with the problem of waste disposal. But a rival technique in heavy chemicals now made its appearance: Brunner and Mond in 1873 formed a partnership for the exploitation of the more efficient Solvay ammonia process in the very centre of the Cheshire salt field.[2] The Tennants opted to continue on the Leblanc method, especially as their by-products were so valuable, and thus lost the position they had held down to this time as the greatest of British chemical manufacturers.

But to recognize the protean nature of the chemical industry through its by-products was to anticipate the future. Charles Macintosh, with a great deal of naphtha left over from his dye manufacture, sought a use for it, and produced, with the aid of George Syme, a sandwich of rubber between layers of cloth, patented it in 1823, and so was born the waterproof clothing industry.[3]

The production of dyestuffs constituted the other great branch of the chemical industry. Britain had a considerable development in tar-distilling, sending raw materials abroad, especially to Germany and

[1] J. F. Chance, *A History of the Firm of Chance Brothers & Co.*, London, 1919; B. E. Threlfall, *The Story of a Hundred Years of Phosphorus-making, 1851–1951; Albright and Wilson Ltd.*, Oldbury, 1951.

[2] J. M. Cohen, *The Life of Ludwig Mond*, London, 1956; H. E. Armstrong, 'The Monds and the Chemical Industry — a Study in Heredity', *Nature*, 1931, pp. 238 ff.

[3] *D.N.B.*, XXXV, 113; G. Macintosh, *A Biographical Memoir of Charles Macintosh*, Glasgow, 1847. For rubber see W. Woodruff, *Rise of the British Rubber Industry*, Liverpool, 1958; P. Schidrowitz and T. R. Dawson eds., *History of the Rubber Industry*, London, 1953.

re-importing the finished colours. In this way British inorganic chemicals production was deprived of the support of an organic sector and so lagged badly behind Germany, with the Germans taking to tar-distilling for themselves. British production of alizarin dyes, based on Perkin's discovery, did very badly. The only British successes were due to immigrant continentals, the German, Ivan Levinstein, and the Alsatian, Charles Dreyfus, but even these were no rivals to the great German concerns. Once more the failure of British manufacturers to take science seriously proved very costly to the nation.

II. THE EXPLOITERS OF ELECTRICITY

The men who sought to develop electricity on a commercial basis were a heterogeneous group. The metal platers already had a long tradition in Birmingham. Men like Elkington knew their markets and materials, and had merely to embody the new process of electrical deposition. Their products were small and ornamental; no long series of operations or repercussion followed upon the electro-plating of their spoons and candelabra and the like.[1]

With electric telegraphy it was quite otherwise.[2] Though wire drawing was an old story, the strength, purity, and behaviour of conductors was not. Insulation provided another range of problems. Not the least of the difficulties was a lack of knowledge of the natural problems of terrain and sea bed. The cable engineers did not propose, like the railwaymen, a surgical operation upon nature, but even to conform to its mysterious submarine behaviour was difficult enough. Operations on the necessary scale demanded reliable production methods for cable: Faraday in conducting his experiments in 1831 had been obliged to coat his wires with sealing wax, or to get covered wire from the spectacle makers and the milliners.

Progress was slow from the operation of the first telegraph in England by Wheatstone and Cooke in 1838, until 1846 when the Electric Telegraph Company was formed, with a capital of £140,000. Thereafter new projects quickly multiplied. The railways demanded better signalling systems, and governments rushed to improve their military communications. In 1853 the Submarine and European Telegraph Company began to lay wires in London to join Parliament, Whitehall, Bucking-

---

[1] R. E. Leader, 'Early History of Electro-Silver-plating', *Journal of the Institute of Metals*, 1919, pp. 305–26.

[2] R. Sabine, *The Electric Telegraph*, London, 1867.

ham Palace, and the West End clubs to the European systems. The Crimea and the Indian Mutiny brought British interest in even remoter communication to a high pitch. Across oceans there were dreams of a great commercial potential. A firm of ships' rope and cable makers thought submarine cable a good line; another firm entered the field through the deposition of rubber as an insulant. Projects multiplied, beginning with attempts to cross the English Channel and other narrow seas, in the early fifties. By the later fifties the Atlantic was being tackled, and even the route to India.[1]

The earliest efforts to lay cables were heroic, but too often failed. Even when simple breakage was avoided, other faults soon multiplied, apparent only when deterioration had had time to do its work. By 1861 over 11,000 miles of submarine cable had been laid, but only about a quarter of it could carry messages. Adventurers who had begun with bland expectations of large profits were often obliged to fall back for their survival on the obsessive urge of a few men who refused to admit failure, long after early optimism had evaporated. Real success came only with the application of scientific method to all aspects of the job. The brothers Siemens dominated this new phase.[2] With their excellent German training in science and technology they had started first as makers of the highly finished control gear (thus avoiding the risks of the project as a whole). But the extraordinary naiveté of the early cable layers caused the Siemens brothers to turn to the fundamental problems involved. In alliance with Kelvin and other great physicists of the day the difficulties that had defeated so many daring and devoted projectors were overcome. Initiative in the new electrical industry, destined to show its full potential in the later decades of the century, passed to firms whose leaders were men not merely of projects, but also of science. By the eighties, the Atlantic and the Pacific, and the principal land routes, were festooned with cables, vibrating with commercial, military, political and personal news and orders. An epic of construction, cutting across societies the railway was yet to touch, had been completed. The copper industry had taken on a new significance.

Such developments had required not merely cable making and laying firms, but firms to sponsor, finance, and operate the new lines. Britain and America stood alone in the fifties in leaving the electric telegraph to private ventures; elsewhere the state in one form or another was in control. But the Post Office, successfully operating its penny post since

[1] See *Report . . . into Construction of Submarines Telegraph Cables*, P.P. vol. LXII, 1860
[2] See J. D. Scott, *Siemens Brothers 1858–1958*, London, 1958.

1840, was casting eyes in the direction of the telegraph. Moreover the problems of competition and duplication of services, similar and often related to those of the railways, were becoming serious. Business men, whatever their general views about competition, were, by the sixties, so committed to telegraphic communication over land and sea that they demanded that the soundness and cheapness of the service be guaranteed. The press too, outraged by the monopoly of news transmission to the provinces, was hostile to the companies. They had responded vigorously to the new mode of news gathering, and were very resentful of the attempt by the companies to exploit them. Even J. L. Ricardo, founder and chairman of the great Electric and International Telegraph Company, advocated nationalization, though his successor, Robert Grimston, was equally convinced in the opposite sense.

The Post Office acquired the telegraph system in 1869. It was a costly operation, requiring £10·5 million. The price was thought by many to be much too generous, and raised new problems as to the manner in which such a bundle of assets should be valued. The telegraph account soon produced a deficit, to be met by subsidy from the letter post. In the full flush of Victorian liberalism the state had acquired a great public utility. The debate about public ownership, though it hardly gained in clarity, took on a new meaning and a new vigour.[1] Meanwhile, the Post Office could settle down to the operation of the domestic system, leaving the private companies to span the seas.

12. VARIOUS MEN OF PARTS

In addition to those who exercised their initiative in the leading sectors of trade and industry, there was a great outburst of secondary initiative responding to the consumption urges of the recipients of rising incomes: the needs of leisure, of food, of homes. The taking of holidays at the seaside provoked a host of maritime housewives to become petty entrepreneurs in the sale of hospitality; hotels too went up in increasing numbers. Thomas Cook, a peripatetic Baptist Bible reader and missionary, inspired by his experience in hiring a special train to a temperance rally in 1841, moved on to sponsor his famous tours, which by the seventies were making the Continent and the Holy Land accessible to new elements in society.[2] The provision of new foods – Cadbury, Fry, and Rowntree with their cocoa and chocolate, Colman

---

[1] See below, Chapter 9, section 7, p. 362.
[2] *D.N.B.*, Supplement ii, 55.

with his mustard, Keiller with his marmalade, Donkin, Hall, and Gamble with their 'embalmed provisions', attested by no less a person than the Duke of Wellington, the factory production of biscuits, John Player with his brilliant ideas for extending the sale of tobacco, incorporated in packaging and brand names, all were responding to the urge to vary consumption.[1] The cattle plague of the sixties, together with the knowledge that in Argentine animals were slaughtered for their hides only, provoked a burst of initiative, including Baron Liebig's famous extract of meat, and culminating, after the laws of heat engines had been generalized, in refrigeration ships.[2] Alexis Soyer, chef and chief of the kitchens of the Reform Club, turned his thoughts during the Irish famine and the Crimean War to the problem of cooking very large quantities of food in an efficient and palatable manner.[3] He made an immense contribution not only to the treatment of crises, but to bringing the economies of scale and planning into the preparation of food for public consumption, and to the improvement of domestic management.[4] Brewing, after 1830, underwent great changes, for the ending of the heavy duty on beer and of the restrictive licensing of public houses, together with the breaking down of local limits to marketing, caused by the railways, made possible the rise of the immense enterprises of Burton-on-Trent.

Industrialists renovated the home. Floor coverings for fairly modest incomes were provided from the forties by the carpet makers, and from the fifties by the linoleum manufacturers; the wallpaper makers provided further embellishment.[5] Producers of brass and other ornamental work were busy.[6] Furniture manufacturers, though to no great degree mechanized before 1860, much increased their output; men with large businesses had arrived by the seventies.[7] Paper makers, printers and publishers provided books for the study and the living

[1] Iolo A. Williams, *The Firm of Cadbury, 1831–1931*, London, 1931; Ann Vernon, *A Quaker Business Man. The Life of Joseph Rowntree, 1836–1925*, London, 1958; Helen Caroline Colman, *Jeremiah James Colman, A Memoir*, London, 1905; Charles C. Maxwell, 'The Confectionery and Marmalade Trade of Dundee', *R.B.A.*, 1867, section F, p. 145; P. Mathias, *The Brewing Industry in England, 1700–1830*, Cambridge, 1959.

[2] 'South American Meat', *Econ.*, 27 October 1866, p. 1254.

[3] Helen Morris, *Portrait of a Chef; the Life of Alexis Soyer*, London, 1938.

[4] Alexis Soyer, *The Modern Housewife or Ménagère*, London, 1849.

[5] Augustus Muir, *Nairns of Kirkcaldy, 1847–1956*, Cambridge, 1956; A. V. Sugden, *A History of English Wallpaper, 1506–1914*, London, 1925.

[6] R. H. Best, *Brass Chandelier*, London, 1940.

[7] Elizabeth Aslin, *Nineteenth Century English Furniture*, London, 1962; *Autobiography of Benjamin North*, Aylesbury, 1882.

room.[1] Toys for young and old, both on new principles and on traditional ones enlivened family life: in 1854 George Swann Nottage founded the London Stereoscopic Company for the exploitation of the invention to which Sir David Brewster essentially contributed. Seeing in three dimensions soon became even more popular than the kaleidoscope of a generation earlier: Nottage amassed a fortune and became Lord Mayor of London.[2] It is impossible to catalogue all the responses to the demand for consumer goods.[3]

So the extraordinary elaboration of business initiatives went on. The successful destroyed the function of some men and created new opportunities for others. The element of vitality is striking as each of these men, in his own particular situation, sought to define for himself an opportunity within which he might venture to bring together labour and capital. In considering the opportunities that were missed we ought to do full justice to the enormous number that were taken, and to remember how great a challenge it was that a society, over some seventy years, should be called upon to find and make effective so many entrepreneurs.

The successful man of business needed characteristics that were not plentiful in the England of 1815. Such men, an essential requirement for expansion, were forthcoming, through trial and error: men who could make sound estimates of the elements of cost and of the behaviour of demand, who could act upon these on the correct scale, who could integrate what they had bought or hired into an efficient enterprise, who knew how to regulate their relations with their workers, combining discipline with equity, who were capable of varying their actions to take account of general prosperity or depression or the intervening states.

[1] D. C. Coleman, *The British Paper Industry, 1495–1860*, Oxford, 1958. For printers and publishers see the Business Archives Council Handlist; also A. E. Musson, 'Newspaper Printing in the Industrial Revolution', *Econ. H.R.*, 1958; Ruari McLean, *Victorian Book Design and Colour Printing*, London, 1963.

[2] Gernsheim, *History of Photography*, pp. 191, 192.

[3] For consumers' problems see Chapter 8, section 8. For some of those who catered for them see *Royal Album of Arts and Industries in Great Britain*, London, 1887.

CHAPTER FIVE

Food Producers and Rent Receivers

THE production of food had since time immemorial been the chief activity of man; moreover the terms upon which men of one kind or another had access to the land, and the relationship thus evolved between men, were the principal elements in the social and political system.[1] An agrarian society can attain to a very high level of coherence if the stresses engendered by nature or human inquisitiveness into nature are not too great, and especially if it is possible to evolve a philosophy of society which unifies and gratifies those willing to conform, and imposes sanctions on those who reject their place in the system. British landowners and tenant farmers were blessed with a most fertile soil and a favourable climate capable of supporting an agriculture as intensive as any in the world.[2]

When the French Wars ended in 1815 the landowners were still immensely powerful, and, by means of the tenant relationship that bound them to the farmers, had attained a strong solidarity. But there were by this time substantial rival interests — traders and industrialists whose outlook might be put, in varying degrees, in opposition to that of the receivers of rents and the producers of food.

The wars had hurried on this sense of difference. Agricultural rents had reached unprecedented heights, raising the pointed question whether, in national emergency, the interest of the landowner, that paragon of patriotism, was in direct conflict with that of society as a whole. The enclosure movement, then in its last phase, called attention

[1] See below, Chapter 8, section 1. The standard work on agricultural history is still R. E. Prothero (Lord Ernle), *English Farming Past and Present* (new 6th ed., with Introduction by G. E. Fussell and O. R. McGregor), London, 1961. For an excellent brief treatment of the first half of the nineteenth century see J. D. Chambers, *The Workshop of the World*, Oxford, 1961, chapter 3; also L. Drescher, 'The Development of Agricultural Production in Gt. Britain & Ireland from the Early 19th century', *M.S.*, 1955.

[2] For the quality of life in the countryside see E. W. Bovill, *English Country Life, 1780–1830*, Oxford, 1962; F. M. L. Thompson, *English Landed Society in the Nineteenth Century*, London, 1963.

to the larger landowners and their sponsorship of a reallocation of the land.[1] In spite of rising rents, the tenant farmers had also done well as the price of food continued to rise in spite of increasing production, and had cheerfully entered into new rent contracts. The margin of cultivation was vigorously extended and the enclosure movement received its crowning impetus.

With peace came very serious agricultural depression.[2] Rents frequently had to be allowed to go into abeyance, for tenants could no longer pay at the agreed level. Poor Law rates rose. The landed interest pushed through the Corn Law of 1815. Its immediate aim was to stop the decline in farm incomes by excluding foreign produce. But at a deeper level there was the idea that society must not be exposed to the uncontrolled impact of the free international movement of goods. For this would mean that the industrial and commercial sector in Britain could take the world for its source of material and its market for manufactures. On this basis it would expand so greatly that agriculture and all concerned with it would be obliged to accept both a diminution of standing in the national life and a new degree of precariousness. But though the landed interests were able to secure the Corn Law, they were obliged to stomach the return to gold, in spite of its deflationary effects, because the commercial and industrial groups insisted that if British trade was to revive, the currency must return to its pre-war footing.[3]

The receivers of rents had been concentrating the ownership of land in their own hands, from at least the seventeenth century down to the outbreak of the French Wars. The movement seems to have been somewhat arrested until 1815; thereafter it was resumed with rather more vigour, with the larger estates leading the trend. The old yeoman class, seriously impaired by 1780, further diminished, leaving the field to landlord and tenant farmer.[4] The devices of primogeniture, entail,

---

[1] J. G. Hunt, 'Landownership and Enclosure, 1750–1830', *Econ. H.R.*, 1959; W. H. R. Curtler, *The Enclosure and Redistribution of our Land*, Oxford, 1921; W. H. Chaloner, 'Bibliography of Recent Work on Enclosure, the Open Fields, and Related Topics', *Ag. H.R.*, 1954.

[2] G. E. Fussell and M. Compton, 'Agricultural Adjustments after the Napoleonic Wars', *E.H.*, 1939; H. G. Hunt, 'Agricultural Rent in South-East England, 1788–1825', *Ag. H.R.*, 1959; L. P. Adams, *Agricultural Depression and Farm Relief in England 1813–52*, London, 1932.

[3] See above, Chapter 2, section 2, p. 8.

[4] A. H. Johnson, *The Disappearance of the Small Landowner*, Oxford, 1909; E. Davies, 'The Small Landowner, 1780–1832 in the Light of the Land Tax Assessments', *Econ. H.R.*, 1927.

and marriage settlement were used by the landed families to consolidate their position.[1] Just as industrialists developed ever larger units as the century proceeded, so too did the landowners, beginning much earlier. Eventually they produced something near a 'land monopoly'. It did not of necessity follow that larger estates meant bigger farms, but the trend was in the direction of larger producing units.

The landlord was thus a capitalist and entrepreneur of a rather curious kind.[2] The successive heads of a family could increase the family stake in the land by means of an elaborate set of legal and matrimonial tactics.[3] The eldest son succeeded, but often to a life interest only, so that though he could increase the estate, he usually could not seriously impair it without overcoming formidable and often impossible difficulties. In this way an ever narrowing group of hereditary claimants upon production was able to improve its position even as the economy was being industrialized. It is not possible to draw any precise distinction between the 'great estate' headed by a nobleman, and the holdings of a more wealthy member of the gentry, but broadly a magnate might be taken as one whose estates exceeded 10,000 acres; about 24 per cent of the area of England was so accounted for by the seventies.

Often, however, the landowner or his agent performed valuable functions. An intelligent owner could do much to sponsor good husbandry among his tenants, and would be especially anxious to prevent damage to his great asset, the land. In the direct cultivation of part of the estate such a man might give excellent leadership. He might well provide capital for improvement, drawn from his industrial undertakings or from the sale of land to the railways. At least in some significant cases the incentive to improve the land and to exploit mines came from the challenge of indebtedness incurred by a predecessor.

The agent made his chief impact through the great estates, large enough to provide full-time employment for such a man, and too large for the landowner to manage on his own. The agent increasingly

---

[1] For the working of the system, especially as related to debt, see David Spring, 'English Landownership in the 19th century: A Critical Note', *Econ. H.R.*, 1957, p. 472; Joshua Williams, *The Settlement of Real Estates*, London, 1879.

[2] For an excellent summary of the discussion of the landowner see F. M. L. Thomson, 'English Great Estates in the 19th Century, 1790–1914', *First International Conference of Economic History, Contributions and Communications*, 1960, p. 385. See below, Chapter 8, section 2.

[3] See David and Eileen Spring, 'The Fall of the Grenvilles, 1844–1848', *Huntington Library Quarterly*, 1956, p. 165.

displaced the solicitor in this role, and in 1834 the founding of the Land Agents Society marked the professionalization of yet another economic function, this time in the heart of traditional England. Agents did well for their masters, bringing their undoubted talents to bear in a situation that was favourable from the forties onward.[1] The new race of agents were in fact managers and executives on a very large scale, making decisions affecting all aspects of the life of the estate and the region.

The well-being of the rent receiver was not necessarily in conflict with that of the tenant food producer. So long as the landlord did not make excessive claims, the tenant might well be glad that he was exempted from the need to tie up his modest capital in land purchase, but could put all his resources into farming operations. This community of interest was especially strong when times were good, though the landowner might resent the social presumption it might induce in farmers. Moreover the farmers could appreciate the stability which the system gave to the countryside, together with leadership in local and regional government. There was also great convenience in having an understood social hierarchy.

Not all farmers were equally embarrassed after the war. Indeed the old distinction between Lowland England south of the Tees–Exe line, and the northern area, showed itself in this matter. The north and west had always been more pastoral; farmers there were less hard hit by the fall in cereal prices that had devastated the south. The rural population was less dense upon the ground, so that unemployed labourers were fewer. Even more important, it was in the north that new jobs were being created by the new industries. Caird showed in map form how the southern limit of coal, though not exactly conforming to the Tees–Exe line, coincided with a difference in wages.[2] Industrialization not only kept rural unemployment down, but pushed wages, including those of farm labourers, up. But the growing cities also sustained farm prices, creating a rising demand for meat, dairy products, and vegetables.[3]

The effects of the collapse of cereal agriculture continued to be felt until the mid-thirties. Farms were without tenants, fertilizers were

---

[1] For one such see David Spring, 'Ralph Sneyd; Tory Country Gentleman', *Bulletin of the John Rylands Library*, March, 1956, p. 553.

[2] Clapham, op. cit., vol. I, p. 466; also James Caird, *English Agriculture in 1850–51*, London, 1851. See below, p. 186.

[3] For the effect on rural amenities see W. G. Hoskins, *The Making of the English Landscape*, London, 1955.

more sparsely applied, drainage and other capital improvements fell away, livestock was drastically reduced. Yeomen who had over-reached themselves left the land for the towns. The relations between landlord and tenant were often bad. The latter lacked the capital with which to make improvements; moreover he was often fearful that if he did invest his labour and resources in the land, the owner, when the tenancy expired, would exercise his right to expel him without compensation.

Between farmer and labourer there were bitter feelings, as the employer sought to pass on his difficulties. The labourer had gained little during the war; thereafter his condition positively deteriorated. Nor could he seek to improve it by moving elsewhere, for the Settlement Laws prevented this.[1] The condition of the southern agricultural labourers was notorious; mentally and physically they were inferior to their counterparts in the north. In 1830 they erupted in riots and rick-burning.[2]

But even in the south there were some gains. As had occurred so often in earlier generations the purge of farmers removed many who were inefficient and created opportunities for new men to try their hand. Rent contracts were often revised voluntarily by the landlord. The population was rising and with it the demand for food. In 1836 the Tithe Commutation Act partly relieved one of the great grievances of the farmers by setting a formula for the size of the tithe, placing the matter on a monetary footing rather than in kind.[3] The farmers of the thirties, thus relieved of some of their burdens, now improved their methods, using the root crops and grasses, interest in which had formerly been largely confined to the wealthy experimenters, the high farmers, to improve their animal husbandry; this in turn, through the greater supply of fertilizer, made possible the cultivation of formerly idle lighter soils.[4]

By the thirties prosperity was beginning to return to agriculture. The general area of arable cultivation had been reduced to a more suitable size. The selective process had produced new initiatives. The tenants, so hard pressed for capital after the war, found their position eased. High farming, with its new ideas, was ceasing to be confined to the élite; new methods and new crops were coming in. Selective stock

---

[1] See below, Chapter 9, section 3, p. 335.
[2] See below, Chapter 9, section 4, p. 338.
[3] See H. C. Prince, 'The Tithe Surveys of the Mid-Nineteenth Century', *Ag. H.R.*, 1959, p. 14.
[4] See J. D. Chambers, op. cit., p. 70.

breeding was becoming more general.[1] The raising of potatoes on small plots became more and more important to the labourer, to the concern of those who sensed the precariousness thus produced. The Royal Agricultural Society was founded in 1838 and chartered in 1840; from it came a new leadership.

It was into this upward trend that the fiercest economic controversy of the century was injected: that over the repeal of the Corn Laws.[2] The landowners and farmers of the south were greatly concerned that the degree of recovery they had achieved would be imperilled by imports of cereals from the great European grain areas of Prussia, Poland, and southern Russia. They mobilized all the strength they could against repeal. But to no avail. For all their still powerful position in both Houses of Parliament the cry of 'free food' was not to be resisted.

The expected calamity did not follow. Instead British farmers accepted in their actions the slogan 'efficiency is better than protection'.[3] Their confidence in their ability to meet the continental challenge rose as their response extended. Good weather produced good harvests in most seasons, yet not such as to cause glut.[4] Prices generally were rising and with them the price of food. Farmers found, in fact, that prices moved well ahead of rents.

New implements, mechanization, and power were increasing in the countryside.[5] By the forties the village smith, the ploughwright and the carpenter were yielding the production of tools and gear to a new generation of manufacturers and engineers. The mobile steam-engine was common after 1850, in threshing and other stationary operations.[6] The system of cable cultivation was introduced in 1859 by John Fowler of Leeds, whereby ploughs and harrows were drawn across the fields on cables. It soon spread, especially on large farms after 1860. But because cable cultivation involved expensive equipment and large uniform areas it did not become so general a use for the steam-engine

---

[1] R. Trow-Smith, *A History of British Livestock Husbandry, 1700–1900*, London, 1959.

[2] See below, Chapter 9, section 7, pp. 353-358; also Briggs, *A. of I.*, chapter 6, section 4.

[3] James Caird, *High Farming under Liberal Covenants, the best Substitute for Protection*, London, 1848; *The Landed Interest and The Supply of Food*, London, 1878.

[4] Systematic agricultural statistics begin in 1867 with the first *Agricultural Returns of Great Britain*. For the general history of farming statistics see J. T. Coppock, 'The Statistical Assessment of British Agriculture', *Ag. H.R.*, 1956, p. 4; W. E. Minchinton, 'Agricultural Returns and the Government during the Napoleonic Wars', *Ag. H.R.*, 1953, p. 29.

[5] Prof. Wrightson, 'Agricultural Machinery', in G. P. Bevan, ed., *British Manufacturing Industries*, London, 1876, vol. X.

[6] Clark C. Spence, *God Speed the Plough: The Coming of Steam Cultivation in Great Britain*, Urbana, 1960.

as did stationary tasks. McCormick in America, using ideas from Britain, perfected his reaper, which along with other American farm machinery, created a great stir at the Exhibition of 1851. The manufacture of mineral superphosphates was begun by John Bennett Lawes in 1842; drainage could be much improved as John Reade's invention of a tile drainpipe in 1843 went into machine production, making possible a new attack on heavy soil husbandry. Specialist drainage contractors, like Josiah Parkes, appeared, using Reade's pipe.[1] In 1846 Peel made available £4 million for drainage; a further, equal, sum followed in 1856. The agricultural press took on its modern form headed by the *Farmer and Stock-breeder* in 1843.[2] The railways made their contribution to agriculture as to industry, reducing the general transport costs of farming, greatly speeding up deliveries, and obviating the costly system of driving cattle on the hoof to the great urban markets.

Even the deep south and the eastern counties made some movement away from their very heavy concentration on cereals, turning to more sheep and cattle raising. This made the restoration of the soil easier. In addition to the supply of new chemical fertilizer a great deal of nitrate was imported, especially in the form of guano from the Peruvian islands. Indeed home agricultural output, like industry, became increasingly dependent upon imports, not merely of fertilizers, but of oil cakes and meals for animals. Agriculture, no less than industry, was becoming dependent upon imports.

But though there was relative prosperity in the countryside there was not freedom from trouble. By the sixties the fundamental instability of cereal agriculture was becoming more apparent. A bad wheat crop, as in 1867, sent prices very high, the result was an increase in acreage sown, which frequently, as in 1868, coincided with good weather, so that total output could vary from 9·3 million quarters to 16·4 in successive years, with a consequent violent fall in price.[3] The rinderpest killed at its peak some 20,000 cattle per week in the sixties, calling into being the Cattle Plague Section of the Home Office in 1865, later to become a new Board of Agriculture. Thus the state entered directly into husbandry.

The concentration of land ownership was proceeding steadily, and

---

[1] Obituary, *Proc. Institution of Civil Engineers*, 1872, p. 233.

[2] F. A. Buttress, *Agricultural Periodicals of the British Isles, 1681–1900, and their Location*, Cambridge, 1950. For improvement generally, see R. Trow-Smith, *English Husbandry*, London, 1951.

[3] James Caird, 'On the Agricultural Statistics of the United Kingdom', *J.S.S.*, 1869, p. 62.

had, by the seventies, gone very far indeed. A great stir was created in 1874 by the *Return of Owners of Land*, a report made at the request of the House of Lords.[1] The rent receivers had intended, in calling for the Return, to rebutt the charges that were increasingly being made by Bright and other radicals. It demonstrated, on the contrary, that they were short of the truth. With a population of some 31 million in the U.K. some 7,400 persons owned about one-half of all the land. Some 41 per cent of the area of England was in estates of 3,000 acres and upwards. Probably no more than 12 per cent of the cultivable land of England was owned by those farmers who worked it.[2]

Protests against the concentration of landownership had been loud in the forties; they gained renewed strength in the seventies and eighties. They took several forms: that such a monopoly was in itself bad, and positively that it made the possibility of peasant proprietorships, or small holdings, a system favoured by philosophical radicals, increasingly remote.[3] The case was argued for 'free trade in land,' uninhibited by legal obstacles, so that individual initiative could operate in agriculture as in industry.[4] The German worker, it was pointed out, was much less interested in rushing to the towns than was his British counterpart, because he was anxious and able to acquire a small piece of land.[5] A further effect of concentration was to delay until near the end of the century the development of a real national market in land.[6]

The tenants had been making improvements steadily; the question of protecting their interest in these became more pressing. For though after about 1830 the respective responsibilities of landlord and tenants were fairly clear, with the costs of true improvement lying on the landlord, maintenance costs were shared. In difficult times the landlord would often accept his agent's suggestion that instead of lowering rents, the farmer might be aided by a programme of improvement carried out by the landlord; the tenant could thus hardly fail to feel that in a sense part of the improvement was his. The price of land was being further run up by newly arrived industrialists seeking to acquire

---

[1] *Owners of Land. England and Wales, Return for 1872-3*, Cmd. 1097; J. Bateman, *The Great Landowners of Great Britain and Ireland*, London, definitive edition 1883.

[2] Clapham, op. cit., vol. II, pp. 253, 261.

[3] Henry Fawcett, *The Economic Position of the British Labourer*, London, 1865, pp. 10, 16.

[4] G. Shaw Lefevre, *English and Irish Land Questions*, London, 1881, chapter I.

[5] B. Samuelson, M.P., *Report on the State of Industry and Manufactures on the Continent*, 1867, p. 52, quoted in *Econ. C.H. and R. of 1867*, 1868, p. 51.

[6] See discussion by F. M. L. Thompson, 'The Land Market in the Nineteenth Century' *O.E.P.*, 1957, p. 285.

estates for prestige reasons. Parliament was bound to intervene to secure the tenants' rights, and after much discussion did so.

In 1877 the Settled Estates Act allowed the tenant to sell his lease to another without it ceasing at his death, unless the agreement specifically provided otherwise. In 1883 the Agricultural Holdings Act made compensation for improvements compulsory. The decline in the political power of the landlords was further marked by the County Council Act of 1888, ending the old system of government by Justices of the Peace, establishing local Councils, and enfranchising the country labourer. The diminution of the landlord's role was by no means all gain for the countryside, for there were many who sought to serve both the land and their tenants; an honorable place among these was held by some of the retired manufacturers, newly come to the country.[1]

As the tenant sought remedies against the landlord, so the farm labourers sought to have recourse against the farmer. But obscurantism was at work. Attempts were made to get a clause into the Census Act of 1860 to provide information about farm cottages and to obtain returns of farm wages; both were refused for England but adopted for Scotland.[2] Yet it seems clear that agricultural wages had risen little if at all during the period of prosperity. The farm workers, because of their dispersal, illiteracy, and traditional conservatism, were among the last to gain from expansion.[3] By the seventies they were depressed to such a point, especially as industrial workers' earnings were improving, that at long last an effective movement toward organization took place. Joseph Arch, a Warwickshire hedger and ditcher and Methodist lay preacher, launched the National Agricultural Labourers' Union in 1872.[4] Its growth was astonishing to its sponsors and frightening to its enemies. It had some 100,000 members within a year, with some 50,000 agricultural workers enrolled in other unions. The town unions gave support; many of the middle class, headed by John Stuart Mill, showed their sympathy morally and financially. At last successful pressure could be exerted upon the farmers to raise wages.

As this urban blight spread to the countryside there was bound to be strong resistance. Farmers and landowners were as one in their hostility; the church in the countryside showed itself a no less doughty

[1] Chambers, op. cit., p. 82.

[2] James Caird, op. cit., 1869, p. 75.

[3] For the agricultural worker see below, Chapter 7, sections 4 and 9, pp. 230, 231, 271.

[4] Joseph Arch, *Joseph Arch, the Story of his Life*, told by himself, London, 1898. Evidence before *R.C. on Depressed Condition of the Agricultural Interests*, Final Report, 1882.

defender of rural order, paternalism, and the proper relationship of classes. Farmers' Associations were quickly formed.

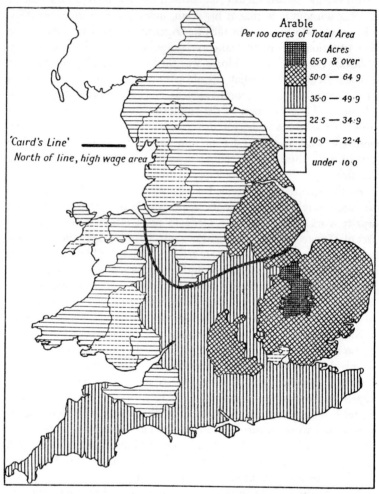

ARABLE ACREAGE IN 1870

Source: Wilfred Smith, An Economic Geography of Great Britain, London 1949, p. 56; James Caird, English Agriculture in 1815–1851, London, 1851, p. 512.

But the contest was never really joined, for the coming of rural depression revolutionized the situation. Between 1873 and 1894 agricultural prices in general fell by about a third, with wheat prices declining more drastically than the rest.[1] The farmers in response to this common danger found a solidarity and determination far beyond what a mere association could induce, and in instinctive concert bore heavily downward on wages, enforcing their position with lock-outs.[2] The precarious basis of the agricultural labourers' union was revealed: its membership fell rapidly away and the countryside returned to the peace of paternalism.

The difficulties of the wool raisers and the cereal farmers were very real.[3] The former suffered the effects of cheap Australian wool. The latter sought and obtained lower rents either by agreement or default. From this the gentry probably suffered more than did the owners of the great estates. The capital value of agricultural land dropped drastically. Over the whole of England and Wales rents fell by some 24 per cent between the seventies and the early nineties; in some areas, especially the arable east, the fall was much greater, bringing landlords' incomes down by some 30 per cent.[4] The heavy clay areas, where draining investment was highest, were hardest hit. Home-grown cereals could no longer compete; by the late seventies Britain was importing some 70 per cent of her consumption. The acreage under wheat fell from 3·6 million in 1874 to 2·3 in 1887, and by 1900 was 1·8 million. But the relative contraction was greater in the north, where corn production was marginal, than in the south. Indeed the north was confirmed in its pastoralism. Some areas at least suffered not at all from depression, and some might well have gained as they contributed to a substantial rise in the cattle population.[5]

Now, if ever, it might be expected that the landed interest would once more seek to use its still considerable political power to re-establish protection. There was no real movement of the kind. It was too late to ask that a now industrialized Britain should forgo the cheapness of foreign corn. For many landed families depended for their income upon industrial prosperity no less than upon agriculture. The

---

[1] E. M. Ojala, *Agriculture and Economic Progress*, Oxford, 1952, p. 146.

[2] Frederick Clifford, *The Agricultural Lock-Out of 1874*, Edinburgh, 1875; A. Clayden, *The Revolt of the Field*, London, 1874.

[3] See Mancur Olson, Jr. and Curtis C. Harris, Jr. 'Free Trade in "Corn"; A Statistical study of the Prices and Production of Wheat in Great Britain from 1873 to 1914', *The Quarterly Journal of Economics*, 1959, p. 145; Pierre Besse, *La Crise et l'Evolution de l'Agriculture en Angleterre de 1875 à Nos Jours*, Paris, 1910; *Royal Commission on Agricultural Depression*, 1882.

[4] H. A. Rhee, *Rent of Agricultural Land in England and Wales, 1870–1939*, 1946.

[5] T. W. Fletcher, 'Lancashire Livestock Farming during the Great Depression', *Ag. H.R.*, 1961, p. 17; R. P. Stearns, 'Agricultural Adaptation in England', *Agric. Hist.*, 1932, p. 91.

only hope for Britain was to participate in a growing world economy and not to seek to cut herself off from the areas abroad that she herself had done so much to fructify. The situation of the land-owners, with their demonstrable 'land monopoly', provoked little or no sympathy.

At last an important sector of British agriculture had succumbed to overseas competition.[1] Farming, in common with industry, was challenged to move out of relatively simple lines of production represented by cereal cultivation into new products where skill and science were more important. The farmers and landowners of the south were ill equipped to make this essential response. As tenants departed from their holdings, many of the great owners, in spite of the exhortations of others, declined to cultivate on their own account.[2] To change, with vigour, and on a sufficient scale, to dairy herds, poultry, orchards, and market gardens was too much for the lords of the great acres and their tenants. Even the pastoral farmers of the north had their difficulties, as frozen meat from abroad began to make its appearance.

Yet the new situation did bring its changes. Traditionalists though southern British farmers were, and vulnerable though their intensive cereal cultivation might be to the extensive methods of the new countries, the British wheat producer worked a more versatile land than did the prairie farmer under semi-arable conditions. The proportion of grassland rose, though too often there was simple reversion to natural pasture with little attempt to use improved grass mixtures. Once more many of the less efficient cultivators were ousted. Scots dairymen went south and did well by laying down to grass old arable farms in Hertfordshire and Essex. Labour intensive production began to give way increasingly to mechanization.[3] The absolute number of workers on the land fell. The reaper and binder were gradually adopted, methods of seeding and threshing were improved. Chemical fertilizers instead of natural nitrates were increasing; with their advent, crops did not depend on stock, and the old Norfolk four-course rotation, so valuable in its day as integrating animals and arable, was being superseded. But for all these changes it is doubtful whether adaptation was as rapid in these decades as it had been in the years of prosperity, for confidence and capital had received heavy blows.

[1] For general discussion see T. W. Fletcher, 'The Great Depression of English Agriculture, 1873–1896', *Econ. H.R.*, 1961.

[2] For the decline in tenants' investment see J. R. Bellerby, 'Farm Occupiers' Capital in the United Kingdom before 1939', *The Farm Economist*, 1952–54, p. 257.

[3] For numbers in agriculture, see J. A. Mollett, 'The Size of Farm Staffs in England and Wales in 1851 and 1941', *The Farm Economist*, 1948–51, p. 150.

CHAPTER SIX

The Suppliers and Manipulators of Capital

THE tenant farmers were much concerned with problems of capital provision; so, too, to an even greater degree, were the industrialists. Indeed the renovation of the economy of Britain during the nineteenth century may be regarded, in one sense, as an act of continuous mobilization of capital. It was necessary that a considerable portion of the mounting product be withheld from consumption, to be continuously applied to the formation of new equipment, embodying new ideas. In order that compounding growth might be sustained, this proportion had to be consistently higher than in pre-industrial times, to become equivalent to something like 10 per cent of the national income.

The men who presided over this process were the bankers. They, like the men who borrowed from them and deposited with them, were men of business. They too had to learn, formulate, and practise appropriate rules. But they also required imagination. 'A large bank', remarked Bagehot, 'is exactly the place where a vain and shallow person in authority, if he be a man of gravity and method, as such men often are, may do infinite evil in no long time, and before he is detected.'[1] The Governor and Court of the Bank of England had to learn the role of a central controlling bank.[2] They had to find a continuous adjustment between three sets of duties, through a period of immense changes: to their shareholders, to the government, and to the economy as a whole. The wide variety of other bankers had to learn to conduct their lesser affairs: how to reconcile profitability (which depended upon

[1] Walter Bagehot, *Lombard Street, a Description of the Money Market*, London, new ed., 1919, p. 214.

[2] E. Victor Morgan, *The Theory and Practice of Central Banking 1797–1913*, Cambridge, 1943; Sir John Clapham, *The Bank of England: a History 1694–1914*, Cambridge, 1944; W. M. Acres, *The Bank of England from Within, 1694–1900*, Oxford, 1931; Elmer Wood, *English Theories of Central Banking Control, 1819–58*, Harvard, 1939.

the extension of their loans) with safety (which involved keeping loans in a safe proportion to ready assets).

The banks were a unique element in the national life, not only because of the essential function of finance which they performed. It was becoming apparent that they and their operations might be reduced to a set of general principles so integrated as to constitute a system of thought. This system was fundamental to the understanding of the growth and stability of the economy, both internally and in its external relations. No other of man's institutional creations had this characteristic to the same degree. The monetary system came closer than did any other element in the national life to supporting reasoning analogous to the laws of physical science. Many men in the first half of the century were intrigued as they watched a set of business enterprises, created pragmatically, to a great extent in their own lifetime, yielding, under successive inquiries, a set of concepts and laws. That there might be radical differences of opinion about their nature did not impair their fascination. Indeed the immense concentration of debate on the monetary system, in the hope that if a full understanding of its working could be achieved it would be unnecessary for the state to look further, much less to intervene, served to obscure other elements in the economy that reacted only over very long periods, if at all, to changes in the monetary and credit structure.[1]

The system was presided over by the Governor and Court of the Bank of England. The question of the shareholders' interest in the Bank, so important in earlier years, though it never wholly disappeared, was by 1815 largely subordinated to the other responsibilities of the Court. The Bank could not act as a simple profit maximizer. But there remained the question, was the Bank an instrument and projection of government, regulated by Charter and Act of Parliament, and occasionally by direct governmental intervention, or was it, on the contrary, an independent institution run by private traders, yet exercising controlling authority over the great range of bodies of a similar or related kind that constituted the money market of the City? Down to the mid-century, the Bank lived on the edge of the political arena, and could be confident that it would be dragged into it whenever, in difficult times, it found itself bound to force a considerable number of traders into liquidation. Only after 1844 did it make any real progress in the direction of becoming non-political. But the fact that the Governor was appointed by fellow members of a private cor-

[1] See below, Chapter 10, section 9.

poration, and not by the political party in power, as in the case of the Bank of France, made it easier to keep banking out of politics.[1]

The London bankers, especially in the City, constituted a kind of middle term between the Bank and the host of country bankers that had sprung up since the mid-eighteenth century. They acted as agents for them, providing advances, advice, and information. The bankers of the newer West End, with their aristocratic connections, were slower to enter into such business. Many country bankers, attracted by the prospects of London, had themselves invaded the City, and some had established themselves there among the most powerful.

As London and country banking grew, a national system of credit mobilization came into being — the bankers in the eastern, agricultural, counties collecting savings in the form of deposits, building up credits with their London agents, through whom loans were made available to northern industrialists through the discount of their bills of exchange — the famous 'inland bills'.[2] In addition, there emerged, from the late eighteenth century onward, a highly specialized class of bill brokers — men who circulated among the London banks, buying and selling personal paper promises to pay.[3] These transactions in bills of exchange, in addition to their role in the provision of credit, were very important as a means of making payments, especially inter-regionally.

Many bankers simply evolved into their position through transactions arising from other types of business, trading as iron masters, mine owners, or textile manufacturers. Some were first prompted in the direction of banking by the need to supply a local currency — issuing their own notes and tokens. Often it was found that specialization in the banking function was more profitable than the original business. For with good management and good fortune, the gains in country banking could be very great. So, too, were the risks, for these provincials were often without considered rules of conduct.

The number of country bankers fell rapidly after 1815, with hundreds of failures. But it recovered again in the twenties. The country bankers could be highly vulnerable to crisis. In part this was due to the state of the law as it prescribed the privileges of the Bank of England. By earlier Acts confirmed in 1808, the right to issue notes, in all cases

---

[1] Bagehot, op. cit., p. 217.

[2] L. S. Pressnell, *Country Banking in the Industrial Revolution*, Oxford, 1956.

[3] W. T. C. King, *History of the London Discount Market: with an Introduction by T. E. Gregory*, London, 1936; Ellis T. Powell, *The Evolution of the Money Market, 1385–1915*, London, 1915. For the special role of bills in Lancashire see T. S. Ashton, 'The Bill of Exchange and Private Banks in Lancashire, 1790–1830', *Econ. H.R.*, 1945.

save that of the Bank itself, was limited to bodies composed of not more than six partners. This rule was generally taken to mean that in England and Wales banking concerns must involve no more than six partners. In consequence, joint-stock banking was a Bank of England monopoly, so that many country banks were kept smaller than they might have been, with consequent limitations on their reserves. On the other hand, the law held all such partners liable for the debts of the bank to the full extent of their assets.

The mechanisms that linked the Bank of England, the London houses, the English country bankers, and the Scots and Irish bankers, were approaching their modern form. All bankers undertook to redeem their notes on demand, in gold or in Bank of England notes. The Bank itself, in turn, would always pay out a fixed weight of gold ($£3$ 17s. $10\frac{1}{2}$d. per ounce) for each one pound worth of its notes presented. This meant that the notes issued by the country banks, the Bank's notes, and gold, were all fixed in relation to one another, so long as full convertibility was practised by all bankers. From this equivalence flowed consequences of the highest importance.

The Bank of England, at the centre of the system, made its loans, largely through the discount or rediscount of bills of exchange. That is to say, it was prepared to buy the promises of business men, from other bankers or from bill brokers, paying something below their mature full value, taking this difference as the price of their service. So long as bankers in London and in the country generally were confident that the Bank of England would buy bills from them if it should be necessary, they would continue, subject to certain tests of their own, to discount, and thus make credit available for business.

The leading theme of central banking history from 1815 to 1914 was the search for a principle capable of serving as a guide to the control of money and credit. In theory there were three leading possibilities. The Bank of England might attempt a judgement about the general situation of the business community, in terms of its proximity to full employment and consequent vulnerability to an inflationary price rise, or in terms of underemployment and consequent falling incomes. But under nineteenth-century conditions, with inadequate data and under-developed theory this principle was too difficult to operate. Moreover the Directors of the Bank had no desire to become involved in the complexities of monetary management. They yearned for a simple rule, which would allow them to discharge their function as directors in much the same way as they conducted their own private businesses.

The Bank fell back upon the two remaining possibilities, both based upon rules of thumb. During the Bank Restriction period (1797–1821) the Bank had defended itself against the charge of excessive discounts by arguing, as Adam Smith had done, the famous real bills doctrine. So long as the banks discounted only such bills as represented real transactions, and at short maturities, there could never be excessive credit, for each instrument had a corresponding element of real wealth associated with it, and the credit supply was closely tied to the legitimate needs of industry. This doctrine was vigorously assailed by the Bullion Committee in 1810.[1] It was then demonstrated that with general rising prices it was impossible to be sure that the supply of bill credit was always such that each trader could always meet his bills on maturity.

This left one further alternative. Could the Bank find the clue to the general management of credit in some aspect of its own situation? Could affairs be so managed that a set of simple rules would ensure the liquidity of the Bank in all situations, and also, indirectly, minimize disturbances in the economy as a whole? Banking legislation down to and including the famous Bank Charter Act of 1844 may be summed up as an attempt to establish liquidity procedures for the Bank which would also be valid for the economy as a whole.

As, in the upswing, the Bank extended its purchases of government securities, or its discounts, two other aspects of its business also expanded: deposits and note issue. Deposits of the public with the Bank of England increased, just as they did with other bankers, for those who had gained credits held a substantial proportion of them in this form, or passed them on to others, increasing their deposits in turn. Such an extension involved vulnerability for the Bank of England, for it meant that sudden demands for gold in exchange for notes might arise, from deposit holders who were pressed to meet their obligations in coin or bullion.

Greater business activity meant an increase in the note issue of the banks, including the Bank of England. This in turn caused prices to rise as the supply of money in the hands of purchasers increased faster than did the supply of goods and services. But any failure of confidence in the soundness of the expansion caused country bank notes to be returned to the country bankers, and Bank of England notes to be demanded instead. If the crisis became severe enough, even the Bank's notes would be returned, and gold demanded. If the supply of gold in

[1] *S.C. on the High Price of Gold Bullion*, 1810.

the hands of the Bank was inadequate, a stoppage of payments, and general breakdown, must ensue. The crux of the system, therefore, by 1821, had become the relationships between the gold holdings of the Bank of England and its outstanding obligations in the form of notes.

The Bank was thus seeking to use its own liquidity position as a barometer giving guidance concerning the entire economy. This, of course, gave rise to much discussion as to whether the balance sheet of the Bank was a true analogue. Should the general supply of credit to the economy as a whole be extended or contracted on considerations stemming from the ratio of reserves to obligations in the central bank?

The debate was the more difficult to interpret as considerations arising from international transactions were inevitably introduced. The gold supply available to the Bank in Britain might well be lessened by external drain. This might be more or less continuous, for example, to the markets of the East where British exports could never match British purchases. Or the drain might be more sudden. A great deal of grain might be imported to make up for crop failure, and because Poland, Prussia, Russia, and Hungary were servile societies unable to respond to British manufactures, only gold was acceptable. Political disturbance, as in 1819, would also send hot money abroad. Or thirdly, in boom, imports might rise more quickly than exports. Gold losses abroad, from whatever cause, and at whatever conjuncture they occurred, obliged the Bank to take action affecting all businessmen, through credit restriction: this affected not only domestic lending, but also the whole complex of world lending as London's position as the centre of international finance was confirmed.

The Bank might try, as an alternative to reducing discounts directly, to deal with the situation by raising the rate of interest in Britain. This it could do by raising Bank Rate: the price at which it was prepared to lend, when called upon to discharge its role as lender of last resort. This served as notice to all, that future borrowing if available at all, was going to be more costly. The effect was likely to be twofold. Businessmen felt this discouraging influence and slackened the pace of their activities, allowing the upward pressure in prices to be relieved. This stabilizing or lowering of prices at home in turn encouraged exports and discouraged imports, staunching the flow of gold abroad. In addition, a higher rate of interest in the City of London might well attract loans from abroad to the City, further reducing the outward gold flow. If both businessmen and foreign lenders had been sensitive to small changes in the Bank Rate, it would have been a much more effective instrument than it sometimes was, but often the Bank found that it was necessary to administer a severe shock to the City in order to get an adequate response. This meant brusque action: either a steep increase in Bank Rate, or a very severe attitude towards discounts, or both.

All this meant that convertibility of the note issue at a fixed ratio to gold was subject to the criticism that the whole technique was too crude: it allowed deterioration to go much further than it should have done, and it involved measures that were severely damaging to the economy and to society when enforced liquidation finally became necessary. The suffering and loss involved in credit curtailment in a banking crisis was often very severe, reaching far beyond what was necessary to remove unsound undertakings. The argument was long and vigorous; the upshot was a determination to return to full convertibility at the pre-1797 standard; this was achieved in 1821.[1]

The crisis of 1825 revived the banking debate. It was suddenly demonstrated that convertibility of the note issue to gold, achieved with such difficulty in 1821, was not a sufficient condition to preclude commercial collapse. Speculation reached an extraordinary pitch. The Bank was slow to raise its interest rate to dampen the boom. Eventually there was a panic demand for liquidity, so that not even the best securities, including those of the government, the East India Company, and the Bank itself, could be sold without grave loss. Some sixty financial houses in London failed, together with some thirty-six country banks.

By the Bank Act of 1826 Parliament sought to remedy several defects in the banking system at a single blow.[2] The small notes of the Bank of England were thought to promote price instability and speculation; there were to be no notes under £5. The country bankers of the new industrial areas were to be permitted to adopt the joint-stock principle: they were to be freed from the six partner rule. But the concession was limited: it applied only to banks outside a radius of sixty-five miles from London, leaving large-scale joint-stock banking in the south-east a monopoly of the Bank of England. Nevertheless it was now possible to begin to put provincial banking onto a large-scale basis. In addition the Bank was empowered to open its own branches in any

---

[1] *Reports . . . On the Expediency of Resuming Cash Payments*, 1819.

[2] For discussion and legislation see T. E. Gregory, *Select Statutes Documents and Reports Relating to British Banking 1832–1928*, London, 1929.

part of England.[1] This provision was intended to hasten the spread of Bank of England notes, in order to promote a uniform stable issue.

In the early thirties there was intense inquiry into national institutions. Banking was not exempt. The Bank Act was due to expire in 1833 and, in spite of the fact that the Reform Bill commotion delayed consideration, a Committee of inquiry reported late in 1832. Its findings were inconclusive, a result that was not surprising from a Committee deliberately chosen to include all points of view.

The note issue was still regarded as the prime consideration. By 1832 the Directors of the Bank were unanimous that, as laid down in the Bullion Report of 1810, note issue should contract or expand in response to changes in the foreign exchanges or in the price paid in notes for bullion. Thus if the exchanges moved against Britain, this meant that imports were in excess of exports in value terms and it was time to reduce credit, in order to stop an external drain of bullion. Or, if the price of bullion at home was rising relative to notes, this indicated internal drain and therefore, again, excess credit facilities. The Governor of the Bank, Horsley Palmer, came forward with his famous 'rule', to formalize the proper relationship between note issue and liquid reserve: the Bank should keep a ratio of one-third of its assets in bullion and coin against its obligations payable on demand.[2] There was a growing appreciation by the Bank of England that it must accept the role of a central bank, responsible for the stability of the system as a whole. This supervisory role meant severe curtailment of its competition with other banks for the discount of commercial bills, and the acceptance of the responsibility to act, in crisis, as a lender of last resort, sustaining those enterprises that were basically sound but threatened by cumulative panic.

The Bank Act of 1833 was the next step. The notes of the Bank, for the first time, were made legal tender. This meant that holders of Bank of England notes (except of course the Bank itself) could pass them on in discharge of debts in excess of £5, and not be obliged to obtain gold for them in order to do so. It further meant that embarrassed country bankers did not need to send panic requests to the Bank of England for gold, but could be supplied with notes, and discharge their obligations therewith. The legal tender status, so long sought by the Bank for its notes, could thus serve to ease internal drains. Secondly, the Bank on its discounting of Bills of less than three months term was freed from

---

[1] R. O. Roberts, 'Bank of England Branch Discounting, 1826–59', *Econa.*, 1958.
[2] For Palmer see A. B. Cramp, *Opinion on Bank Rate, 1822–60*, London, 1962.

the old Usury Law limiting the rate of interest to 5 per cent.[1] It was now possible, in times of credit shortage, for the price of short-term funds to go as high as the market was prepared to bid.

But the Bank in turn had to make concessions. Public opinion now demanded at least some disclosure of its affairs: the result was a clause in the Act requiring the Bank to inform the Chancellor of the Exchequer monthly of the amount of its bullion, the size of its note circulation, and the total of its deposits. This return was to be published in averaged form, in the *London Gazette*, three months in arrears. The joint-stock banks made a further important encroachment. After 1833 they were permitted to operate within the sixty-five mile radius, and thus set themselves up in the City itself. But any bank doing so was required to abandon its note issue, this leaving more scope for the Bank of England in this regard. The joint-stock banks found this a good bargain, for they had been developing the new technique of cheque issuing, which was soon to eclipse notes as a vehicle of business. The Act certainly caused the joint-stock principle to be further extended.

But banking crises came again in 1836 and 1839.[2] There were, as was usual in crisis, plenty of grounds upon which the Bank could claim to have been the victim of special circumstance. Prosperity was very real between 1833 and the culmination of the first railway boom in 1836. The number of joint-stock banking companies in England had been increased in these three years by some seventy-two new foundations. They had added to inflationary pressure by excessive re-discounts. In 1836 gold flowed outward to the United States as American securities were sold in London to provide gold for President Jackson's banking reforms. Ireland too was a heavy drain due to the failure of the Agricultural and Commercial Bank. Lancashire was a centre of difficulty because of heavy involvement with American and Irish investments. The American houses in London were also gravely embarrassed. Bad harvests in 1838 meant gold exports to pay for corn. Late in 1838 America, France, and Belgium all hastily withdrew gold from London because of the threatened breakdown of their over-extended credit systems. Only by borrowing from the Bank of France, a necessity especially shocking to the public mind, was the situation saved. The effects were felt as late as 1843; some sixty-three country banks suspended payments.

It was painfully obvious that the banking world was still a long way from understanding itself. The Directors of the Bank were under

[1] See Chapter 9, section 5, p. 344. [2] See above, Chapter 2, sections 3 and 4, pp. 14, 17.

heavy fire. Out of this situation, full of intellectual confusion and bitter recrimination, came the long debate leading to the most celebrated of the Bank Acts, that of 1844.

Control of the note issue was still uppermost in men's minds. It was pointed out that though many banks had abandoned their issues, by failure, or by invading London, the issue was still not unified under the control of a single bank. There was a fair element of agreement that a single issuer should be the objective. But on the problem, how should this monopolist behave, there was deep division.

The Banking School and the Currency School, both employing arguments with long pedigrees, joined in a great full-dress debate.[1] The former said, in effect, 'Trust the Bankers'. Demand convertibility of the note issue, but beyond that leave discretion to those in charge of banking policy; they will not, the argument ran, be able to put excessive paper afloat. Do not impair the scope and flexibility of the Bank's actions by an insistence upon rigid rules. The Currency school, on the other hand, denied that the bankers should be left so free of control. They said, in effect, that the issue of notes by the Bank of England should be backed pound for pound, by gold reserves. Paper currency would thus be convertible at all times, with the result, so it was argued, that liquidity crises could not occur. The Bank, with its discretion thus limited, would find itself less likely to make errors. The essential need was for 'a regulation based upon principle'.[2]

Samuel Jones Loyd, later Lord Overstone, was the great protagonist on the Currency side.[3] Thomas Tooke led for the Banking school.[4] Tooke had begun during the war as a bullionist — stressing the need for full convertibility, helping to carry the Bank, against its will, toward the resumption of 1821. This he did in the interests of restoring an automatic monetary system, in which the state need have no part, yet which was subject to the binding rule of convertibility. Then his views

[1] For the background see R. S. Sayers, 'Ricardo's Views on Monetary Questions', in T. S. Ashton and R. S. Sayers, eds. *Papers on English Monetary History*, Oxford, 1953; J. K. Horsefield, 'The Origins of the Bank Charter Act, 1844', ibid. p. 109; A. E. Feavearyear, *The Pound Sterling*, 2nd ed., revised by E. Victor Morgan, London, 1963; K. Niebyl, *Studies in the Classical Theories of Money*, Oxford, 1946; Jacob Viner, *Studies in the Theory of International Trade*, New York and London, 1937, chapter V.

[2] Gregory, op. cit., vol. I, p. xxi.

[3] L. A. Helms, *The Contributions of Lord Overstone to the Theory of Currency and Banking*, Urbana, 1939; Lord Overstone, *Tracts and other Publications on Metallic and Paper Currencies*, London, 1858.

[4] See Alvin H. Hanson, *Monetary Theory and Fiscal Policy*, New York, 1949, p. 83; T. E. Gregory, *An Introduction to Tooke and Newmarch's a History of Prices . . .* London, 1928.

began to change: his great opponent, Overstone, referred to him as being like the artist Turner, having his 'periods'.[1] It was still important to Tooke there should be convertibility. But was it really necessary, as the Currency school insisted, to oblige the Bank to back its note issue, pound for pound? Was it not enough to require convertibility, allowing the Bank to use its own discretion as to how far the note issue might safely exceed gold holdings? For Tooke believed that the Bank would not serve the economy to the best effect if it thought only in terms of the note issue. He had come to the opinion that it was not the quantity of money that was of primary importance, but the volume of expenditures, itself related, in turn, to the size of incomes. Thus it was to incomes, rather than to prices, that the controllers of the monetary system, the Bank, should look. This new emphasis opened a much broader inquiry than had been associated with the quantity theory of money, for it involved consideration of the factors governing aggregate outlays, and through them, output and employment. Moreover it meant that the level of prices did not depend upon the quantity of money, but the reverse, 'the amount of circulating medium is a consequence of prices'.[2] This line of reasoning opened the way to attack upon 'The preponderant, almost the exclusive theory . . . which refers all the phenomena of high prices from 1793 until 1819, and of the comparatively low prices since 1819, to alterations in the system of our currency . . .'[3] But Tooke, for all his insight, could not consolidate his theory sufficiently to overthrow the Currency school. Indeed the 'income' approach was only to come into its own with Keynes.

The Currency school won the ear of Sir Robert Peel and the government. A memorandum was drafted by William Cotton, Governor of the Bank, and J. B. Heath, the Deputy Governor, embodying all the basic features that were incorporated in the Act. It now appears that the Act of 1844, far from being imposed upon the Bank, was its preferred solution to the problem of monetary management. It prescribed that the balance sheet of the Bank should be published in two parts, reflecting a distinction of function within the Bank between the Issue Department and the Banking Department. The account of the former showed the number of notes outstanding, and against this total the amount covered by holding of gold (and to a minor degree, silver) and the amount issued against holdings of government securities. The

---

[1] *D.N.B.*, LVII, p. 47.
[2] Thomas Tooke, *An Enquiry into the Currency Principle*, London, 1844, p. 123.
[3] Ibid., pp. 71–2.

nation could thus make a weekly inspection of the Bank's behaviour in terms of note issue and gold cover. The Banking Department was to be entirely independent, freed from the special responsibilities devolving upon a bank of issue.

The crux of the Act was that the amount not covered by gold holdings was absolutely limited. It could not exceed £14 million. This 'uncovered' portion was known as the fiduciary issue. It could, however, be extended to compensate for contractions in the issues of other banks: in the case of any discontinued issue the Bank of England might increase its fiduciary issue by a sum not exceeding two-thirds of the notes withdrawn. Other banks continuing to issue were forbidden to exceed their average issue of the twelve weeks preceding 27 April 1844. No new issues were to be made. So the way was prepared for centralizing this function with the Bank.

The hope behind the fiduciary rule was that the Bank would continuously and closely adjust its lending behaviour, keeping its note issue closely related to its metallic reserves, so that arbitrary movements in prices and extravagant boom and crisis would be precluded. This hope was not to be realized. Crises came again in 1847, 1857, and 1866: on each occasion authority for abandoning the limitation of the fiduciary issue had to be given by the government to the Bank, in order to restore confidence, though actual over-issue occurred only in 1857. On the other hand there is little doubt that the Act secured that the ratio of the Bank's treasure to its note issue was much more favourable than it would otherwise have been.

The drastic simplification of thought behind the Act of 1844 depended a good deal upon the Quantity Theory of Money, held by Ricardo and others. It asserted that, in the short run, an increase in prices followed automatically on an increase in the money supply, with consequent effects on exports and imports and therefore the foreign exchanges. It seemed to follow that the note issue should be strictly controlled. But this was to ignore other elements in the situation, as Thomas Tooke pointed out. These were to increase in importance with the passage of time. The growth of deposits and the use of cheques were capable of having a real effect upon prices. After the end of the Usury Laws in 1833 and 1854 the Bank developed a policy of operating upon interest rates as a means of controlling the economy that could often be much more effective than variations in the money supply.[1] The rate of exchange could vary from natural

[1] For the usury laws see below, Chapter 9, pp. 335, 336, 344, 359.

causes, like irregularities in harvests. The free movement of gold and the operation of fixed parities only really produced stable exchanges, as evidenced by the rate for American dollars, after 1875; previously there were wide short-term fluctuations, associated with variations in the rate of expansion of the American economy (sterling going to a premium in American booms).[1] But for all its limitations, the Act of 1844 was the basic document in British banking right down to 1914. Robert Lowe's proposals of 1873 in his 'elastic issue' Bill, failed; so did the suggestion that minimum reserve regulations on the American plan should be enforced by law on the commercial banks.

Because of the unreality of the assumptions upon which it rested, the historical role of the Act of 1844 was quite otherwise than that intended by its drafters. It did not, in fact, provide the guidance needed for the conduct of central banking in the subsequent decades: for this the Bank relied on a more flexible approach, and was aided in its policies by mounting prosperity. The significance of the Act of 1844 lay in the way in which it served to put an end to any serious prospect of state interference in the monetary system. Apart from the fiduciary issue rule, the state prescribed nothing about the Bank's behaviour in the manipulation of money and credit. Neither side of the debate was wholly satisfied with the Act. There were sporadic debates, led by irreconcilables like George Rae, leader of the country bankers, Tooke, and William Newmarch. But most Englishmen of business, for some forty or fifty years, regarded the banking debate as exhausted, and accepted, in a vague way, the Act of 1844 as definitive.[2] Thus, largely freed from public concern, the Bank Directors gradually adjusted their policies to new conditions.[3] For the discussion of 1844 had, in a sense, caused a movement backward in banking thought, over-emphasizing the note issue. But by the later sixties the Bank had developed earlier techniques so that it knew how to curb the money market when it became over-exuberant, and, if this proved impossible, was able to preside effectively over the process of liquidation when speculation had gone too far. 'Open market operations', experimented with a good deal earlier, became established practice after 1873.[4] This term described the purchase or sale, in the open market, of government

[1] L. E. Davis and J. R. T. Hughes, 'A Dollar Sterling Exchange, 1803–1895', *Econ. H.R.*, 1960.

[2] George Rae, *The Country Banker: his Clients, Care and Wo k*, London, 1885.

[3] P. Barrett Whale, 'A Retrospective View of the Bank Charter Act of 1844', in T. S. Ashton and R. S. Sayers, eds. op. cit., p. 127.

[4] Clapham, *Bank of England*, vol. II, pp. 294–8.

securities, chiefly Consols. By purchasing securities the Bank created greater balances with itself held by individuals or corporations, including banks, and so made credit expansion easier; conversely by the sale of Consols, it diminished the balances held with itself, and so obliged borrowers to come to it, thus making the rate at which the Bank was prepared to lend (Bank Rate) effective as ruling the market. In this way the Bank learned to use the official debt as a means of controlling the volume of credit and money. If necessary it could create sufficient stringency to force the market 'into the Bank'. The idea that the Bank should act as the lender of last resort, when there was monetary stringency, so minimizing the damage of a panic rush for liquidity, had been expressed as early as 1797; it was made explicit in the seventies.[1]

City and country were first linked on a significant scale by the financing of slaves and commodity movements, especially in dealings with the Americas and the West Indies — sugar, coffee and cotton. When the outports were admitted to the East India trade after 1813 this was reflected in an extension of finance. A little later came the financing of textiles, metal goods, and industrial products generally. In both these phases, it was the financing of the *movement* of goods, and the holding of stocks, rather than their production, that constituted the bankers' major contribution. This is not to diminish its importance — rather the contrary. Though the bankers provided mainly finance of this kind, and that for only some portion of total transactions, their attitude was crucial to the level of activity. When misgiving appeared, the attempt would be made to gain bankers' approval for a much higher proportion of the volume of commercial paper, so that to restore confidence it was not enough merely to maintain the level of discounts, but it was necessary to extend them. *A fortiori*, if the bankers hardened their attitude, the effect would be much more than proportionate.

Even at short term, much discounting could be done without recourse to bankers, with private promises to pay privately held. At longer term, a landed man, for example, could pledge his acres or his interest in the income from them, continue to work them without consequential impediment, so long as he met interest and amortization, and yet by loaning the proceeds he could place a projector in a position

---

[1] Sir Francis Baring, *Observations on . . . the Bank of England*, London, 1797, p. 22; Walter Bagehot, op. cit., chapter VII; R. S. Sayers, *Central Banking after Bagehot*, Oxford, 1957; R. G. Hawtrey, *A Century of Bank Rate*, London, 1938.

to acquire control over other productive factors. Shipbuilders undertook construction, accepting the bonds of those who placed the orders; iron masters supplied rails to railway builders on the same basis. But sooner or later bills would be drawn by those requiring negotiable instruments, and at this point the cardinal issue would be, what estimate would the bankers make of such bills: on what terms would they buy them?

With the coming of the railway, especially from 1840 onward, the relationship between City and country began to alter.[1] Short-term bills of exchange still remained very important. But other ways of mobilizing capital and strengthening the central position of the London money market in the economy of Britain were rapidly extending. Lawyers with clients' funds to place had long known how to do this with advantage in the City; the funnelling of Nonconformist funds to the money market by solicitors throughout the country increased in importance. The old insularity of the City was being modified. In the early part of the century it had been dominated by matters that were of relatively little direct concern to the rest of the country—government debt, bullion movements, the great trading companies. Funds from the provinces were now flowing in to participate in these and other activities.

Effects also appeared in the opposite direction; the City reached out into the country looking for profitable investments. The railways provided the great opportunity, though not in their first phase of expansion. The early initiative in this greatest of all bursts of capital investment came from provincial men of business seeking to exploit local or regional profit opportunities. But though the railway boom of the thirties was largely a provincial affair, during the second, in the forties, the City financiers seized the opportunity with a vigour that soon became excessive. The Stock Exchange quickly adapted itself to the new opportunity.[2]

Though railway shares became one of the principal elements on the Stock Exchange, the City financier did not become intimately involved with the finance of the industrial north in any other substantial direction, at least until the last decades of the century.[3] This

[1] For regional banking see references in Clapham, op. cit., vol. II, p. 349, n. 4.

[2] E. Victor Morgan and W. A. Thomas, *The Stock Exchange: Its History and Functions*, London, 1962. For the securities dealt in see *Fenn on the Funds*, London, successive editions, 16th ed., 1898.

[3] Harold Pollins, 'The Marketing of Railway Shares in the First Half of the Nineteenth Century, *Econ. H.R.*, 1954; 'The Finances of the Liverpool and Manchester Railway', ibid., 1952; S. A. Broadbridge, 'The Early Capital Market: The Lancashire and Yorkshire Railway', ibid., 1955.

was in part due to the fact that limited liability, with its great improvement in the marketability of shares, was not effective over much of industry before that time. Railways seemed a fairly simple proposition to assimilate; the diverse and emergent sectors of the industrial economy were much more difficult to construe and to make attractive to City lenders. Moreover the City had always been a pretty liquid place, with the greatest gains available to those who could buy and sell fairly quickly; industrial investment where each venture had an air of uniqueness was quite a different matter and took a good deal of getting used to. The insurance companies, growing in importance with the expansion of the economy, tended by their nature to be conservative lenders, looking with distrust and bewilderment upon industry.

Industry, in fact, until at least the last quarter of the century, had to rely upon traditional means for the supply of capital: forming partnerships into which the members placed the necessary means from their own resources, borrowing from the landed interest and indeed sharing industrial initiative with these traditional leaders, and ploughing back. This latter could take two forms: the firm might use its profits for its own growth, where the field of activity was such as to promise continued and even increasing gains, or its constituent owners might take their gains and apply them in an altogether new direction. It was this last possibility that opened the way to the company promoter. For though some groups of like-minded business men could come together from different sectors of the economy to establish a new enterprise, in general the divergencies and dispersions of interest and energies were so great that it was better to form a new concern with its own identity, and very often with a hired manager. The railways led the way in this delegation of managerial power. They were also an important force in the formation of provincial stock exchanges in the great cities from the thirties onward.[1]

Though the business world was finding its own resources and allocating them over the area of opportunity, the bankers were still crucial to the process. They could aid particular projects by supporting their credit, but their most important function was still to control the total supply of short-term credit made available very largely for the movement of goods and the holding of stocks. In controlling the expansion of the system in terms of their own liquidity position they controlled the whole, both in quantitative and in qualitative terms. For the bankers' judgement as to what direction was sound for expansion

[1] For other London markets see above, Chapter 4, section 1, p. 106.

and what was not was an important determinant of which projects among many were pursued. Moreover the very process of expansion itself, bringing buoyant prices and new profits, both probably moving in advance of wages, served to alter the distribution of incomes and wealth in the direction of the men of initiative.

In the late fifties the system for the discounting of bills was undergoing important changes. The leading bill-brokers, including Overend and Gurney Limited, established themselves as discount companies, buying bills to hold for their own account and for resale to bankers, guaranteeing all bills that passed through their hands. They traded on short-term borrowed money, underwriting much of the commercial paper in the market, conducting a business that was profitable but often precarious, as the Overend crash of 1866 showed. These houses, of which the Union and the National were the greatest by the mid-eighties, made possible a very highly developed short-term money market, aided by more numerous small bill-brokers.[1]

If the City was perhaps too cautious in participating in the domestic industrial sector, it was excessively generous in its foreign loans. It had always had a cosmopolitan bias. The extension of trade and the enormous traffic in semi-tropical produce had confirmed it. Bills arising from innumerable foreign transactions called for discounting facilities in the provinces and often for rediscounting in the City. If the government wished to wage war it required the means of payment in the currency of the theatre of war; even more important the home government was well aware by the early nineteenth century that the City stood between it and the means of making war, or for that matter the carrying out of any policy that required money. Foreign governments were no less conscious of this strategic position of the City and increasingly through the century sought loans in London. Speculation in land in the colonial empire and in foreign countries made further demands on the banking system. For down at least to the eighties Britain was ahead of the United States in the development of her credit mechanism, and far ahead of France and Germany. London was the undisputed centre of the world's transactions.[2]

In meeting such foreign appeals lay the greatest of all sources of bankers' gains. The great merchant bankers of the City, with Rothschild at their head, and Baring not far behind, took an especial interest

[1] Clapham, op. cit., vol. II, pp. 353, 354.
[2] Harvey H. Segal and Matthew Simon, 'British Foreign Capital Issues, 1865–1894', *J. Econ. H.*, 1961.

in the provision of funds for governments at home and abroad.[1] Foreign banks came to London in increasing numbers, playing a large part in the exchange market from the seventies onward. In this way a leaning toward foreign lending developed in the City; London became the centre of international finance on a new scale. But too often the loans made were not well grounded in any calculation of real productivity; the great City men had no more intimate knowledge of the states and economies they thus financed than they had of industry in northern Britain. But this was not their task. Floating a loan depended upon estimates of what lenders would advance and to whom; the kind of appeal to which they would respond abroad was too often emotional, including the hopes of speculative gain, rather than rational. There were exceptions to this lack of knowledge; a great house like that of Baring, intimately concerned with the discounting of bills from a particular area, in their case America, might well develop an extensive knowledge of the economy concerned. But too often in the selling of a foreign issue in the City such information was not brought to bear.

Though the London money men were responsible for great movements of funds abroad, large balances also came in. In the second half of the century the system of multilateral trade, so rapidly elaborating, depended for its settlement between countries upon the holding of large sterling balances in London. To the extent that the world was prepared to hold its assets in sterling balances these constituted a continuous loan to Britain. Moreover, with many foreign securities held in London, when the need for large imports of corn arose, the bill could be met, in part at least, by sending these instead of gold, thus minimizing the monetary effect. On the other hand money held in London might be hastily withdrawn if there was a loss of confidence, so that this element of instability was increasing.

Continental and American money men attracted by these opportunities coverged on London, especially after the mid-sixties. They made a formidable community: Morgans, Hambros, Goschens, Erlangers, Raphaels, Oppenheims. These additions to the community of merchant bankers further elaborated the City's international finance: with their aid the formation of syndicates for the exploitation of the world's resources could accelerate.

---

[1] Frederic Morton, *The Rothschilds*, London, 1962; Ralph W. Hidy, *The House of Baring in American Trade and Finance; English Merchant Bankers at Work, 1763–1861*, Harvard, 1949.

While foreigners were moving to London, British bankers were taking up new initiatives in Europe, the Middle East, and South America. There were particularly notable developments in the colonies and the Far East. Colonial banks appeared in Canada and Australia.[1] The rise of British banks 'East of the Cape of Good Hope' produced a whole new system. In 1842 there were only the three Presidency Banks in India (in Bengal, Bombay, and Madras), all forbidden to deal in foreign transactions; east of India there were no banks at all. It was the Agency Houses, arising from the trading of the East India Company's servants and that of the private traders after 1813 that dealt with international business; banking was subsidiary to trade, though often profit or loss depended more on wisdom in exchange dealings than the traffic in commodities. But Eastern trade was about to burst its old bonds: the steamship began to reduce travelling times, the restrictive hand of the East India Company was falling away, and the Western powers, under the Treaty of Nanking in 1842, were forcing China to trade.

The new beginning came with the Oriental Banking Corporation, originating in 1842.[2] After receiving its Royal Charter in 1851 it expanded rapidly over India and the Antipodes. Many of the older Agency Houses were weakened and many fell in 1847. This was the signal for new fouudations. During the fifties there was a rush to start Anglo-Indian banks, amounting almost to a speculative mania. The Chartered Bank of India, Australia and China was founded in 1853.[3] Banking development accelerated again in the sixties as India enjoyed high cotton prices during the American Civil War. Though some failed, in general the extension of banking proved very helpful to trade. The new banks spread their branches to China, and banks of Chinese domicile came into being. Thus the great Exchange Banks began to perfect the foreign exchanges of the East, aided by the great reductions in the lag of information and remittance brought about by the electric telegraph. They relied heavily upon their exchange transactions for their profits; indeed these were often their predominant activities. As at home, they took no part in the finance of long-term investment;

---

[1] A. S. J. Baster, *The Imperial Banks*, London, 1929.

[2] See C. N. Cooke, *The Rise, Progress and Present Condition of Banking in India*, Calcutta, 1863; W. F. Spalding, *Eastern Exchange, Currency and Finance*, London, 1917; A. S. J. Baster, 'The Origins of the British Exchange Banks in China', *E.H.*, 1917; Michael Greenberg, *British Trade and the Opening of China, 1800–1842*, Cambridge, 1951.

[3] Sir Compton Mackenzie, *Realms of Silver: One Hundred Years of Banking in the East*, London, 1954.

their concern was with the movement of goods; they left to the merchant bankers the issue of long-term bonds. To improve their facilities many colonial banks opened branches in London; by 1886 some twenty-eight had done so.[1]

At home the banking system, though showing a continuous evolution so far as the role of the central bank was concerned, changed radically in its other components. By the early thirties the country banks and the London banks were being threatened by newly formed joint-stock concerns, since 1826 freed from the general rule limiting them to six partners, and since 1833 permitted to invade London itself. Soon the London and Westminster Bank, the London Joint Stock Bank, the Union Bank, and others were flourishing in the City and leading an accelerating trend away from partnership banking to the joint-stock principle. For it was clear that concerns with numerous branches were now the best means of collecting savings and making loans. Moreover the joint-stock banks in the fifties began to offer interest on deposits at a rate kept close to that available on ordinary securities; so successful were they that by 1858 over £40 million had been attracted from the securities market and lodged on deposit.[2] By the mid-eighties the new kind of joint-stock bank, with many branches and a strong and accelerating tendency to amalgamation, dominated the scene and increased public misgiving.[3] Even bankers had their doubts. The President of the Institute of Bankers referred to the vast magnitude of the financial undertakings of today — 'like our huge iron-clads — unwieldy, uncertain and liable to unforeseen dangers'.[4] Yet there were real signs by the seventies that, at least with some of these thrusting new institutions, a code of behaviour was developing, though the criminal failure of the City of Glasgow in 1878 revived old fears.[5] By Sir Stafford Northcote's Act of 1879 'reserve liability' became available under which banks might increase their nominal issue of shares, with the shareholders liable, in the event of failure, to pay the difference between the amount paid up and the nominal amount of their subscription; all important banks rushed to take advantage of the measure.[6]

[1] Clapham, op. cit., vol. II, p. 349.
[2] 'Bank Deposits at Interest', *Econ.*, 10 April 1858, p. 391.
[3] Joseph Sykes, *The Amalgamation Movement in English Banking, 1825–1924*, London, 1926.
[4] Thomas Salt, 'Inaugural Address', *Journal Institute of Bankers*, 1892, p. 590.
[5] A. W. Kerr, *History of Banking in Scotland*, 4th ed., London, 1926, p. 25; William Purdy, *The City Life: its Trade and Finances*, London, 1876, p. 28.
[6] Clapham, op. cit., vol. II, p. 352. For some histories of leading banks see R. S. Sayers, *Lloyds Bank in the History of English Banking*, Oxford, 1957; W. F. Crick and J. E. Wadsworth, *A Hundred Years of Joint Stock Banking*, London, 1936; T. E. Gregory and Annette

Other new forms of banking made their appearance. The Savings Banks were founded to encourage thrift among the workers, though not to make loan facilities available for them, for workers' savings were largely advances to the government.[1] The pawnbroker had always loaned to the poor, but on a petty emergency basis; he flourished greatly in the poorer areas of the great cities. Door-to-door canvassers had long been on the scene pressing goods upon uneducated or weak-minded housewives; in the thirties it was the acceptance of German clocks that landed the poor in the small debt courts. By the sixties there were a few experiments in hire purchase by manufacturers and retailers such as the Singer Sewing Machine Company. The same decade also saw the beginning of the investment trust in Britain: companies that used their subscribed capital to acquire shares in other companies, based upon the idea that the individual small investor who would not wish to commit himself to the fate of a single company would be glad to take a share in one with a diversified range of investments.[2] With the great extension of insurance the investment portfolios of the companies became an increasingly important element in the money market, especially in the case of life insurance.[3] By 1860 they had become one of the greatest monied institutions of the day, with some 120 offices for the insurance of lives, and with some 1,500,000 people interested in their solvency, to whom were owed some £250,000,000.[4]

From the early nineteenth century the City was dominated by a financial aristocracy — the investment bankers or issuing houses, headed by Rothschild and Baring. It was from this small community that the Governor and Court of the Bank of England were chosen, for the view was held that they could properly preside over the financial mechanism, whereas the bankers whose main concern was with deposits and discounts (the clearing-house bankers) would fail to see the issues in their true light, and were therefore not eligible for membership of the Court.

These were the men whom Disraeli described as 'the mighty loan mongers, on whose fiat the fate of kings and empires sometimes

[1] H. Oliver Horne, *A History of Savings Banks*, Oxford, 1947; Henry W. Wolff, *People's Banks: a Record of Social and Economic Success*, rev. ed., London, 1896.

[2] T. J. Grayson, *Investment Trusts*, London, 1928, p. 2.

[3] See P. G. M. Dickson, op. cit., esp. last two chapters, also above, p. 105.

[4] 'Novelties in Life Insurance', *Econ.*, 22 December 1860, p. 1423.

Henderson, *Westminster Bank through a Century*, Oxford, 1936; H. Withers, *The National Provincial Bank, 1833–1933*, London, 1933; J. A. S. L. Leighton-Boyce, *Smiths the Bankers, 1658–1958*, London, 1958.

depend'. The politicians of England, unlike those of Germany and France, struggled throughout the nineteenth century to convince themselves that finance of this sort was a matter distinct and separate from the functions of government, and that its practitioners were men of business whose actions and decisions could be distinguished from those taken in the political field. The City men, for their part, expected the government to accept their guidance in monetary matters and not, as in 1854 and 1884, to seek to borrow at less than the rate thought proper by the City. On the other hand syndicates launching new bond issues knew how to encourage investors by sending up the price by their own purchases.[1]

The fear of a 'money power' goes far back before the nineteenth century; now it was to grow to new proportions. Wartime flotations by the government had produced loan mongers on a new scale. Dislike of these men often coincided with antisemitic feelings; Victoria reflected the attitude of many of her subjects in declining to ennoble Rothschild.[2] Moreover it was difficult to reconcile the cosmopolitanism of the money men with the demands of patriotism: during the Crimean War Britain honoured past financial obligations to the enemy, Russia.[3] The liberals, with their fear of monopoly in all its forms, were aware of its possibility in the supply of money and credit; Gladstone complained in 1854 that 'The government itself was not to be a substantive power in matters of finance, but was to leave the money power supreme and unquestioned'.[4] It was partly in order to provide the government with an independent source of borrowing that he established the Post Office Savings Bank in 1861.

As lending abroad extended there was much fear that the nation would become involved in the enforcement upon foreign creditors of the payment of their debts, a prospect abhorrent to those who condemned foreign adventure, especially when it was likely to take the form of using the tax-payers' money to enforce contracts made in the City. By the seventies the concession hunters in Egypt, Turkey, China, and elsewhere were certainly leading Britain into situations that might place the interests of the City above those of the nation. The promoters of new ventures were becoming increasingly plausible and adroit, with

[1] Jenks, op. cit., pp. 272, 276.
[2] Philip Guedalla, *The Queen and Mr. Gladstone*, London, 1933, vol. I, pp. 199, 207.
[3] See F. W. Fetter, 'The Russian Loan of 1855: A Postscript', *Econa.*, 1961; Olive Anderson, 'The Russian Loan of 1855: An Example of Economic Liberalism', *Econa.*, 1960, and 'A Comment', *Econa.*, 1961.
[4] John Morley, *The Life of William Ewart Gladstone*, London, 1903, vol. I, p. 651.

the Stock Exchange by the eighties producing a frightening array of scandals, upon which journalists learned to play. Throughout the century the idea was potent that the great City men always arranged their affairs in such a way that when crisis came, whatever might happen to those with fixed commitments, like industrialists and workers, the money men would weather the storm, and enforce such curtailments and liquidations upon the rest as were necessary.

In this there was much truth, though somewhat unfairly expressed, for it was indeed the function of the banking system and its ancillaries to enforce a purge when expansion had gone too far. Nor was public opinion altogether fair toward the City in other respects. The kind of finance that was provided for foreigners did a good deal to develop plantation societies, in South America and the West Indies. British bankers helped to make possible the extension of such societies, with immense inequalities of wealth, largely dependent upon Britain for commercial services, and for a market for their foodstuffs and raw materials. The City in this way contributed much to the cheapness of such goods, benefiting the British consumer. But the populations concerned were without the means of evolving toward stable self-government. The benefits of this relationship were complacently received in Britain; only in the eighties were its implications pointed out by the new generation of socialist critics.

In discharging their functions most of the men of the City were honest, but in a strict and limited way. The precise and full discharge of bargains was, in spite of outbursts of fraud from time to time, the practice of the City; indeed this was a *sine qua non* for success. There was also a strict code about rumour and the passing of information. But beyond these rather narrow rules, the City had no moral conscience. Each dealer had to rely upon his own commercial judgement; if it failed him he must fail.

Closely related to the absence of moral responsibility was the repudiation of any analytical mission.[1] The men of the City accepted no public responsibility for understanding and explaining the result of the totality of their actions; indeed many felt strongly that it was wrong to expect them to. Their job, and the key to business success, was to foresee and act upon the short-period course of events in particular markets. But in spite of this the City had, by the fifties, a general view on policy, believing in sound money and the automatic gold standard. This had its appeal for two reasons. It worked with reasonable

---

[1] S. G. Checkland, 'The Mind of the City, 1870–1914', *O.E.P.*, 1957.

effectiveness; even more important it kept the state and its minions out of the money market.

The efficiency of the City must be judged at several levels. Bad projects could certainly gain a footing, and good ones could be mishandled. There were many financial houses that collapsed; indeed every crisis brought its crop of failures, usually including some of the largest and apparently most substantial, some of which were known to be unsound long before they had exhausted all the expedients by means of which time could be bought and others more widely and deeply embroiled. Overend Gurney, the 'great house at the corner' had, long before its collapse in 1866, been known as an asylum for bad securities. But many houses had a long and successful continuous existence, building up strong reserves and wisely declining to push opportunities too far.

As to the policymaking of the City, the Bank of England did not always behave wisely, nor apply the lessons of past experience in reacting to the state of the economy. It made 'a long series of great errors' in the management of its reserves till after 1857.[1] After that date it showed very considerable skill in dealing with crisis when it came. On the other hand the approach to crisis was sometimes unwisely managed. Moreover, after 1866 the Bank largely withdrew as a competitor in the discount market. Its discount rate in normal times was 'high and dry' above the market rate, which meant that Bank Rate no longer served the City as a guide to interest rates except in crisis.[2]

When monetary debate was resumed in the seventies, with the price level in decline after a generation of upward movement, no clarification of principles came from the City. Yet there were plenty of new problems calling for solution: the difficulties in trade to the East, the debate over bi-metallism bogging down in confusion, the great increase in the supply of money held in London on a precarious short-term basis likely to be withdrawn when political or other rumour was afoot.[3] Perhaps it is unfair to expect the operators of the system, now long since committed to the automatic operation of the gold standard, and engrossed in short-term anticipations, to question its fundamental arrangements.

---

[1] Bagehot, op. cit., p. 220.
[2] *Bankers' Magazine*, 1890, p. 389.
[3] *Final Report, R.C. on Gold and Silver*, 1888, Historical Section.

PART THREE

The Elements of Society

THE initiatives of the men of invention, the men of business, the food producers, and the capital suppliers went far beyond their intentions. In involved interaction, they revolutionized the lives of the members of the labour force. They did no less for the initiators themselves, their families, and the orders of society in which authority lay.

CHAPTER SEVEN

The Lower Orders

I. MEN, WOMEN AND JOBS

THE first three censuses, 1801, 1811, 1821, sought to distinguish between the number of persons supported by the four great branches of human activity: agriculture, trade, manufactures, and the learned professions.[1] By 1831 it was felt that this primary distinction was not enough, and the employment categories were elaborated in that year, and again in 1841. But it is not until the census of 1851 that we are on ground that is at all safe, and which allows us to consider in comparative terms the many sectors of which the economy was then composed. Even then the pitfalls are still numerous: even thereafter there was a good deal of shifting between categories. The Registrars General and others interested in social measurement as a preliminary to social diagnosis had to fight hard and long in order to obtain the right to ask the questions that interested them; moreover they too, like the men of business, had to learn how to deal with their job.

The only matter upon which comment is possible before the mid-century is the proportion of the working population engaged in farming occupations. This fell continuously from about 35 to 16 per cent between 1801 and 1851. The absolute number engaged in agriculture only began to fall after 1851.[2] But in another primary sense the distribution of the population did not change greatly between 1851 and 1881. It fell into three almost equal parts. The principal earning group, men of over fifteen years, constituted slightly over 30 per cent. Adult women comprised almost exactly a third. The children under fifteen years were a little more than a third, rising from 35·4 in 1851 to 36·5 per cent in 1881.[3]

---

[1] See Clive Day, 'The Distribution of Industrial Occupations in England, 1841–61', *Transactions of the Connecticut Academy of Arts and Sciences*, New Haven, 1927.

[2] James Caird, 'On the Agricultural Statistics of the United Kingdom', *J.S.S.*, 1869.

[3] C. Booth, 'Occupations of the People of the U.K., 1801–1881', *J.S.S.*, 1886.

Of the adult women about one-fifth were in employment: the rest were dependants, if such a term can be properly applied to a group mainly composed of housewives. Those employed were divided between industry (mainly textiles), dressmaking and millinery, and domestic service. The number of textile factory girls, after a virtual doubling in the forties, hardly increased in the sixties and seventies. There was a considerable redistribution of women away from the factories to the houses as domestic service grew and as opportunities for women in textile employment declined relative to the female population. In the clothing trades there was a similar increase in the demand for female labour, exceeding a doubling in the forties, followed by a more modest rate of increase thereafter. Domestic service was the great alternative for these women also, for retail trade had hardly begun, even by the eighties, to demand their services, except perhaps in terms of small millinery and dressmaking establishments.

By 1881 more than one in seven of the general working population was in domestic service, totalling over 1·8 million persons, with women greatly predominating.[1] The great mass of women thus employed worked indoors, either as skivvies in small households under a female member of the bourgeoisie, or as servants in larger establishments under butlers and housekeepers. In this way a large proportion of the rising number of females could be taken into employment and, as extensions of the wealthier households, provide consumers' services. The number of male servants indoors diminished sharply in the fifties, and failed thereafter to keep pace as the number of indoor females mounted; but outdoors the number of men increased rapidly as, with the growth in the number and extent of the estates of persons with large incomes, the cult of hunting spread, and especially as the compulsion for more elaborate private carriages became stronger.

This very large category of people in service embraced wide differences in rewards and manners of life. Those who ruled backstairs, among whom males had a large place, were managers of large-scale establishments, separated from their employers by a social gap now almost inconceivable, which it was in their interests to widen. Butler and housekeeper, often man and wife, were able to develop for themselves a mixture of deference paid to employers and authority demanded from underlings. The mass of the females, on the other hand, was largely composed of younger women who though without prospect

[1] R. H. Hooker, 'On Forty Years Industrial Changes in England and Wales', *T.M.S.S.*, 1897–98.

of independence while in service, had a lively hope of marrying and leaving employment. Much good could come of a period in disciplined service, in the formation of habits of order and cleanliness. But there were also negative considerations. Illegitimate births were frequent among servant girls, and their experience in service might result in either excessive respect for their betters or in ill-considered revolt and revulsion from discipline.

The general distribution of men between jobs showed much more dramatic changes than in the case of women. In 1841 agriculture employed some 1·3 million people, overwhelmingly men; the peak of 1·7 was reached in 1851. Then came accelerating decline. By 1881 the absolute number had fallen to almost the same level as in 1841, and as a percentage of the labour force from nearly 21 to 11·5. In 1851 more than one employed worker in five had been in agriculture; by 1881 there was little better than one in ten. This change in proportion between producers and consumers of food was most remarkable.[1] The composition of farming labour too had changed. The least skilled element (some 0·9 million men in 1841), after increasing in the fifties, fell continuously to 0·8 million in 1881. The number of farmers and their relatives, after growing in the forties, also fell away. But nurserymen and gardeners, and breeders of horses and cattle, became more numerous.

The total of those employed in manufacturing seems to have followed a similar up and down course. In the forties, though the figures are subject to considerable qualification, it appears that not only was there a very big increase in manufacturing employment from 1·8 million to 2·7, but this more than kept pace with a very general simultaneous move from the category of dependants (especially female) to that of the employed. But in the sixties the manufacturing share of the workers began to decline, and from a peak of 32·7 per cent of the total labour force in 1851 fell away until in 1881 it was 30·7. As England further advanced as an industrial society, manufacturing began to lose ground as a proportion of the whole, though not to anything like the same degree as did agriculture.

Within manufacturing there were striking changes. Textiles and the making of clothing represented the major cases of failure to maintain the rate of growth of employment present in the economy as a whole.

[1] See John Saville, *Rural Depopulation in England and Wales, 1851–1951*, London, 1957, chapters 1 and 2; S. W. E. Vince, *The Rural Population of England and Wales, 1801–1951*, London Ph.D., thesis, I.T.A., no. 477.

This was due in no small measure to the economies of the machine as they were brought to bear upon the making of cloth and its assembly. From employing nearly two-thirds of all in industry in 1851, cloth and clothing requirements fell within thirty years to little more than half. Both, in spite of their relative decline, were increasingly becoming women's trades. Textiles and especially the clothing trades, became the chief source of auxiliary incomes to married households, just as domestic service sustained the unmarried females. In textiles, men had comprised a good deal more than half the workers in 1841; by 1881 they were hardly more than one-third. The clothing trades swung from a two-to-one preponderance of males, to the reverse situation. The new demand by women for clothes made by others accelerated as more women began to enjoy incomes that permitted them to follow fashion changes, however partially, but by this time the making of clothing was being mechanized with the advent of the sewing-machine.[1]

Within the manufacturing sector, in spite of its relative decline, there were industries that were gaining manpower. Their variety shows how the diversity of the economy was advancing. Men working in metals, including engineers and shipbuilding workers, showed the greatest increase, from 15 per cent of the manufacturing population in 1851 to some 23 per cent in 1881. These men in general had a higher level of skill and education than those in other industries. The shift from fibres to metals is further emphasized when mining is considered, for it too was on the increase.

With both agriculture and manufactures diminishing their net shares of the labour force, there must have been other elements that were gaining. These were to be found chiefly in merchanting and finance (the white collared clerks doubled in the sixties and again in the seventies, an astonishing rate of growth), in transport (a new race of railway workers had sprung into being), in domestic service, and in the professions.[2] In short, the elements in the economy providing services of one kind or another were expanding both absolutely and relatively. For as the volume of real goods increased and diversified, the number of those standing between immediate producer and final consumer grew. The number of workers in the building trades also advanced

---

[1] See above, Chapter 2, section 5, pp. 25, 26.

[2] P. W. Kingsford, *Railway Labour, 1830–70*, London, Ph.D. thesis, I.T.A., 1951–52, no. 888; G. C. Halverson, *The Development of Labour Relations in the British Railways since 1860*, London Ph.D. thesis, I.T.A., 1951–52, no. 887. David Lockwood, *The Blackcoated Worker*, London, 1958, chapter 1.

notably, for an increasing force of men was now required to provide buildings to house men, machines, and institutions.

But we must not imagine that these changes meant that the unskilled labourer was diminishing everywhere. Indeed, as the unskilled of the countryside moved townward the number of general labourers in the cities grew rapidly, especially in the sixties. The number of those whose skill was rudimentary, as in the docks, was also on the increase.

2. THE WORKER AS ENTREPRENEUR

Robert Applegarth, a leading trade union spokesman, remarked in the sixties that the task of the unions was to ameliorate the condition of the workers at that level at which they found themselves, for few had any real prospect of moving out of the category of hired labour.[1] The opposite opinion was expressed by radical employers, namely that there was a continuous ladder of opportunity of which all might avail themselves.

In some sectors, especially in the earlier decades of the century, the latter view had some reality. In the earliest stage of textile mechanization a weaver might well blossom as a factory owner. Though enterprise in the Newcastle coal area was in the hands of patricians, in south Staffordshire it was possible to rise from the ranks to ownership. In engineering and the metal trades, especially in the great period of emergence, down perhaps to the mid-century, a man of initiative and intelligence could often launch his own concern.[2] In the building trades it was certainly true that artisans could become masters; indeed it might be argued that ease of entry upon entrepreneurial status was excessive.[3]

After 1825, in textiles, the opportunities for the small man to found an enterprise were rapidly diminishing. One of the few ways in which a family in industrial Lancashire or Yorkshire could seek self-improvement outside textiles was through opening a small shop. Too often this ended in disaster. Yet there was a great number of these petty undertakings, a high proportion of their owners landing in the small debt courts. The small shop might also be the form in which former domestic servants sought independence, frequently with the same results.[4] In coal mining after the mid-century it was virtually impossible

[1] Asa Briggs, *Victorian People*, London, 1954, chapter 7.
[2] See above, Chapter 4, sections 4 and 6.
[3] See above, Chapter 4, section 9, pp. 165, 166.
[4] See H. G. Wells, *Experiment in Autobiography*, London, 1934, chapter 2.

for the collier by his exertions to rise to the head of an enterprise. Nor could he do much by way of shopkeeping. But there was a form of delegation of managerial function, the butty system, under which the man who carried authority with his fellows could organize groups of them and work on a sub-contract basis.[1] In iron founding too there were such sub-contractors. Typically these men worked only to spend, without thought of accumulating capital. There was much exploitation of worker by worker under the system, though the butties also knew how to put pressure on the masters. Sub-contracting was common also in ship construction and the building trades. In railway contracting gangers working under a leader were typical.[2] The spinning operative employed his 'piecer' at a wage one-half of his own. In most trades and manufactures, down to the forties and later, children were hired by the workmen, and were entirely under their control.[3] In agriculture, in the eastern counties especially, gangs of labourers were organized under leaders.[4]

The commercial clerk, if he was well connected, could hope to become a partner, but the host of ledger keepers and correspondence copiers, the sons of working-class families aspiring to respectability, had little chance to dazzle their employers. The servants in wealthier households certainly did not lack initiative of a kind, for they had securely established the tradition that they should be tipped for the slightest service.[5] A wide range of ingenuity appeared among the lowest levels of society. Mayhew described with clinical precision the host of men and women in London who were both entrepreneurs and evaders, doing all kinds of jobs, very often forms of scavenging and pest control: collectors of rags, bones, fats, and refuse of all kinds.[6] In many fields the operation of the Factory Acts, so necessary in order to improve conditions of work, had hastened the elimination of the small man in industry. In the socially useful sense the competitive market system was, by the fifties, drawing very little upon the initiative of the working class.

Nor did the system of education help the worker to exercise initiative. Even the enlightened employers who campaigned for technical

---

[1] A. J. Taylor, 'The Sub-contract System in the British Coal Industry', in L. S. Pressnell, ed., *Studies in the Industrial Revolution*, London, 1960.

[2] See R. A. Lewis, 'Edwin Chadwick and the Railway Labourers', *Econ. H.R.*, 1950.

[3] D. Schloss, *Methods of Industrial Remuneration*, London, 1892.

[4] *R.C. on Employment of Children in Agriculture*, 1868–69.

[5] Dorothy Wise, ed., *Diary of William Taylor, Footman, 1837*, London, 1962.

[6] Henry Mayhew, *London Labour and the London Poor*, London, 1851–62.

education did not usually see it as a means of embuing employees with the urge to participate in ownership and in business policy making. Workers were aware that their opportunities for higher earnings were tied to their physical vigour, and thus diminished from their early twenties, if not sooner. Indeed so heavy were the disabilities upon their personal initiatives that when the manual workers thought to challenge the system they turned naturally to state action.

A few of the middle class realized how confined the initiative of the worker had become in England. Especially conscious of this were those who had been to America where the release of the energies of humble men had been so great. The English operative could not embark his savings in business, but was obliged to be content with the miserable return of the Savings Bank. It was hardly surprising that his interest in the wealth-creating process should flag, and that well-meaning middle-class improvers, with their reading rooms, baths, wash-houses, temperance coffee-houses, and prizes for houses and gardens, should be disappointed in the reactions of the beneficiaries.

But there was a rapidly growing class of foremen and managers of one kind or another, often of quite humble origin, who rose to posts of great responsibility; not infrequently the whole enterprise for which they worked came to rest on their acumen. The men of inherited wealth and position, now derivable from industry as well as from agriculture, often developed other tastes and interests. Before the spread of the limited liability company controlled by directors, the managerial element within the firm could hardly aspire to become its masters. But as owner-managers yielded place to directors elected by shareholders, often recruited within the firm, the former servant of the business might well find himself in a new position of power, especially when a further generation had passed. In this way leadership from among the lower orders of society could, from the seventies onward, take a new form in place of the butty, the sub-contractor, and the gang leader. To the more militant workers these managers, with their defence of the interests of the concern, might well appear traitors to their class. Indeed they were likely to be assimilated to the lower middle class, and, from there, perhaps in a subsequent generation, to move further upward.[1] Perhaps it is fairer to regard them as the means whereby the free enterprise community was being obliged to draw more generally upon its reserves of ability.

Although the worker found it difficult to rise to responsible status in

---

[1] See below, Chapter 8, section 5, p. 303.

most of private industry, he was able to provide some institutions of his own, of which the cooperative movement and the Friendly Societies were the chief examples.[1] In shopkeeping workers were able to score a considerable success through the cooperative principle, aided by the corruption that afflicted so much of commercial retail trade, itself too often practised by smallish men who would not escape from their difficulties by legitimate means. On the side of production, success was more modest. In cotton spinning the 'Oldham Limiteds' made a promising beginning, but this was not sustained.[2] Though the workers, using committee organization and the cooperative principle, could occupy a few areas of production where the competitive system had broken down, they were wholly ineffectual at the growing points of the economy. The Friendly Societies, after difficult times in the fifties, had a membership of some four millions by the seventies, who benefited greatly in security from sickness and poverty from this form of self-help.[3]

3. BIRTH, MARRIAGE AND DEATH

The general trend of population growth though clear enough in out-line, was the result of many forces operating upon the birth, marriage, and death behaviour of these many groups of men and women, so diverse in the way in which they lived and sought the means to live.[4]

Fertility varied markedly between various occupational categories.[5] The highest birth rates appeared among the agricultural labourers, the miners, the less skilled building workers, the general labourers, and the submerged of the great cities, especially in the East End of London. With the exception of the last group, these were the sections of the community in which physical exertion was greatest and skill least. But other aspects of their lives varied widely, including the level and stability of their incomes.

The procreative power of the farm workers was strong, and was little affected either by ideas or techniques concerned with birth

[1] See below, Chapter 9, section 8, pp. 366, 367.

[2] See above, Chapter 4, section 3, p. 130.

[3] 'The Difficulties of Friendly Societies', *Econ.*, 19 November 1859, p. 1288; P. H. J. H. Gosden, *The Friendly Societies in England, 1815-1875*, Manchester, 1961; *Report of the R.C. on Friendly Societies*, 1871-74.

[4] See above, Chapter 2, sections 6 and 9. For much interesting material see William Farr, *Vital Statistics*, N. A. Humphreys, ed., London, 1885.

[5] T. H. C. Stevenson, 'Fertility of Various Social Class . . .', *J.R.S.S.*, 1920; J. W. Innes, *Class Fertility Trends in England and Wales, 1876-1934*, Princeton, 1938.

control: it was normal to marry and beget children. Moreover the countryside was continuously disembarrassed of its excess numbers by the movement to the towns and abroad. Housing shortage seemed to have little effect upon the birth rate in the countryside, even in the case of the close parish where control lay with one or two landowners who could strictly regulate the supply of cottages, a device operated down to the Union Chargeability Act of 1865. This high birth rate was reinforced by a lower than average child mortality rate. But though the farm worker might continue to be prolific, his loss of relative importance in the population as a whole was bound to reduce the nation's rate of increase.

The miners, like the farm workers, reached their peak earnings early, with the result that the age of marriage was low and the period of child-bearing was long. Mining, however, was not the kind of occupation in which the worker enjoyed alternatives. Miners were by the nature of their job ill suited for other employment; their children too were bred to their ways. But in contrast to agriculture, mining was an expanding industry. It seized eagerly upon the sons of miners over successive generations, making it possible for them to marry young and thus aid the process of expansion.

In building many of the rapidly growing labour force were of a low order of skill. But their earnings were often protected by effective combination, both among themselves, and among their masters who could thus keep the rewards of the industry higher than might have been the case if unregulated competition had prevailed. This could often occur in spite of the great number of small enterprises in building.

The general labourer represented that range of society that had no effective means of combining for the protection of earnings. Some could do well, where tempting piece-rates were offered to accelerate the tasks to which machinery could not be applied (as with railway navvying), but on the whole the unskilled worker was on a casual, ill-paid basis. Yet his numbers increased at a startling rate. At the bottom end of the skill-less scale came the slum-dweller. For him all elements of life were adverse: poor housing, low incomes, and unemployment destroyed the potential for improvement in successive generations. Yet the result was not to reduce the birth rate; on the contrary. In terms of the death rate, however, these factors were effective, carrying off slum children in great numbers, especially the very young.

Just as textiles lost ground in terms of the growth of employment, so textile workers failed to maintain their fertility. In the first decade of the century they had helped to swell the expansion; by the mid-century they were beginning to lose their power to propagate; in the third quarter of the century its decline was marked. It was in the textile areas of Lancashire and Yorkshire that female employment was highest. The effect of this upon child-bearing was much stressed by observers, for the physical problems of raising a family were made much more difficult. Increasing money earnings, especially in a house with a dual income, might further reduce the disposition to have children, especially as the earning capacity of the latter was reduced by the Factory Acts.

The skilled artisans also showed signs of producing smaller families, though for different reasons. With such men the timing of earnings could be important, for virtually alone among the workers they did not attain maximum rewards until after a considerable apprenticeship and educational delay, thus causing postponement of marriage and the adoption of control measures within it. There is evidence of the spread of birth control in the towns of the north from 1870 onward. The new education and the increased aggregation in towns made contraceptive information and sale easier to arrange. The inhibitions involved in the older view of family life were weakening, especially as the church lost ground. It was the skilled operatives of the wool towns, like Halifax and Huddersfield, and the shoe workers of Northampton, who received the particular attention of neo-Malthusian propagandists like Annie Besant and Charles Bradlaugh. This factor became more powerful as the need for skill and technical education was increasingly felt.

The death rate was constant for thirty years down to 1870. The first reduction in the early part of the century had been achieved in relatively easy conditions. But the whole pace and scale of change from the forties onward was greater. Even to keep the rate constant required that amelioration should go forward. Urban death rates were well above the national rate of 22·4 in the early forties: at least a dozen of the largest towns had rates of 26 or more.

The new challenge was not the traditional one of food supply, but pestilence. Formerly the two had been closely related in that food shortage both reduced resistance and made people indiscriminate about the condition of the things they ate. But now pestilence had taken on an independent existence. Great numbers of people in close proximity with no official intervention to serve their common needs meant trouble, even though real wages might be rising. A full private purse was no insurance against impotent public bodies.

In the late thirties, for 100 persons per thousand who died in the home counties, 164 died in London. The great problems were to cleanse and purge: to do this involved a clean and copious water supply, and efficient drainage.[1] At last, in the seventies, equipoise between deterioration and improvement was effectively disturbed and the death rate, stationary for thirty years, resumed its downward trend. Thus, coincident with the redundancy of labour power appeared the new capacity to preserve it. Yet infectious diseases were still responsible for one-third of the total deaths at all ages, with respiratory, nervous, and digestive diseases a formidable second.[2]

The great reduction in urban death rates, especially among infants, beginning in the seventies, was apparent particularly among those who had suffered most from the earlier deterioration of urban conditions. But it was among these receivers of low incomes and occupiers of inferior houses that there was least response in terms of diminishing births. The effect was to induce rapid increase among families with the poorest prospects. For those of the middle classes whose philosophy was liberal this posed a most serious problem.[3]

4. THE WORKERS' REWARDS

One of the most obvious but difficult questions posed by the nineteenth-century expansion of the British economy is that of the share enjoyed by the workers. There are two problems; that of real wages and that of the distribution of income between classes. If we compare the real wages of the worker with his own past position, we avoid ethical questions of fairness but we do not dispose of them; we simply ask: were the workers and their families becoming better or worse off in material terms compared with their own past? Or more precisely, did they receive more or less in terms of real goods per unit of effort, taking into account the availability of employment? This of course leaves quite open the more profound questions were they more or less happy, though it contributes to an answer.

The distribution problem, on the other hand, involves first estimating what was available for distribution, then putting the workers' share alongside that of the capitalists and landowners. Were wages or

---

[1] See below, Chapter 7, section 8, pp. 255, 256; *Report of R.C. on Water for the Metropolis and other Large Towns*, 1868–69.

[2] W. P. D. Logan, 'Mortality in England and Wales from 1848–1947', *P.S.*, 1950.

[3] See below, Chapter 10, section 2, p. 398.

profits and rents rising (or falling) the more rapidly? Here we must consider not merely the distributive shares, but the uses to which the profit and rent share was put. If the latter was wholly ploughed back into the growth of the economy the concept of 'exploitation' is much affected thereby; if it was wastefully spent by the capitalist and landowner upon their own pleasures, the argument that the worker was being damaged is stronger, though even here it is not conclusive.

The manner in which the national income was divided between workers and owners is very difficult to determine. There are, however, a few pointers. In 1860 in agriculture it seems probable that landowners received rent payments in excess of the total wages bill, with an additional share for farmers' profits, so that the three were in the ratio of rent 2 : labourers' wages 2 : profits 1.[1] For the economy as a whole between 1860 and 1880 Bowley's calculations suggested that the workers' share of the national income fell from 47 to 42 per cent.[2] In the seventies *The Economist* was prepared not only to demonstrate that the wages share had not increased in proportion to that of interest, rent, and profits, but also to justify its failure to do so.[3] The worker, it was said, had no natural right to any particular aggregate share, much less to one that progressively increased. But from the eighties at least it would seem that the depression caused some redistribution in favour of the wage earner.[4] The reduced return on capital due to its accumulation, and diminished rents due to a reduction of cereal agriculture, helped to bring this about. But even with this general redistribution there was increasing concern in the eighties about the principles that underlay the distribution of the national product.[5]

Much of the great gains accruing to capital found its way into new productive investment at home or abroad, and so could contribute to the workers' well-being. But a good deal was wasted, especially abroad, and a good deal went in supporting the rising standards of the owning class. Those who protested against the smallness of the wage share, the waste and the ostentation, argued for a diminution in the income of capital owners; those who stressed the need to keep the

[1] Frederic Purdy, 'On the Earnings of Agricultural Labourers in England and Wales, 1860', *J.S.S.*, 1861, p. 355.

[2] A. L. Bowley, 'Changes in Average Wages (Nominal and Real) in the United Kingdom between 1860 and 1891', *J.R.S.S.*, 1895.

[3] 'How Far Have our Working Classes Benefited by the Increase of Our Wealth?' *Econ.*, 24 January 1874, p. 95.

[4] See above, Chapter 2, section 8, p. 56.

[5] See *Report of the Industrial Remuneration Conference*, London, 1885.

consumption of the general population well within productive capacity in order to provide for new investment, were opposed to any intervention in the distributive process, arguing that only large incomes based on profits and rent could make possible the formation of new capital.[1] It is fair to say of many liberal apologists that, market forces having produced a redistribution in favour of the workers, they accepted and even applauded it. But the arguments seldom really met, for the real facts of the case were not available, nor was there an adequate theoretical framework in terms of which to understand them.

No less intractable than the distributive problem was that of real wages. Three lines of attack are available. Strict quantifaction would be the ideal: an authentic money wage index, and an authentic price index, the latter divided by the former to give real wages. This kind of calculation is necessary both for the main groups of workers, and in a unified form for workers taken as a whole. But the difficulties in the way of such measurements are very great, and perhaps insoluble.

As an alternative, an attempt can be made to avoid the index number problem by taking the physical quantities of certain commodities consumed – tea, sugar, coffee, meat, each expressed per head of the population.[2] These results can be compared with a third approach: the comments and estimates of contemporaries. For though these are usually highly subjective, taken together they are often illuminating. Finally, there is the corpus of general economic theory which consists in working out the logical implications of given assumptions – this can be used sometimes to correct the picture by demonstrating which estimates are mutually consistent and which are not, and sometimes to fill the gaps in data provisionally by defining a residual. The essence of the historian's task is to attempt a general synthesis of all these types of measurement. He cannot escape the riddles of quantification, nor ignore the reactions of those who lived through the period, nor wholly reject the abstract relationships of theory. But he must not be ruled by any of the three, for none rests upon an adequate basis.

Immediately after the end of the war, with falling prices, it is likely that those workers who remained in employment made some gains. From the twenties to the early thirties real wages in urban occupations

[1] See below, Chapter 10, section 7.
[2] George Henry Wood, 'The Investigation of Retail Prices', *J.R.S.S.*, 1902; 'Some Statistics relating to Working Class Progress since 1860', *J.R.S.S.*, 1899, p. 648; A. Sauerbeck, 'On Prices of Commodities and the Precious Metals', *J.S.S.*, 1886.

in the United Kingdom increased, though by no means dramatically, with big differences between towns.[1] The effect upon the national situation was dampened by the lesser gains of rural workers because their advances were both less and pertained to the larger element in the population. Even urban wages after a brief period of improvement, 1833–35, were not sustained in the later thirties.[2] Over all, if 1840 is taken as the base for urban real wages the trend would seem to have been an index of 94 in 1816, 108 in 1831, and 100 in 1840.[3] This result is obtained using a retail price index; a wholesale one gives almost no change. For the countryman calculation is almost impossible, with very great regional differences.

The forties are just as intransigent. Wood's urban index using retail prices rises from 100 in 1840 to 110 in 1850; a wholesale price index gives a real wages index of 137. The railway boom culminating in 1845 sent wages up, even for unskilled men from agriculture, but with the end of the boom, accompanied by potato famine and high food prices, there was serious general distress. Tucker, describing the London artisan, regarded 1847 as 'the last example of a real famine or an approach to one in time of peace'.[4] There seems little doubt, however, that wages were recovering by the end of the forties and early fifties. But it would seem that such gains as the workers had been able to make over the decade were not very striking. Moreover, such gains in formal real wages as did occur came in a context of deteriorating town life, bringing social losses not embodied in any index.

Little advance appeared in the fifties. It is not, in fact, until the middle or later sixties that any really convincing aggregative evidence of rising real wages appears. Even then the matter was in some doubt. Fawcett, one of the leading economists of the seventies, argued that, notwithstanding the immense increase in the national wealth, many of the labouring classes were little, if at all, better off than they had been

---

[1] See above, Chapter 2, section 3, p. 11, for references; also Gayer *et al.*, op. cit., vol. I, pp. 167, 208, 238.

[2] Gayer, ibid., p. 273.

[3] George Henry Wood, 'The Course of Average Wages between 1790 and 1860', *R.B.A.*, Section F, 1899, p. 829; 'The Course of Average Wages between 1790 and 1860', *E.J.*, 1899, p. 588. Arthur L. Bowley, *Wages in the United Kingdom in the Nineteenth Century. Notes for the use of Students*, Cambridge, 1900. In Deane and Cole, op. cit., pp. 25–27, it is suggested that real wages in the U.K. underwent 'an unprecedentedly rapid improvement in the second quarter' [of the century]. This improvement seems overstated. For a vigorous exchange see E. J. Hobsbawm and R. M. Hartwell, 'The Standard of Living during the Industrial Revolution: A Discussion', *Econ. H.R.*, 1963.

[4] Rufus J. Tucker, 'Real Wages of Artisans in London, 1729–1935', *Journal of the American Statistical Association*, 1936.

twenty years before.[1] Cairnes could find no convincing evidence of advance.[2] Taking the whole period from the forties to the later sixties it seems likely that there was a continuous trend in betterment, but so slight, and so often reversed in bad times, that the statisticians found real demonstration very difficult. From the sixties the trend was markedly upward to 1900, though 1875–79 appears as a minor setback.[3] (The seventies were a turning point in the relation between money wages and prices; down to the seventies the former rose faster than the latter, thereafter they fell more slowly.[4]) Real wages in the United Kingdom rose some 20 per cent between 1860 and the early seventies, thereafter the rise continued with little interruption, so that taking the whole period 1860–91 the increase was of the order of 60 per cent.[5]

All this is not to deny that great gains were being made much earlier by some groups of workers. But until well into the sixties it would seem that the number of those whose condition had improved, multiplied by the degree of their improvement, did not greatly exceed the number whose condition had deteriorated, multiplied by the degree of decline.

Thereafter, gains clearly predominated over losses. But it was still possible, and indeed likely, that the absolute number of those whose lot deteriorated was equal to or greater than it had been in the early decades of the century, for the overall growth in numbers permitted this sector of the population to diminish relatively, yet to increase absolutely. Moreover with the growth in the absolute size of the areas of dilapidation and dereliction, in the East End of London and in the provincial cities, the degree of degeneration possible within them might well have increased.[6]

The trend of wages as a whole depended upon two factors: the rise in productivity within each of the multiplicity of lines of output of

---

[1] Henry Fawcett, 'The Effect of an Increased Production of Wealth on Wages', *F.R.*, 1874.

[2] 'How Far Have our Working Classes Benefited by the Increase of our Wealth?' *Econ.*, 24 January 1874, p. 93.

[3] A. L. Bowley, op. cit., *J.R.S.S.*, 1895, p. 248.

[4] Rostow, *British Economy*, p. 95; also Robert Giffen, 'The Progress of the Working Classes in the Last Half of the Century', reprinted in *Essays in Finance*, 2nd ed. 2nd Series, London, 1886.

[5] A. L. Bowley, 'Comparison of the Rates of Increase in Wages in the United States and in Great Britain, 1860–1891', *E.J.*, 1895, p. 382; also George H. Wood, 'Real Wages and the Standard of Comfort since 1850', *J.R.S.S.*, 1909, p. 99.

[6] R. Giffen, *The Growth of Capital*, London, 1889, p. 113.

which the economy was composed, and the changes going on in the distribution of the labour force between them. As to the latter it is clear that there was, in general, a substantial shift of workers into jobs demanding greater skill.[1] This meant that wages of all workers taken as a whole could increase even if the wages of each category of workers fell somewhat, for the proportion of those in the better paid jobs was rising. The growth of engineering, shipbuilding, the metal trades, coal mining, building, and of the more specialized branches of agriculture, all contributed to the shift to higher rates. The rise of the service industries provided a further area in which many workers could improve themselves.

Increasing productivity per man was present to greater or lesser degree in almost all lines of output. But the extent to which this benefited the worker depended upon his bargaining power.[2] Where trade unionism was weak his position turned upon the question of the supply of labour. There were important lines in which the worker was relatively redundant — especially in agriculture and in the labouring elements attached to many trades. In spite of the enormous migration from the farms to industrial areas the large size of rural families replaced the losses for a long time; moreover the countryman often preferred to accept the conditions of his life, which though imposing a pretty severe limit in general, none the less left scope for ingenuity and even for luck. His income consisted of so many elements, and often included a basic minimum even in the worst of times in the form of the chance to maintain himself and family from a small holding, that he might well prefer to stay where he was. All those receipts that baffle the wages statistician comprised an income which, though it did not rise much with the growth of the industrial sector, had attractions that are hidden in any attempt at monetary aggregation: weekly money wages and payments for piece-work, both for himself and his family, harvest wages, often at twice normal rates, payments in kind of food, cider and other things, gleanings made by the wife and children, pigs, poultry, common rights, fuel, a cottage with a piece of ground, perhaps some domestic industry like lace or glove making. There were also important regional differences: in the sixties, so far as calculation is possible, farm workers in Dorset and Devon received barely one-half of the

---

[1] See above, Chapter 7, section 1, p. 218.

[2] For studies of wages in various industries see A. L. Bowley and G. H. Wood, 'The Statistics of Wages in the United Kingdom during the Nineteenth Century', *J.R.S.S.*, 1899 to 1906 inc.; for building, E. H. Phelps-Brown and Sheila V. Hopkins, 'Seven Centuries of Consumables, compared with Builders' Wage-rates', *Econa.*, 1956.

income of a northern worker, perhaps seven shillings against fourteen shillings per week.[1] In general, the latter pulled steadily ahead. The farm worker and the general labourer shared two disabilities: fecundity, and the competition for jobs by the flood of Irish immigrants. But whereas the farm labourer kept his own rewards down partly because of a non-commercial preference for his traditional way of life, the general labourer received low rewards because he had no alternative. He was the man who depended solely upon his money wage, and this was wholly determined by market forces, for not until the eighties could he make an effective beginning in trade-union organization.[2] In boom times the unskilled worker might well make gains, but these would be more modest and later in coming than in the case of a man with a specialized job.

Even the latter did not enjoy a steady increase; on the contrary it was in boom times that wages rose significantly, doing so first, fastest, and farthest in those trades that constituted the bottlenecks at the time. The colliers were a highly paid group, but they suffered the widest vacillations in wages between good and bad times. The iron puddlers were in a similar position.

The mechanics of Birmingham, the artisans of Sheffield, the engineers of Manchester and elsewhere, the craftsmen engaged in shipbuilding, the potters of Staffordshire, the skilled members of the building trades, the printers, and makers of printing machinery, were the urban industrial élite during most of the nineteenth century.[3] Their wages were both high and relatively stable, especially in the downswing. Their relations with their employers, as exemplified in Birmingham, were described by Cobden in 1857 as far more 'healthy and natural in a moral and political sense' than where a capitalist oligarchy was present, as in Manchester, sharply distinguished from the workers. It should also be remembered that the length of the working day was falling, a gain not reflected in wage statistics; hours worked were probably reduced by a fifth between 1836 and 1886 in textiles, engineering and house-building.[4]

[1] Purdy, op. cit., p. 344. For a summary of agricultural workers' money wages see A. L. Bowley, 'The Statistics of Wages in the United Kingdom during the last Hundred Years (part IV). Agricultural Wages', *J.R.S.S.*, 1899, p. 555; Arthur Wilson Fox, 'Agricultural Wages in England and Wales during the last Fifty Years', *J.R.S.S.*, 1903.
[2] See below, Chapter 9, section 8, p. 371.
[3] Sidney Pollard, 'Wages and Earnings in the Sheffield Trades, 1851–1914', *Y.B.*, 1954; 'Real Earnings in Sheffield, 1851–1914', ibid., 1957; M. and J. B. Jefferys, 'The Wages, Hours and Trade Customs of the Skilled Engineer in 1861', *Econ. H.R.*, 1947.
[4] A. L. Bowley, *Wages and Income in the United Kingdom since 1860*, Cambridge, 1937, p. 25.

The advantage enjoyed by the skilled workers because of their market scarcity was reinforced by their ability to organize effectively. It is likely that down to the mid-century they failed to extract all they might have done from their employers, and so failed to extend the gap between themselves and the unskilled.[1] The skilled may well have thought mainly in terms of the conventional differential that was proper between them and the unskilled, and, satisfied if this was maintained, failed to press for even higher payments. But this attitude was changing; the skilled worker, free from this inhibiting convention, and much more tightly organized in his union, learned to press for what he could get. From the fifties to the eighties wage differentials between skilled and unskilled grew steadily. It seems probable that in the sixties the ratio of average wages between skilled, semi-skilled and unskilled was of the order of $5 : 3 \cdot 3 : 2 \cdot 4$, or, in absolute terms, the skilled man earned between £60 and £73 per annum, the semi-skilled between £46 and £52, and the agricultural and unskilled labourer between £20 and £41.[2]

But there were signs that new factors were in operation, likely to reverse the trend of separation. The Education Acts of 1870 and 1876 (the latter making elementary education compulsory) were great strides toward universal literacy, and removed some of the difference in educational opportunity.[3] The distinction between skilled and unskilled was being obscured with the rise of the semi-skilled machine operator. As early as 1867 there were, of the 'higher skilled' some 1·3 million workers, of the 'lower skilled' some 5 million, and of the unskilled and agricultural some 4·5 million.

5. THE WORKER AND HIS WIFE AS SPENDERS

There was some justification for those employers in the early nineteenth century who feared the effects of higher wages. They were dealing with men and women whose consumption patterns had always been both narrow and conventionalized; too often an increase in wages served nothing but demoralization. It could be applied, in part, to an extended consumption of customary articles, but thereafter it tended to run away in gin and dissipation, or to cause the incentive to work to

[1] E. J. Hobsbawm, in Asa Briggs and John Saville, eds., *Essays in Labour History*, London, 1960, chapter 3.
[2] R. Dudley Baxter, *The National Income of the United Kingdom*, London, 1868; 'The National Income of the United Kingdom — Wages, Profits, Rents', *C.R.*, 1868, p. 93.
[3] See below, Chapter 7, section 8, pp. 257, 258.

fail. A sanitary reformer in a fit of discouragement might well remark that the drainage of the kingdom might be dealt with in ten years with ten millions of money but a generation would be needed to alter living habits.[1] The boom of the seventies produced a frightening burst of debauchery even among classes that had enjoyed better incomes for a generation. As to the very poor, many observers agreed with Giffen that 'to give them more to spend would be a curse'.[2] A profound sociological problem is lost sight of if our minds run simply in terms of varying injections of money into the family purse, and varying movements in the prices of consumables.

The matrix from which the labourer came was the traditional agricultural one. This had given him a framework of more or less settled desiderata within which to make his petty decisions. In the new towns and manufacturing districts he and his wife were called upon to deal with the problem of spending in an unprecedented form; they were challenged to find the skill and discipline to discount present benefits for the sake of the future. If they failed in this, whatever might be the relationship between prices and money wages, little real gain was possible.

The difficulties were especially great where people gathered in groups within which there was little real leadership. In the mining areas, in many mill towns, in the slums congested by the unskilled, from all of which the middle class largely held themselves aloof, mutual imitation had often given rise to a fixed frame of convention to which all tended to conform, and which was hostile to improvement. Drink often became a symbol of virility. Chadwick was outraged by the sums spent upon funerals by all classes of society, but especially by those least able to afford such extravagance.[3] A typical adult labourer's funeral in the London of the forties cost from four to five pounds, at least the equivalent of as many weeks' wages even before the costs of the festivities or of the mourning were considered. This, among people with no savings, resulted in crushing burdens on widowhood and the congestion of the small debts courts by prosecutions laid by the undertakers.

Gains could probably be made most effectively in the direction of diet. For food is the most immediate of needs; it can also be bought

[1] W. L. Sargent, 'Certain Results and Defects of the Reports of the Registrar-General', *J.S.S.*, 1864, p. 202.

[2] Giffen, *Essays*, vol. II, p. 94.

[3] *Report on the Sanitary Condition of the Labouring Population of Great Britain. Supplementary Report . . . on Interment in Towns*, 1843, p. 78.

in small quantities, with increases in consumption moving in step with higher wages. This meant that the tendency was strong to take improvements in real wages out in food rather than in shelter. Observers, having heard of the deplorable condition of the workers, were often puzzled at the splendid physique of iron puddlers and coal miners: the explanation was of course that these were a selection of men who ate well, and who could thus thrive physically in the most degrading conditions. But not all workers enjoyed such earnings: malnutrition was so distinct a feature of the textile towns that rickets were known on the Continent as the 'English disease'. For these reasons caution is necessary in assessing the meaning of figures constructed to show changes in the consumption of meat, tea, tobacco, cocoa, sugar, and other food items, for even where the trend with these is upward, their meaning is very limited. As the point of satiety in food was approached, there followed a great deal of waste through diminishing efficiency in food purchasing.[1] But there were changes over time. On the whole, beer and cheese gave way to tea, cocoa, sugar, and fruits as real incomes rose, with food diminishing as a proportion of total expenditure, and clothing, fuel and light, and sundries gaining.

Failure to use rising wages to improve the standard of social life took its most dramatic form in drunkenness, which meant, of course, vulnerability to the tommy shop, the money lender, and the landlord. Many employers of miners, dock labourers, builders, and bricklayers paid wages in beer.[2] Excessive drink meant also that the wife found it impossible to defend her petty savings, her only security, against her husband.[3] Even more important the poor woman had no effective defence against her husband's physical demands and so was obliged to deal with unwanted pregnancies. In 1830 the government acted so as to increase drunkenness greatly: the tax was taken off beer, and under the Beerhouse Act any householder assessed to the poor rate might open his house as a beershop, free of licence or control, on payment of two guineas. The consequences were appalling. No less than 31,000 new beer-sellers came into existence; the trade remained outside the jurisdiction of the magistrates down to 1869.[4] Like the truck system, adulteration of food served to diminish spending power and to damage

[1] For some discussion of changes in family budgets see Tucker, op. cit., p. 74.

[2] George W. Hilton, 'The British Truck System in the Nineteenth Century', *J.P.E.*, 1957.

[3] 'Poor Men's Wives', *J.S.S.*, 1869, p. 115.

[4] George B. Wilson, *Alcohol and the Nation*, London, 1940, pp. 99, 100.

health.[1] Cooperative shopkeeping was, in part, an attempt to avoid toxic Prussian blue in tea, plaster of Paris in bread, red lead in pepper and mahogany sawdust in coffee. Real success in spending depended largely upon the element of housing. For a real rise in consumption standards to occur, it was necessary that the spending pattern of the family be raised by a sort of dead lift, by taking the home out of the context of overcrowding and cumulative degradation, and placing it in a new setting wherein money might be spent on an entirely new range of amenities, through better sanitation, greater comfort, personal privacy, and the acquisition of books and other possessions. But only the family itself could provide the means. This meant allocating the budget toward a high rent payment, or saving over time in order to make a house purchase. Those of the middle class who placed so much emphasis upon the need for working-class thrift have sometimes been accused of urging on the workers a habit that was almost an end in itself. But the real case was that saving should make possible this essential change of scene so that all subsequent spending could be made more effective in the promotion of a better life, or, negatively, that death benefits should be available so that a dying bread-winner might not drag his family into debt. So too with the temperance enthusiasts, for though they sometimes showed signs of monomania, there was some truth in their claim that alcoholism prolonged business depression because it made impossible an effective demand for the comforts of life and was destructive of all trades but one.

But many other circumstances militated against good budgeting. Divided jurisdiction might have serious results. Often the head of the family kept what he thought appropriate for himself, leaving only the residue for his wife to manage. The middle-class employer kept his property out of female purview. This had some sort of rationale among the middle-class males, for they were practised spenders in their daily business, and indeed highly developed discounters of future returns. But the labourer had no such experience; indeed the main spending experience of his family lay with the females. Early marriages further impaired spending judgement. Where the combined ages of the parties was often no more than thirty-four years ('boom' marriages) and both sprang from homes where example was not strong, there

---

[1] See A. H. Hassal, *Adulterations Detected*, London, 1861, for descriptions of adulterations and methods of discovery; J. Burnett, *The History of Food Adulteration in Great Britain in the Nineteenth Century, with Special Reference to Bread, Tea and Beer*, London Ph.D. thesis, I.T.A., 1957–58, no. 661.

could hardly be a high standard. Even more disastrous was the employment of wives in mines and factories. Where the wages of the wife are counted in with that of the family in arriving at real wages, the picture is palpably false. The cost of this double labour was very high, for it meant that, with no true housewife, the efficiency of the family as spenders was greatly reduced.[1] The irregularity of earnings militated drastically against the building up of a long-term view of the spending problem; too often the worker tended to occupy a room the rent of which he could safely meet in bad times. The result was a housing provision related to his lowest rather than his average earnings, with the rest largely wasted. Pawn shops were plentiful — visible sign of the vulnerability of the worker to instability of employment.

It is difficult to assign the trend with respect to the efficiency of spending. It would seem likely that this, like so many aspects of the national life, was at its worst in the thirties and early forties. The state intervened with the Truck Act of 1831 to stop the intentional exploitation of the workers through their purchases, but no means of enforcement were provided. Though there was some amelioration in some trades, truck continued to be a serious problem, as was revealed by the inquiry of 1871, and was by no means dead in the eighties, in spite of a Truck Act of 1887, with the initiative and expense of prosecution still lying with the workmen.[2]

Yet in spite of inefficiency and low standards, new houses were built, and especially among the more skilled artisans the level of comfort could improve considerably. The old manual knack demanded by textile machinery had been quite consistent with drunkenness and debauchery; the new tasks demanded by the Stephensons, Fairbairn, Nasmyth, and the rest of the great engineers called for a clear mind, precision, and initiative.[3] Many young wives had learned something of how to plan their homes during their period of domestic service. Female employment declined so that there were fewer homes in which there was no real budgeter. The ending of underground working for women and children in 1842 made possible the improvement of miners' homes. In the Glamorganshire coalfield many wives, raised on the hill farms, brought to their marriages high standards and cleanliness with which they successfully withstood the trend toward debasement to

---

[1] Margaret Hewitt, *Wives and Mothers in Victorian Industry*, London, 1958.

[2] *Report of the R.C. on Truck*, 1871; G. W. Hilton, *The Truck System Including a History of the British Truck Acts, 1465–1960*, Cambridge, 1960.

[3] 'Mr Nasmyth on the Intellectual Influence of Machinery', *Econ.*, 7 November 1868, p. 1272.

which husbands and fathers were prone. The teetotalling Calvinists were there and elsewhere fighting for betterment. There were also the Sunday schools and the evening schools where self-knowledge might dawn.

An important element in better budgeting was the urge for respectability, increasingly manifest from the forties onwards. The working family engaged in the struggle with a no-margin budget and yet anxious to improve its condition, was often subject to painful stresses. The best safeguard against slipping into lower standards was to choose an objective; for this purpose the aim was respectability. This concept was not merely dependent upon imitation of the middle class, but embraced also much of the old self respect and independence of the pre-industrial craftsman. Fundamentally there was the feeling, noted by observers of the working class, of dignity and confidence among these men, some in very humble situations, because they were possessed of a skill which was in itself a source of pride, and for which there was a strong social demand. For all the rebuffs and rejections suffered by the less fortunate, at least this element among the workers could not only maintain the self-esteem enjoyed in pre-industrial society, but might even increase it, provided the skill in question was still called for.

Negatively, respectability meant establishing a gap between the aspiring family and the great mass who were incapable of any real effort at betterment. In this sense the working man and wife who wished for a better life for their children were engaged in a continuous struggle to draw away from other workers. There was an awful chasm lying between those who could maintain a stable home, spend their money according to some reasonable plan, avoid drunkenness, and plan the children's future, at least to some degree, and those who could do none of these things, but relapsed into various degrees of fecklessness and fatalism. Like the middle-class entrepreneurs who employed them, each such family had to constitute itself a single and separate entity consciously seeking to distinguish itself from the mass. To stay on the right side of the respectability gap they needed to reinforce their determination by all the means available. The outlook thus formed persisted when they were at last successful in moving to an artisan suburb.

As had always been the case with humble men, skill and craftsmanship were the basis upon which personal and family self-respect rested. But many attitudes concerned with dress and the furnishing of the home were derived from the middle class.[1] Such ideas as referred to the

[1] See below, Chapter 8, section 8.

working of society as a whole often came from middle-class intellectuals, especially John Stuart Mill, Carlyle, and Ruskin. Only such men seemed able to provide a systematic view of the working of the society and the economy; it is not surprising that as the respectable workers gained in political consciousness after the fifties they turned to the Liberal Party. But with the masses to whom high-minded radical individualism had less appeal, it might well be different. Disraeli believed that the lower the franchise went in the social scale, the better for the Tory Party, for thus it could minimize the influence of the respectable among the workers who sought social improvement.[1]

Perhaps down to the seventies or eighties, the artisans approximated to the activist element in the middle class, distinguishing themselves from the worker who took no thought of his position. But with the passage of time this was changing, at least so far as ideas and politics were concerned, for the liberal alliance, with its emphasis upon individuality rather than solidarity, became less acceptable to the intelligent skilled men, as class differences became more obvious, and to the trade union movement as, from the eighties onward, industrial unions for the unskilled began to rival the craft organizations.

The distinction between those who owned sufficient property to be called capitalists and those who possessed little more than their labour certainly did not exhaust the differences within society. For a man's responsiveness to pleas for altering society by political action might well be more fundamentally related to his view of himself and his family relative to those within his own experience and observation, than to more highly generalized distinctions between classes.

6. THE HOUSING PROBLEM

All the large towns underwent an expansion even more impressive than that of the population generally. It was very rapid in the 1811-21 decade, with rates of percentage increase between 22·63 for Sheffield and 41·92 for Manchester. But the peak rise for the great provincial centres of the midlands and the north came in the following decade. In the twenties the rate of increase of Liverpool, Manchester, Birmingham, Leeds, Sheffield, Bradford, and Salford was extraordinary — none below 40 per cent, and most over 45 per cent.[2] The thirties brought a

---

[1] 'The Last Crisis of the Reform Bill', *Econ.*, 4 May 1867, p. 493.
[2] R. Price Williams, 'On the Increase of Population in England and Wales', *J.S.S.*, 1880, p. 468.

sharp decline in all the foregoing rates, except for Liverpool and Bradford, but even so the growth of towns with over 20,000 inhabitants was very much greater than that of the country as a whole. Further, the towns of the north-east coast, including Newcastle, Hull, and Sunderland, hitherto lagging, responded to the coming of the railway, showing their greatest rate of increase in the thirties.

Though changes in civic boundaries complicated the picture, certain trends were apparent. By the fifties the lustiest striplings of the urban expansion of the twenties had almost ceased to grow. Liverpool increased in the forties by only 4·07 per cent, Manchester by 6·81 per cent. But in the metal and engineering towns the rates were still high: Birmingham 22·27 per cent, Sheffield 22·44 per cent, Wolverhampton 21·83 per cent.[1] By 1861 overcrowding was at its worst in the towns that had led the industrial revolution, in Manchester and Liverpool, and in London. Leeds, Sheffield, and Birmingham, where population was still buoyant, were below national average in overcrowding.

The concentration of the housing problem in Liverpool, Manchester, and London, reflects two general circumstances. The great Lancashire cities were the focus of the first great burst of productivity. They were the first to feel the impact of a wave of people in the midst of an unprepared nation. Moreover, with cities as with countries, there is a tendency for men to overrun their opportunities, thus producing social situations characterized by cumulative degradation which can only be reversed by time, or by the strongest measures of control. But London and Liverpool shared another problem. Both were magnets to attract and retain a disproportionate share of population increase. The economic expansion of the nation hastened the long trend of increase in the metropolis. Liverpool was the unwilling recipient of many who, having reached an English city, could not or would not go further; the greatest of the seaports except London, it was the greatest of the sumps, especially for the unfortunate Irish immigrants.

As more persons converged on the towns the supply of housing inevitably failed to increase proportionately.[2] For it was easier, on a step by step basis, simply to divide into smaller and smaller units, with families sometimes occupying merely a portion of a room. With such

---

[1] W. L. Sargant, 'Birmingham and other large Towns', *J.S.S.*, 1866, p. 104.
[2] *Two Reports of the R.C. for Inquiring into the State of Large Towns & Populous Districts, 1844–45.* E. Chadwick, *Report of the Sanitary Conditions of the Labouring Population, 1842.* See also James Hole, *The Homes of the Working Classes, with suggestions*, London, 1866; J. F. C. Harrison, *Social Reform in Victorian Leeds: The Work of James Hole, 1820–1895* The Thoresby Society, Leeds, 1954.

a premium upon shelter, and with no standards to maintain, it was all too tempting to take in a lodger, adding further to the disruption of the family. Those who were commercially interested in town housing found that there were greater gains in a policy of overcrowding than in one of reconstruction. Moreover, in fairness to such landlords, it was too much to ask that they should take such a new and daring initiative. They were essentially men and women of petty, pre-industrial mind, who knew how to make gains where an intense demand for a facility in short supply had arisen, but they were incapable of the initiative necessary to place the matter on a new basis. Slum landlords have never been slum clearers.

Even in the countryside of pre-industrial times, the supply of houses had never been forthcoming on an adequate basis when population was increasing rapidly. Under the new conditions rural overcrowding became greater than ever, acting as an additional factor impelling people from the land. The problem was serious in the mining areas.[1] Nor would many of the victims of the system have applauded any insistence on higher standards, for if these had meant higher rents, the result would have been either to render many homeless altogether, or to encroach on their miserable budgets for food. Statistical measurement for housing shortage is very difficult, if not impossible, not least because of the problems of defining the basic terms of family and house.[2] So too is the measurement of rents, costs, and output.[3]

The challenge of an adequate housing programme was very complex. There were all the problems of design: what kind of house would be suitable for such occupants, what kind of facilities? Anything that was more than barely minimal would mean a cost of construction that was prohibitive. Indeed house-owning artisans were sometimes found opposing a sanitation programme for fear of increased local taxation. Pipes and fittings were usually very poor, and buried in the plaster so that water companies had to be very careful when

[1] H. Richards and P. Lewis, 'House Building in the South Wales Coalfield, 1851–1913, *M.S.*, 1956.

[2] For such attempts see R. H. I. Palgrave, 'On the House Accommodation of England and Wales, with Special Reference to the Census of 1871', *J.S.S.*, 1869, p. 411; H. Robinson, *Economics of Building*, London, 1939, p. 109. For discussion of the number of occupants per house see Hoffmann, op. cit., p. 77, and Clapham, op. cit., vol. II, pp. 489–496.

[3] See H. W. Singer, 'An Index of Urban Land Rates and House Rents in England and Wales, 1845–1913', *Econometrica*, 1941; K. Maiwald, 'An Index of Building Costs in the U.K., 1845–1938', *Econ. H.R.*, 1954. See A. K. Cairncross and B. Webber, 'Fluctuations in Building in Great Britain, 1785–1849', *Econ. H.R.*, 1956.

raising the pressure.[1] In some areas the building operatives were charged, and with some justice, with keeping down recruitment through apprenticeship, thus keeping their own wages up, but curtailing the housing supply.[2] Skimping was all too general — in slump to get the cost down in order to sell at all, and in boom to speed completion before prices broke.[3]

Attempts were made by various philanthropists, including Prince Albert, to design houses for workers, but their ideas were seldom taken up. If economies were to be achieved it was necessary that the construction of a considerable number of units be undertaken simultaneously. The financial system did very little to aid the northern industrialist himself until well into the second half of the century; the provision of long-term funds for systematic house construction was out of the question. Then there was the matter of site. It was impossible to rebuild in the old areas, for to pull down the housing there was to render a large part of the population homeless for an interval, and even when reconstruction was complete only a proportion could be contained. Clearances did occur, but for other reasons. Some such projects — dock extensions, railway yards, and stations — could offer a prospect of return even greater than slum rents.[4] Street widening was often essential in order to allow the increased traffic to move. Such clearances invariably had the effect of sending rents even higher in the adjacent areas. But too often there was no new ground upon which planned construction could take place. Large areas were pre-empted for middle-class housing; planning schemes of this kind were practicable and profitable, for the growth of a recognized convention of what was appropriate to the various non-plebeian orders of society had been going on steadily at least from the early decades of the century when bourgeois families moved to suburbs. This low density construction covered many acres, and formed something of a cordon round parts of the growing town, especially to the west. In addition there was industry, continuously extending, and often leap-frogging the congested slums to open space to the east, and so constituting another kind of barrier to rehousing. Shaftesbury described the London workers as like people in a besieged town, running to and fro.

[1] Thomas Hawksley, Presidential Address. *Proceedings, Institution of Civil Engineers,* 1872, p. 344.
[2] James Chadwick, *J.S.S.*, 1860, p. 20.
[3] *Report of the R.C. on Depression of Trade and Industry,* 2nd Report, 1886, Evidence of Albert Shiers, p. 45.
[4] H. J. Dyos, 'Railways and Housing in Victorian London', *J. Tpt. H.*, 1955–56.

Eventually the workers began to encroach both west and east. To the east the owners of factories made little complaint, for if extensions to the works proved necessary it was fairly easy to dispose of the interstitial houses. To the west the situation was different — the workers certainly encroached, and indeed the typical slum by the sixties was often composed of the surrendered houses of merchants or industrialists, so proudly occupied three or four decades earlier. For a time there might be attempts at solidarity among the middle class, to resist encroachment, but often their hearts were not in it. The more successful, conscious of neighbouring deterioration, were hearing calls from even farther out of the city to new areas with a prestige that was rising. The workers thus spread into housing intended for others, which was often of such a nature, in continuous terraces, that a very high density could be produced quickly, with inevitable deterioration.

There were indeed some areas developed for artisan dwellings, like the Hulme district of Manchester.[1] Occasionally cooperative building societies, sponsored by determined artisans, struggled through the morass of obstacles to create a housing estate, in Birmingham or Sheffield. Living conditions for cotton workers were certainly better in the smaller towns than in the large. But by and large, housing for the workers was a residual affair in all senses: they got what was left over, in terms of initiative, design, finance, and site. To have inverted the priorities, to begin with the concept of workers' housing as taking precedence was alien to the spirit of the age, and indeed may well be alien to any society seeking rapid industrialization.

There were a few philanthropists who tried to provide an escape from the degradation of bad housing. Southwood Smith and other doctors sought in the forties to demonstrate the wisdom of re-housing if only to avoid the costs of epidemics.[2] In 1844 the Society for Improving the Conditions of the Working Classes began operations; in 1845 the Metropolitan Association for Improving the Dwellings of the Industrious Classes was founded. George Peabody, a highly successful American merchant, became much concerned, and on his death in 1869 left half a million pounds to pursue the work begun in his lifetime of constructing model tenements. Through the Improved Industrial Dwellings Company Sir Sydney Waterlow demonstrated that with sufficient middle-class money and sustained devotion, workers' housing

---

[1] *R.C. on Children's Employment, Second Report*, 1843, vol. II, p. 848; W. Abram, 'The Social Conditions and Political Prospects of the Lancashire Workmen', *F.R.*, 1868, p. 429.

[2] Thomas Southwood Smith, *Results of Sanitary Improvement*, London, 1854.

could eventually be made to pay its way. Others, like the Baroness Burdett-Coutts, followed such examples. Octavia Hill, responding to the same conditions that had moved her grandfather in the forties, took the initiative in the seventies.[1] But the model dwelling house movement with its emphasis upon the need to make housing conform to commercial criteria, with rents that covered costs, could not, in the prevailing conditions, do more than touch the problem. It did have the effect of bringing the question to public notice and calling attention to the many and complex obstacles to effective rehousing. The Artisans' Dwelling Act of 1875 was the first real acceptance of public responsibility.[2]

It was Lord Shaftesbury's opinion that overcrowding in London had become progressively worse between the inquiries of 1844 and 1884–5.[3] The fearful embrace of the city meant that all must live within hailing distance of the prospect of work; none dared breach this centripetal attraction. Especially bound were the unskilled dockers, building labourers, costermongers, and the like. Women and girls added to its force, for they too were obliged to seek employment near to home. Once involved in debt, such families dared not move, for their credit would be gone; they were obliged to stay near those petty shopkeepers who knew them. The troubled mass was swollen by contractors, who, finding the Cockney too debilitated for heavy work, advertised and brought in lusty countrymen. When movement did take place, it was not from areas of high pressure to lower, but from one slum to another, for it was bound by the search for work.

There were certain advantages from this crowding, but they were dubious. Food was a good deal cheaper in the centre of the town than in the suburbs, due to the competition of the very costermongers who swelled the population. Earnings might be higher than elsewhere, but certainly did not keep pace with rents. In the eighties, in the poorer districts, almost one-half of family income went in rent, with the meanest accommodation in return. The house jobbers, the house farmers, and the house knackers had become formidable middlemen. Against all this Giffen posed the argument, based on a false conception of the availability of choice, that 'it is clear . . . that the concentration of men in cities is due to the fact that cities, on the whole, weigh in the

---

[1] Octavia Hill, *Homes of the London Poor*, London, 1875; William Thomson Hill, *Octavia Hill*, London, 1956; E. Moberly Bell, *Octavia Hill*, London, 1942.
[2] See below, Chapter 9, section 9, p. 374.
[3] See *R.C. on Housing of the Working Classes*, 1884–85, p. 7.

balance against the country'.[1] This was to ignore the possibility that the advantages of the town loomed disproportionately large to the immigrant, but when he had acquired the experience for a more balanced judgement, he simply could not reverse his action in urbanizing, by returning to the countryside. Once in the city his chance of acquiring his own house was slight, for, quite apart from the problem of saving, the leasehold system and the legal costs and difficulties of buying small parcels of land made it impossible for the majority of men to undertake the great venture while young and strong; as age and tiredness came they no longer had the will.

7. THE CONDITIONS OF WORKING

As the manner in which men, women and children supported themselves changed, a new kind of vulnerability appeared. Agrarian society had always offered a fair degree of tolerance, in the sense that though the farm workers might be exploited through the tenure of land, there was a kind of limit imposed by nature upon the deterioration of his conditions of work. In the factory, with humans integrated with powered machines in a single process, pace and context were no longer tied to the traditional tempo and conditions of nature, but were determined by employers and managers. The latter had little time or inclination to consider this new aspect of their self-acquired responsibilities. Employers never had done so; such preoccupations would scarcely dominate the minds of the new kind of employer, engrossed as he was, in that portion of his time that was devoted to business, with questions of raw materials, factory organization, finance, and sales. None of these latter would look after themselves. They were clamorous, for the survival of the enterprise depended upon them; the workers on the other hand could not raise a concerted voice — indeed they too, especially in cotton, were involved in the struggle between firms, as competitive reductions of prices in all markets became the order of the day.

Such consideration as the factory master gave to the condition of the workers might well make him less rather than more solicitous. In practical terms there was the necessity to enforce factory discipline upon a labour force that was often unwilling to conform to the new routines. The early owner of a factory was a member of the generation of innovators who thrived partly because the ancient system of

[1] Giffen, op. cit., vol. II, p. 84.

protective labour regulation dating from Tudor times was in abeyance – an excessive concern with the worker might well seem to be in conflict with the growth and prosperity of the enterprise. This showed with particular clarity over the question of shortening hours, for the employer, with his capital investment, quite naturally fell into the habit of mind of regarding it as his paramount concern, with the worker as hirable and expendable.[1]

Nor were the factory owners the only group who failed to appreciate the situation they were creating. Parliament itself, with its preponderance of landed men, had little knowledge of the new industrial sector that was extending so rapidly. The road to political reputation lay along paths quite different from the relief of domestic squalor. Moreover, the landed men had no wish to undertake a crusade against elements in the national life that appeared to have no great direct significance for them: only when the industrialists began to show aggressive tendencies against the Corn Law did the landed interest begin to develop a sense of responsibility toward the benighted industrial worker. Even then many a landed gentleman, like Sir James Graham, a prominent member of three governments, was fearful of public investigations of ills to which there appeared to be no remedy not worse than the evil itself.[2]

Grave decline of working conditions in many industries was the inevitable result. Even those employers who owned large concerns, and aimed at a paternalistic relationship between themselves and their workers, could not escape the blight when market competition was strong. The only way to avoid deteriorating conditions was to follow Owen's example, adding principles of scientific management to cottage building and school provision. But few were capable of this, for many did not wish to be continuously engrossed in their mills, but preferred to turn to other things. Some such employers, like the first Sir Robert Peel with his Health and Morals of Apprentices Act of 1802 sought to impose equal conditions upon all spinners of cottons in order to stop competitive deterioration.[3] Thus the state entered the field first to protect the most vulnerable of all groups – the pitiful pauper children,

[1] See above, Chapter 4, section 3, pp. 123, 124.

[2] J. L. and B. Hammond, *Lord Shaftesbury*, London, 1924, p. 70.

[3] J. T. Ward, *The Factory Movement, 1830–1855*, London, 1962. For a bibliography of statutes and reports see B. L. Hutchins and A. Spencer, *A History of Factory Legislation*, 3rd ed., London, 1926; also M. W. Thomas, *The Early Factory Legislation*, London, 1948. For a convenient summary of the legislation, together with an attempt to assess its effect on wages, see George Henry Wood, 'Factory Legislation, considered with reference to the Wages, etc. of the Operatives Protected thereby', *J.R.S.S.*, 1902.

charged on the parish rates often at birth, but pushed into factory employment as soon as possible by economical Poor Law guardians. They were not to be worked at nights, their maximum was to be twelve hours a day; and they were to receive rudimentary education. So far had sentiments of responsibility for these defenceless and parentless children carried Parliament. The rest of the workers in the mill would also benefit insofar as the Act was effective, for it did prescribe general conditions of ventilation and cleanliness for all mills employing over twenty persons. But other canons of proper legislation had to be observed: there was to be no prying into the concerns of free men. As a result there was no provision for inspection or enforcement. Even if the master was proceeded against, the penalties for infringement were so slight as hardly to weigh in a commercial calculation.

All this greatly annoyed Robert Owen, who had no inhibitions about manipulating society in the direction of his own ideas. In his own business he had shown the way to commercial profits without physical and moral degradation, and agitated with great vigour that the power of the state should be used to establish a set of general rules the effect of which would be not only to define minimal conditions for the workers, but more positively to force employers into better principles of industrial management. As a result of his agitation Owen obtained from the second Sir Robert Peel a committee to inquire into child labour in cotton factories. Two more committees were called before a Bill was drafted. But both Lords and Commons set themselves to whittle it down so that the Act of 1819 was a shadow of Owen's programme, with all provision for inspection deleted. It applied to cotton factories only. It is hardly surprising that Owen abandoned the idea of legislative reform as a way of meeting the social challenges of industrial society.[1] But the Act did prohibit the employment in cotton mills of children under nine, and limited the hours of all under sixteen years to twelve a day. Later, in 1825 and 1831 the age covered by restricted hours was raised to eighteen, with night-work prohibited until twenty-one.

The anomaly of a system of regulation without enforcement grew with the years. So too did the abuses. Agitation was renewed in 1830, led by Lord Ashley, Oastler, Sadler, Fielden, and Wood.[2] The latter three of these were factory owners. For the real dimensions and the

---

[1] See below, Chapter 9, section 6, p. 345.
[2] A. P. W. Robson, *The Factory Controversy, 1830–1853*, London Ph.D. thesis, I.T.A., 1957–58, no. 346. J. L. and B. Hammond, *Lord Shaftesbury*, London, 1924; C. H. Driver, *Tory Radical: the Life of Richard Oastler*, Oxford, 1946; J. Fielden. *The Curse of the Factory System*, London, 1836.

fearful potential of the problem were becoming apparent to the percipient among the great industrialists. C. Turner Thackrah, a devoted Leeds surgeon, drew attention to the new industrial diseases.[1] The landed gentlemen were now preparing to take a more active interest in the concerns of the industrialists, learning to comment upon the sins of factory exploitation, so much blacker than the countercharge of maintaining high food prices through the Corn Law.

Eventually, in 1833, came the first effective Factory Act. At last an inspectorate, with powers of entry, came into being. There remained the problem of discovering a child's true age; this was solved by an Act of 1837 for the compulsory registration of births. Hours were more tightly regulated: those under thirteen might not work more than forty-eight hours per week, those from thirteen to eighteen were limited to sixty-five. Regular meal times were required, and provision was made for two hours' schooling per day. For the first time some elements of regulation were extended to textiles other than cotton — to silk, linen, and wool.

Perhaps the greatest boon of the 1833 Factory Act was to open a window on industrial conditions through the reports of the inspectors.[2] In the bad years of the early forties commercial competition was keen and industrialists were in grave difficulties. It is hardly surprising that the tale of abuses mounted steadily: poor materials, bad ventilation, high temperatures, long hours, and, especially, the speeding of the machinery to increase the workers' tempo. These conditions inevitably meant that the effective working life was reduced, with disease, especially tuberculosis, thriving in exhausted bodies, causing employers or their managers to discard burnt-out adults and to rely increasingly on children and youngsters. At the other end of their lives, as infants, the workers suffered damage as young mothers exhausted themselves in the mills and so failed in the care and feeding of their children. In justification of the system two abstract ideas were posed: the theory of the workers' liberty to choose how, when, and under what conditions he would work, and the theory that industrial profits, and hence industrial expansion, would cease if hours were curtailed.

The new industrial inhumanity had first come to public attention through the textile mill exposures. But there was another group of

[1] Charles Turner Thackrah, *The Effects of the Principal Arts, Trades and Professions . . . on Health and Longevity*, London, 1831.

[2] See R. K. Webb, 'A Whig Inspector: H. S. Tremenheere', *J.M.H.*, 1955. There are *Reports of Inspectors of Factories*, semi-annually from 1835 to 1877, thereafter *Annual Reports of the Chief Inspector of Factories and Workshops*.

workers whose conditions produced a shock of horror even more intense. The *Report of the Commission on the Employment of Women and Children in Mines and Collieries* of 1842 revealed conditions that demonstrated even more starkly the new depths of human degradation now possible, where the owners of enterprises accepted no responsibility for their workers, where the butty system caused the profit motive to extend to the leading workmen and the management of their gangs, where the worker was incapable of effective bargaining, and often indeed sought to excel within the system that was destroying him.

So frightening were the revelations that no attempt whatever was made to alleviate the condition of women and children below ground: they were simply removed from such occupation by Lord Shaftesbury's Mines and Collieries Act of 1842. No doubt indignation was able to become effective without real resistance because women and children were rapidly becoming uneconomic in this employment: in textiles the reverse was true, where they progressively replaced men.

In the textile trades deplorable facts gradually overbore dubious generalized ideas: a further notable instalment of reform came with the Factory Act of 1844. The earlier solicitude for children was now extended to women so that both worked under the same safeguards. It was hoped by the reformers that this would have the effect of protecting the men also from abuse, without the necessity of explicitly breaching the liberal creed. The children's working day was to be limited to six and a half hours, either in the morning or the afternoon so that they might enjoy a half-day for schooling or recreation. The system of penalties was strengthened. Dangerous machinery had now to be fenced so that bad employers were forced to behave as the better ones had done long since. But there was one retrograde step: the minimal age for entry into the factories, set at nine in 1819, was lowered to eight.

These gains, important as they were, left the question of hours still in a most unsatisfactory state. Much abuse even of women and children was possible by various adaptations of the shift system. The Ten Hours movement, led by Ashley, had been active since 1831, contributing to the pressure producing the 1833 Act. Doherty had succeeded in forming the cotton spinners into a union and in pressing the woollen and worsted workers to do the same. By this time workers' support of the idea of the control of hours was strong. In 1847 the Ten Hours Act for Women and Young Persons was carried by Fielden. The Act, strengthened by further legislation in 1850 and 1853, provided a ten and a half hour day, from 6 a.m. to 6 p.m. with one and a half hours for meals. With the hours of women and children thus regulated the men too were provided for, as millowners were obliged to regulate running on the same principle for all.

There was a hard-fought rearguard action by many masters as trade recovered and the old attractions of overtime working reasserted themselves. In the Manchester district the magistrates refused support to the enforcing inspectors; in Scotland the inspectors refused to act. In 1849 a court ruling showed the law to be ambiguous; the relay system, the cause of so much abuse, came back. The Short-time Committees were remustered and the position secured by a new Act of 1850.

In the coal mines the removal of the women and children was followed by a vigorous extension of working, and with this a great increase in risk. Pit disasters had long been part of mining life; they now became more general than ever. The Coal Mines Inspection Act of 1850 provided for regular inspection of the mines to enforce safety, lighting, and ventilation. Safety was amplified in 1855, and the first code of 'General Rules' was introduced. The Coal Mines Regulation Act of 1860 improved the rules and prohibited the employment of boys less than twelve years old. The right of the miners to appoint checkweighmen to ensure that they were properly credited with their output, was, however, ineffectual, for the employers had the right of dismissal.

By the sixties legislative control had brought great benefits to the textile factories, including calico printing, bleaching, and dyeing. As the variety and range of trades extended some early opponents of control were converted to its necessity. But all other elements of industry, with the exception of coal mining, were entirely unregulated. A new Commission with wide terms of reference was set up in 1862. The result was a series of enactments between 1864 and 1878 defining and enforcing minimal working conditions with respect to safety, ventilation, and sanitation in the metal trades, blast furnaces, printing, paper making, eventually extending to the small workshops. Factory legislation was codified in 1878 when over a hundred different pieces of legislation were brought together in a great consolidating Act.

With the immense increase in coal output all the old problems of mining became more serious. The men and their leaders increasingly pressed for two kinds of protection: against exploitation by the owners, and against accident. The old dispute about checkweighmen was still the source of much bitter feelings; some 1,000 lives were lost annually in accidents, with maimings some considerable multiple of this. The Coal Mines Regulation Act of 1872 brought further relief. The position of checkweighmen was made more secure, colliery managers were required to have State Certificates of their training; miners were given the right to appoint inspectors from among their own number.

Against the growth of regulation the petty producer fought on. There had been a severe toll in many trades where the cost of proper regulation had meant losses, and where new methods made bad conditions uneconomic: by the eighties the paper trade, for example, was in the hands of large concerns that no longer had any use for the child carried to the factory by its parents to work for sixteen hours kneeling down to feed the machine. The small manufacturers in the match trade were extinguished. Many brickyard owners had deliberately kept the number of their workers under fifty so as to be treated as workshops rather than factories and so avoid inspection; this was no longer possible when the brickyards were brought under the Factory Acts in 1878.[1]

In the countryside the conditions of itinerant women and children were regulated by the Gangs Act of 1867; the Agricultural Children's Act of 1873 prohibited employment under eight years of age and made provision for their education.[2]

The merchant seaman, those custodians of Britain's trading lifeline, had always been accustomed to dreadful conditions. Drunkenness among captains and crews, making travellers prefer American ships for their lesser danger of shipwreck, appalling living conditions afloat and the attentions of crimps, pimps, prostitutes, and pettifogging lawyers ashore, were typical of the age of the Navigation Acts, persisting to 1849. The double challenge of competition and steam forced owners to consider the seamen in more human, and in consequence, more economic, terms.[3] But direct legislation was still required where rapacity overbore efficiency: in 1876, by a Merchant Shipping Act, Samuel Plimsoll was successful in prohibiting overloading through the adoption of a marking line on the ship's hull.[4]

The steamship also altered the life of the docker. Because of the new premium upon speed, handling in port was now done under intense

[1] Escott, op. cit., vol. I, pp. 256, 257.
[2] Clapham, op. cit., vol. II, p. 292.
[3] See above, Chapter 4, section 5; Thornton, op. cit., p. 47.
[4] Samuel Plimsoll, *Our Seamen. An Appeal*, London, 1873; A. W. Kirkaldy, *British Shipping*, London, 1941.

pressure. Loading was a semi-skilled job, done by stevedores organized and paid by the employers; unloading was an unskilled job, done by disorganized men, probably the most casual group among the workers and the most exposed to vacillations of trade.[1]

The state by the eighties had come to accept the role of industrial policeman. None of the dangers to industry that haunted the debate from its beginning proved to have any reality. Indeed it seems probable that shortened hours and improved conditions did not lower wages, as some had feared, but rather raised them, through greater efficiency; in foreign competition Britain's position was strengthened rather than otherwise.[2] But the threat to civil liberties did have some meaning, for the growth of control undoubtedly led to the extension of bureaucracy and to the delegation to it by Parliament of many of its powers.[3]

8. THE CONDITIONS OF LIVING

The industrial cities of England came into being almost by inadvertence: they were merely the places where factories, offices, depots, and warehouses were built and to which the new industrial population was attracted.[4] The long tradition of town planning and regulation going back to the middle ages had largely passed into abeyance.[5] In the newer cities it had never been operative. The urban revolution was so great in scope and scale as to obliterate the older sense of responsibility.

Problems accumulated steadily. Down to the forties no real attempt at solution was made. Learning about cities had to be imposed by experience; deterioration had to be well advanced before the necessary response could be evoked, for the immensity of the effort required to regulate an industrial city was so great that it was only forthcoming as

---

[1] W. Page, op. cit., p. 332.

[2] Wood, op. cit., *J.R.S.S.*, 1902, p. 313.

[3] A. V. Dicey, *Lectures on the Relation between Law and Public Opinion in England during the Nineteenth Century*, London, 1905.

[4] For a general view see Asa Briggs, *Victorian Cities*, London, 1963. For particular places see Conrad Gill and Asa Briggs, *History of Birmingham*, Oxford, 1952; Arthur Redford and I. S. Russell, *The History of Local Government in Manchester*, London, 1939–40; T. C. Barker and J. R. Harris, *A Merseyside Town in the Industrial Revolution: St. Helens, 1750–1900*, Liverpool, 1954; Sidney Pollard, *A History of Labour in Sheffield*, Liverpool, 1959; W. H. Chaloner, *The Social and Economic Development of Crewe, 1780–1923*, Manchester, 1950; A. Temple Patterson, *Radical Leicester; a History of Leicester, 1780–1850*, Leicester, 1954; *The British Association Handbooks* provide valuable studies of the cities visited by the Association. For general bibliography, W. H. Chaloner, 'Writings in British Urban History', *Vierteljahrschrift für Sozial- und Wirtschaftsgeschichte*, 1958.

[5] W. Ashworth, *Genesis of Modern British Town Planning*, London, 1954.

an inescapable imperative. Moreover, the form and content of the new kind of city represented the attempts of its myriad of members to find the formula that allowed of the implementation of their individual plans, either as masters or men. Until the nature of the resulting complex could be perceived it was impossible to plan to control it. There was thus a paradox: for the morbid aspect of urban growth to be treated it had to be allowed to develop until it carried irresistible conviction.

By the forties a state of affairs had been reached in which three great dangers threatened. There had long been fear of the great mysterious masses accumulating in the areas that were unknown and unpenetrated by the middle class. This could lead to a politically explosive situation. Even more imminent was the danger of epidemic. The cholera outbreak of 1832 and subsequent visits in the forties brought terror to the towns.[1] Every middle-class family relied upon its servants; they constituted an almost instantaneous channel along which disease travelled from slums to residential squares. In the fifties it was discovered that the Royal apartments at Buckingham Palace were ventilated through the common sewer. Outbreaks of disease were predicted by the opponents of the Great Exhibition; before 1851 it was thought unreasonable that the Queen should visit Liverpool and Manchester, and when she did so in that year it was 'almost a matter of wonder'.[2] Finally, the growth and deterioration of the cities was very damaging to efficiency and therefore an obstacle to the lowering of production costs. British industry and trade were doing well in the third quarter of the century, but it was becoming apparent that the conditions of living of the workers would soon become a seriously limiting factor.

The dangers of dirt were the first to provoke thought and action. The system of natural liberty could provide the food and raw materials required to maintain the population of vast cities; indeed the philosophically minded marvelled how the grand design of the Deity was thus demonstrated. But it could not purge its wastes. It was as though the organs of sustenance were efficient but the organs of elimination were defective.

Men, women, and children, in varying degree, were wearing, breathing and drinking refuse. Old garments moved down the social scale and passed from peer to pauper at its nether end. The air was

---

[1] Asa Briggs, 'Cholera and Society in the Nineteenth Century, *P. and P.*, 1961; Charles Creighton, *A History of Epidemics in Great Britain*, 2 vols., Cambridge, 1891-94.

[2] 'Loyalty in the Workshops', *Econ.*, 18 October 1851, p. 1147.

defiled with industrial and human effluvia. Water-courses became open sewers. Tipping and dumping were uncontrolled; there was a lack of depots for night soil. The sewage system was largely on the surface, courts were unpaved, the movement of air was blocked by crowded buildings. The builder might place the primitive privy where he wished, inside or outside the houses: when indoors the smells in winter were dreadful in houses tightly closed to keep them warm, when outdoors women and children, unwilling to visit them in exposed places, became habitually constipated. Cemeteries gave off noxious smells and polluted the water supplies; tanneries, breweries, dyeing works, chemical plants, slaughter houses, and manure driers were uncontrolled in their disposal of waste matter, as gas, liquid, or solid. The cesspool, 'that magazine of all the contagions' as Farr described it, was still general.[1] The children were the heaviest casualties. In the sixties about twenty-six out of every hundred died under the age of five; in the best districts the number was eighteen, in the worst it was thirty-six. The first great task of the urban improvers was to deal with the toxic refuse of urban life.

Chadwick and his disciples began their attack in the forties.[2] They encountered a confusion of authorities, sometimes ludicrous, often tragic. Delay, confusion, waste, and omission were everywhere, presided over by a welter of Town Councils, Highway Boards, Health Committees, Commissioners of Sewers, Water and Gas Authorities, Improvement Commissioners, Watch Committees, Poor Law Unions, Turnpike Trusts. These bodies, laudable in themselves, relics of the tentativeness of the first decades of industrialization, were now archaic.

At the death of William IV in 1837 the Statute Book contained no general sanitary law, in spite of the fact that for two years, from the autumn of 1831, England, in common with Europe, had passed through its first taste of Asiatic cholera.[3] But the cholera did call attention to the depths to which much of urban life had sunk. At last the reformers could make progress against the inter-epidemical indifference. Edwin Chadwick, installed after 1834 as the Secretary of

---

[1] William Farr, 'Mortality of Children', *J.S.S.*, 1866, p. 9.

[2] B. W. Richardson, *The Health of Nations: a Review of the Works of Edwin Chadwick*, 2 vols., London 1887.

[3] Sir John Simon, *English Sanitary Institution*, London, 1890, p. 167. For earlier years see B. Keith-Lucas, 'Some Influences affecting the Development of Sanitary Legislation in England', *Econ. H.R.*, 1954; E. P. Hennock, 'Urban Sanitary Reform a Generation before Chadwick?' *Econ. H.R.*, 1957; S. E. Finer, *The Life and Times of Sir Edwin Chadwick*, London, 1952.

the new Poor Law Board, was placed in a position to begin systematic inquiry into sanitation.[1] He and other devoted men, including Sir J. P. Kay-Shuttleworth and Dr Thomas Southwood Smith provoked committees of inquiry, culminating in 1842, with the *General Report on the Sanitary Condition of the Labouring Population of Great Britain*. But the scale of the political and administrative implications could not be accepted immediately: a Royal Commission was appointed in 1843, reporting in 1845.[2] The Health of Towns Association and other such voluntary bodies helped the agitation.

At long last came the Public Health Act of 1848. A General Board of Health was established; it and its inspectors of nuisances did good work under Chadwick, but the zest of Chadwick as its chief working member added to the animosity of hostile interests. The Board was reduced in effectiveness in 1854, and wound up in 1858, its attentuated functions passing to the Privy Council. It had done much to deal with epidemics, improve water supplies, clear up the awful cemetery problem, and provide for sewage disposal, all against the powerful resistance of those who fought against additional charges on the rates and the enforcement of improvement from the centre.[3] Fortunately John Simon, the new Medical Officer of the Privy Council, appointed in 1855, was another dedicated man, under whom further progress was made.[4] But he had to struggle with a situation with which Parliament was not really prepared to deal; in the absence of considered policy and without effective powers Simon's task was heartbreaking. An assortment of miscellaneous Acts were passed, affecting common lodging houses, powers to deal with epidemics, burials, various aspects of sanitation, food defilement and adulteration, and venereal diseases.[5]

Between the fifties and the seventies England made substantial progress against the zymotic diseases: fevers, smallpox, cholera, diarrhoea – all due to filth. Deaths from such causes fell by some 23 per cent. But this was not enough. The shamefully high mortality of soldiers in barracks had been much reduced, inspiring contemporaries

[1] *Report on the Sanitary Condition of the Labouring Population*, Poor Law Commissioners, 1842; W. M. Frazer, *A History of English Public Health, 1834–1939*, Baltimore, 1951.
[2] *Select Committee Report on the Health of Towns*, 1840; *First and Second Reports of the R.C. on the State of Large Towns*, 1844, 1845.
[3] See below, Chapter 9, section 7.
[4] Sir John Simon, *Public Health Reports*, Edward Seaton, ed., 2 vols., London, 1887; for his contribution to vaccination, see R. J. Lambert, 'A Victorian National Health Service: State Vaccination 1855–71', *J.R.S.S.*, 1962.
[5] Simon, *English Sanitary Institutions*, chapter XIII.

with the hope that more could be done in the towns.[1] Southwood Smith, Edwin Chadwick, and Florence Nightingale were the driving forces of this second campaign.[2]

By 1869 the confusion of authorities was so conspicuous that a new Royal Commission was appointed; its Report of 1871 revealed the extent of the frustration.[3] It called for a national sanitation and health policy. At long last prompt action was forthcoming; the Local Government Act of 1871 divided the country into sanitary districts and made obligatory the appointment in each of a Medical Officer and an Inspector of Nuisances. The Public Health Act of 1875 provided a code of principles and responsibilities. The real attack on urban filth could begin. But there were immense arrears to be overtaken.

More positively, there were new challenges in the provision of two great articles of public consumption: water and gas.[4] A vast leap was required to move from the well and the water cart to a piped supply. For when taps appeared in the houses, it was no longer a question of supplying the same quantity more conveniently; the consumption of water shot up at an extraordinary rate. Gas, too, was in ever-increasing demand. Private companies for the provision of both gas and water had come into existence, and were highly profitable, not least because their extension of supply lagged behind what was needed. Demonstrations that they had increased output and reduced prices could not dispose of the fact that by the fifties they were a serious clog on urban improvement. Rival companies, with their high overheads, were unwilling to invest more, because each was fearful that the other might cut prices. In the absence of water meters the companies imposed complex and troublesome regulations about installations and use. The worst sufferers were, of course, the workers and the poor. Only a fully planned and efficient service could bring an adequate supply of water and gas within their means. So great was the commitment to the competitive idea, and so little developed was the alternative, that effective resistance was prolonged. But advocacy was powerful also; by 1855 more than a dozen towns in the north were empowered to set up water undertakings. The gas companies held out a good deal longer, in spite of spokesmanship in the Lords in favour of consumers. But before 1870

[1] W. L. Sargant, 'Certain Results and Defects of the Reports of the Registrar General', *J.S.S.*, 1864, p. 175.
[2] Cecil Woodham-Smith, *Florence Nightingale, 1820–1910*, London, 1950.
[3] *R.C. on the Sanitary Laws*, 1871.
[4] Herman Finer, *Municipal Trading: A Study in Public Administration*, London, 1941, chapter 3.

some thirty-three towns had their own gas undertakings.[1] In 1875 Birmingham bought out the private companies and through municipal enterprise showed what could be done with the economies of scale. The Public Health Act of the same year gave powers to urban authorities to supply gas where no private company existed. No powers of compulsory purchase of either water or gas companies were ever given. But ideas were changing. Municipal transport, especially in the form of tramway systems, was coming into being in the seventies. Though its actions were equivocal, Parliament sought to prevent a recurrence of the situation that had arisen over water and gas.[2] In the case of electricity Parliament, as early as 1878, regarded civic management as the best formula.

The new and better water supply aggravated the sewage problem — there was now so much more liquid waste to dispose of.[3] Sir Joseph Bazalgette, servant of the Metropolitan Water Board, provided a mighty but invisible monument to himself by constructing the sewers of London. Other cities tackled the sewage problem: Birmingham's great drainage scheme was begun in 1845. Too often, however, the sewers served to pollute rivers and estuaries.

All the other challenges of a society that was both expanding its numbers and concentrating them in cities were accumulating. The widening of streets and the provision of bridges became pressing needs as the number of people and the volume of goods moving about increased. But each such improvement was an *ad hoc* affair, to be fought through confused jurisdictions and embattled resistance, paying as it went toll to those in possession of lands or rights affected. Not infrequently the town, in order to bear such costs, sold its own patrimony of land, making the problem of municipal improvement so much greater for the following generations. Not until the seventies did the idea of area planning make any converts and these were few. The provision of port facilities was something of an exception: from 1857 Liverpool's greatest collection of equipment was placed in the hands of the Mersey Docks and Harbour Board, a non-profit-making monopoly. By this time the railways had slashed their way into the centre of most cities. Parks sometimes came into the possession of the town through gifts or, occasionally, by purchase.[4] Hospitals

---

[1] Gill and Briggs, *History of Birmingham*, p. 71.
[2] Finer, *Municipal Trading*, p. 52.
[3] C. E. de Rance, *The Water Supply of England and Wales*, London, 1882.
[4] Jacob Larwood, *The Story of the London Parks*, 2 vols., London, 1872.

were built, and police and fire services, each with its many problems.[1] Public education received no financial support from the state until 1833; in that year the reformed Parliament voted the sum of £20,000 for buildings for elementary schools.[2] There had been a good deal of voluntary activity which played a large part in causing this tardy and inadequate recognition of responsibility. The religious revival brought about by Methodism in the later eighteenth century had produced the Sunday School movement; the desire to encourage habits of industry had inspired 'schools of industry', to teach local children simple crafts; there were newly endowed parish schools combining religious instruction and the 'three Rs'.[3] The development of educational theory had been accelerated by the discussions of the eighteenth century and by the experiments of Owen at New Lanark and of a few other factory masters. The most vigorous controversy was between Andrew Bell (1753–1832) and Joseph Lancaster (1778–1838) and their respective protagonists. But though they had their differences on implementation, they shared the same basic principle, that of the 'mutual' or 'monitorial' system, under which a single master taught monitors only, these, in turn, being responsible for the instruction of their schoolfellows. The economy of this system was very attractive to the education enthusiasts. By 1858 the governmental support amounted to about two-thirds of a million.

In that year the Newcastle Commission began its inquiries into the provision of education for the masses.[4] Religious debate, bedevilling the matter from the outset, assumed a new virulence; many people came to favour a purely secular system in order to get on with the task of conveying knowledge and making possible the growth of mind. The effect of the religious controversy on the Commission, however, was quite otherwise — their reaction was to reject the principle of general compulsory education. The Educational Code of 1862 was intended to ensure 'efficient' schooling, with money grants dependent upon satisfactory examination performances by the children. The

---

[1] W. L. Melville Lee, *A History of Police in England*, London, 1901; Charles Reith, *A New Study of Police History*, Edinburgh, 1956; G. V. Blackstone, *A History of the British Fire Service*, London, 1957.

[2] See S. J. Curtis, *History of Education in Great Britain*, 2nd ed., London, 1950; J. W. Adamson, *English Education, 1789–1902*, Cambridge, 1930; Sir Henry Craik, *The State in its Relation to Education*, 3rd ed., London, 1914.

[3] See below, Chapter 8, section 7, p. 309; H. J. Burgess, *The Work of the Established Church in the Education of the People, 1833–1870*, London, Ph.D. thesis, I.T.A., 1953–54, no. 549.

[4] *Report of the R.C. on the State of Popular Education in England*, 1861; Marjorie Cruickshank, *Church and State in English Education, 1870 to the Present Day*, London, 1963.

education grant accordingly fell, as insufficient pupils qualified. The Education Act of 1870, presented by W. E. Forster, provided for elected School Boards to be set up where schools were inadequate.[1]

A dual system was thus made explicit, composed of voluntary and Board Schools. The supporters of religious education, chiefly members and clergy of the Church of England, exerted themselves to increase the voluntary element, in order to preclude the necessity for Board Schools. The dissenters worked through the latter, in which an undenominational protestantism was taught. A further Act of 1876 brought compulsory education to the age of twelve. Though fees were payable until 1891, now for the first time universal literacy became possible. The state, having imposed educational duties on local governments, virtually withdrew, leaving the field to the church and the Boards. In many places the Boards, the object of great working-class interest, responded with vigour, causing a remarkable extension. Participation in such work was one of the first effective opportunities afforded to men of working origin to work for the improvement of their fellows.[2]

Secondary education for the sons of farmers, traders, and manufacturers, and for the brighter and more fortunate sons of the workers, had enjoyed private endowments for a long time, though these were inadequate and unsystematic. Civic responsibility for endowed schools was very ancient, but the new conditions revolutionized the situation. The Taunton Commission reviewed the national provision in 1867–68.[3] It found that those schools having connections with the universities were sufficiently bad, the others were in chaos. But the Endowed Schools Act of 1869 did not carry through the comprehensive reform called for by the Report, failing to treat English secondary education as a whole.

With the struggle for such rudimentary provisions as sanitation and education so long and so greatly resisted, it is hardly surprising that the positive idea of making the cities places of beauty and pleasure was scarcely broached.[4] There were some who thought as planners, including the Owenite socialists, and James Silk Buckingham and Robert

[1] Sir Thomas Wemyss Reid, *Life of the Right Honourable W. E. Forster*, London, 1889.

[2] For the situation in the eighties see *Reports of the R.C. to Inquire into the Elementary Education Acts (England and Wales)*, 1886–88.

[3] *R.C. on Schools not comprised within the Two Recent Commissions on Popular Education, and on Public Schools*, 1867–68. The public schools were investigated by the Clarendon Commission, *R.C. to Inquire into the Revenues and Management of Certain Colleges and Schools*, 1864. It provoked the equally mild Public Schools Act, 1868.

[4] See R. J. Mitchell and M. D. R. Leys, *A History of London Life*, London, 1958.

Pemberton, and later on, General Booth.[1] Some of these men revived ideas from the great European tradition of town planning. But they were utopians: they thought and wrote in terms of new beginnings on an unencumbered site. They constructed ideal solutions which, in obviating all ugliness and conflict, ignored the essential dynamic character of cities in continuous response to changing conditions.[2] The result was a formalism that looked slightly ludicrous in the light of actuality.

Moreover the utopians confined their thinking to a scale that ignored much of the problem. The great extension in the size of cities gave rise to changes that were qualitative as well as quantitative. For to add successive new blocks to the city was to cause the centre itself to change and with it the relationship of the components. Some parts of the city became enormously more valuable, others decayed and were rejected as growth assumed a new direction. Some parts were paradoxical: site values rose with scarcity, but there was no dramatic redevelopment to economize and make efficient use of the ground. Instead there was an intensification of the old use, with very high density slums as the result. Thus appeared the baffling situation, that in the case of landlords the market system could cause the highest level of gain to be associated with the lowest level of human degradation. If the deterioration went on long enough the basic shape of the city might be affected, for the new growth took place in areas chosen so as to avoid the area of dereliction, so that eventually the possibility of redevelopment passed away and the slum was further confirmed. The only hope then for its removal was public intervention. But this called for a great effort of communal will that was not forthcoming.

In the newly forming suburbs developers bought parcels of land as they became available.[3] Usually these were arbitrary in shape, defined by existing main roads, farm access roads, and even field boundaries; the layout of the houses was made to conform to the fragments so created. The city thus became a jig-saw of components, its coherence dependent to a great degree upon luck. Those who lived in this ever-extending perimeter had homes divorced both from their place of

---

[1] James Silk Buckingham, *National Evils and Practical Remedies, with the Plan of a Model Town*, London, 1849; Robert Pemberton, *The Happy Colony*, London, 1854; General William Booth, *In Darkest England and the Way Out*, London, 1890; Herman Ausubel, 'General Booth's Scheme of Social Salvation', *A.H.R.*, 1950/51.

[2] Helen Rosenau, *The Ideal City in its Architectural Evolution*, Boston, 1959, p. 130 *passim*.

[3] H. J. Dyos, *Victorian Suburb: A Study of the Growth of Camberwell*, Leicester, 1961.

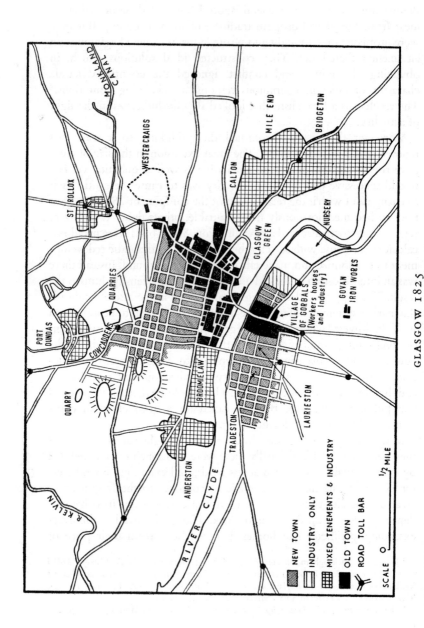

GLASGOW 1825

NEW TOWN
INDUSTRY ONLY
MIXED TENEMENTS & INDUSTRY
OLD TOWN
ROAD TOLL BAR

SCALE 0   ½ MILE

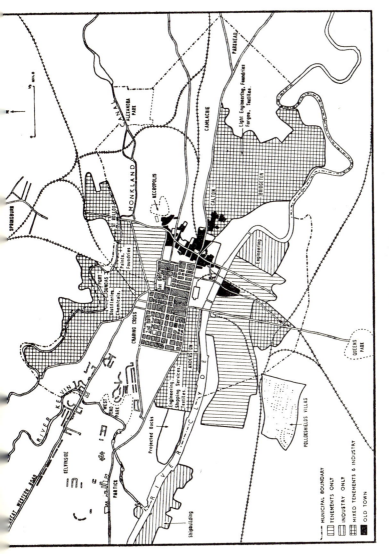

GLASGOW 1875

work and from their places of urban entertainment. This dormitory kind of life was made possible by new transport facilities, in the fifties the omnibus, in the sixties the railways, and in the seventies the beginning of the tramways.

The architect could do little to improve the cities, proliferating without centres and without form. He could exercise himself on single buildings, or at best on a range of terraces.[1] But he had little or no control of the use of space or of the relationship between buildings. Older towns in the south of England had their great churches, market places, castles, and guildhalls upon which to focus; the new cities of the north had little around which to synthesize their meaning. Leeds, Sheffield, Bradford, and Birmingham, the creations of the new age of independence of water travel, had paid the price of this freedom in being denied that greatest of urban amenities, a noble river. Even a monumental town hall was no substitute for a true focus. In London a traditional glory had been defiled; by the forties the Thames had become so noisome that the city turned its back upon its river; it was in Belgravia to the north that wealth went to dwell, and not on the South Bank.[2] It was only in the sixties that the northern side of the Thames was embanked.

Cities, in short, as well as houses, were constructed on the cheap, with virtually no restraints upon the builders except such as were archaic, arising from a cumbersome system of law and land tenure. The conditions of urban living, like the conditions of working, made their contribution to the great task of capital formation, for both were catered for on a minimal scale. Not only were costs kept down; flexibility was maximized in both cases, for both builder and entrepreneur could delimit their problems, excluding from their risk calculations difficult considerations arising from the social interest, especially as it would develop in the longer run, and could thus produce startling growth. In both cases a point was reached at which cheapness and flexibility ceased to be economic because of the impairment of efficiency.

The conditions of work were improved first, beginning in the thirties; the conditions of living were redeemed much more slowly, for in spite of the heroic efforts of the pioneers, the effective national attack upon dirt really dated from the seventies, and positive thinking

[1] Cecil Stewart, *The Stones of Manchester*, London, 1956; John Gloag, *Victorian Taste; Some Social Aspects of Architecture and Industrial Design, from 1820-1900*, London, 1962.
[2] Eric de Mare, *London's Riverside*, London, 1958.

about the urban environment had not made much progress even by the end of the century. But it had made its entry into the political arena.[1]

9. THE IMPACT ON PERSONALITY

Even more fundamental to happiness than monetary rewards and living conditions was the state of mind of men, women and children who needed to be able to find comfort and even pride in their relations with one another and with the productive tasks upon which their livelihood depended. Each was obliged to seek his identity and self-expression in terms of some basic human group, usually the family, and in terms of the larger group or class to which he belonged. Within these he needed a working concept of the function he discharged.

Pre-industrial society had over very many generations solved these problems. The closely knit, though often very extended, family was the fundamental unit within which personalities, loves and loyalties were formed. The general divisions of society were so obvious and irrefragable that most men lived their lives in the class into which they were born, itself related to others by well-established conventions. As to function, most men, women and children worked on the land, directly engaged in the primary task of producing food or raw materials – preparing nature for her yield and gathering and storing the product. Application to these tasks was the condition upon which the continuance of life itself depended – there could be no futility here. Even the village craftsman was half a farmer and half a provider of the simple tools of a farming community.

The new industry disrupted and eventually destroyed this ancient pattern within which the individual could find assimilation. The new structure of society, instead of minimizing individual assertion, put a premium upon it.

The result was grave social damage.[2] A family dwelling in an industrial town found itself not only divorced from nature and from the particular place of its origin, but cut off from other families. To isolation were added serious stresses within the family. Not all its members were engaged, as formerly, as a group, doing essential farming tasks, and sharing the benefits on the basis of a family communism administered by the senior members. Instead, the members of the family might well be employed in different occupations,

---

[1] See below, Chapter 9, section 10, p. 378.
[2] Neil J. Smelser, op. cit., *passim*.

with a premium placed by the wages system upon youth and vigour, so that the rights and authority by which the older generation traditionally maintained their status and self-respect into old age were impaired or destroyed. The new mobility made possible by the railway operated to disperse families as their more enterprising members sought new opportunities. Many of these became lodgers in the houses of others, or servants living in dingy conditions in the lesser middle-class families. So there developed an element of the population with a kind of sub-status, neither heads of families nor true members. Others of the unassimilated were the growing hordes of homeless children, the street-arabs who became a kind of nomad class in the heart of the great cities.[1] The disruption of the family was not arrested by official policy, but, rather, furthered. The Poor Law caused many families to be broken up upon admission to the Poor House so that even for those in this extremity there was no official thought of providing support for personality through the family.

The new conditions of work imposed further psychological strains. There were many men whose skill, once an important element in community life and a source of personal pride, was now wanted by no one. Much of the new employment involved long repetitive hours doing jobs that were meaningless in themselves, a mere portion of a process that ended in a product the worker would not consume. Many tasks now involved the worker in tying himself to a machine powered by a great engine, the whole controlled by managers. The latter would probably be of their own kind, caught up in the same new situation, but distinguished in their ability to manipulate the new system. Divisions of interest thus came within the family as some rose and others did not.

Fissures also appeared between classes of men in the same community, depending upon differences in skill, income, and respectability. Added to this was the growth of distinctions between communities. By the forties the Lancashire collier had sunk to an appalling degradation. For though the family had maintained its solidarity, with man, wife, and children acting as one productive unit, the man taking payment in a lump sum, this had involved a debasement of outlook and living that caused the colliers to be wholly isolated from other sections of society. The mill workers and the colliers maintained, in close juxtaposition, communities that were absolutely divorced. Other families found it difficult if not impossible to assimilate to the rest of society. By

---

[1] Benjamin Waugh, 'Street Children', *C.R.*, Lancashire, 1888.

the seventies the bargees were a sizeable group: no less than 44,000 adults, more than half of them unmarried, with some 72,000 children, about 40,000 illegitimate, were floating up and down the country, innocent of sanitation, religion, or education.[1] Economic breakdowns had not only sent Englishmen abroad; they had brought foreigners in — men and women to whom England offered opportunities for improvement on eastern European conditions that combined oppression, poverty, and military conscription. From 1870 onward came Jewish immigrants, settling especially in the East End of London and the other great cities, with their Yiddish newspapers, *chevras* or conventicles, some to fall victim to, and some to operate, the sweating system.[2]

The natural models upon which conduct might be based withdrew themselves from the sight of those who might have accepted them as mentors. The middle class, who had always dwelt in the centres of the towns at or near their places of business, had taken, by the thirties if not earlier, to abandoning the old setting. The incentives to do so were usually some combination of three factors. There was a desire for landed space, and the ways of the landed. Profits could be realized by urban sales: property values rose steeply in town centres of business (Cobden reckoned in 1832 that the house he had bought for 3,000 guineas in Manchester had become worth 6,000 within some four years.)[3] Finally there was serious deterioration of amenity, for old town centres were often made unsavoury by the masses who had been brought to being by the new business initiative — especially those who had sunk furthest and who constituted the growing mass of idlers, importuners, and disturbers of the peace. The centres of the towns, as places to live, were abandoned to the petty shopkeepers, the publicans, the labourers, the 'dangerous classes', and the police. Not for another fifty years were the English middle class to resume contact with the workers in their sordid streets, and then only in charity.

But for all this, the worker was not crushed.[4] Indeed humanity was able to demonstrate its extraordinary adaptability in a more dramatic form than ever before. For as urban living became typical, men and

[1] Escott, op. cit., vol. I, p. 258.

[2] Lloyd P. Gartner, *The Jewish Immigrant in England, 1870–1914*, London, 1960; V. D. Lipman, *A Century of Social Service, 1859–1959*; *The Jewish Board of Guardians*, London, 1959, chapter II; *Social History of the Jews in England, 1850–1950*, London, 1954.

[3] John Morley, *The Life of Richard Cobden*, London, 1893, p. 22, n. 2.

[4] For some aspects of adjustment see W. A. Abram, 'Social Condition and Political Prospects of the Lancashire Workmen', *F.R.*, 1868, p. 426.

women, taking the city as a datum, developed new attitudes, new entertainments, and new ways of mutual assimilation. Many indeed subsided into dirt and fecklessness, seeking the oblivion of drunkenness. But even these could often find pleasures and loyalties. The vitality of the countryside was a kind of hidden reserve upon which the Englishman at the nether end of the social scale could draw during the worst decades of the first half of the century. A new birth brought its worry, but also brought its joy; as Sam Laycock, the Laureate of the Cotton Famine put it:

> But tho' we've childer two or three,
> We'll mak' a bit o' room for thee
> Bless thi, lad.
> Th'art prettiest bird we have in t'nest
> So hutch up closer to mi breast;
> Aw'm thi dad.[1]

The genetic miracle of the renewal of life continued to manifest itself so that dilapidated parents produced babies capable of responding anew to the challenge of their situation. The children who survived the dreadful infant death rate learned to play on the inches of paving that separated them from the earth, and to develop their games in the enclosing streets and alleyways. Youth renewed its spirits with each generation as errand boys and apprentices took to ribaldry and larking. As kinship groups were established, usually through mothers and grandmothers, there came a new security, confirmed by gathering together for celebrations of christenings, marriages, and funerals.

Many means of assimilation to town life developed, most of them with some kind of roots in the past but transmuted in urban conditions: sport, both to participate in and to watch, the public house, the music hall, oratorios and brass bands, cheap fiction, the communal washhouse, the small shops, the barrows and vendors with their cries bringing neighbours onto the street.

Down to the seventies much of the entertainment of the workers grew up on the basis of traditional sports of the aristocracy.[2] Bareknuckle fist-fighting for a purse was common until 1860, with hardy working lads backed by an aristocratic fancy. So too with rowing: the first 'oars' to compete before admiring crowds were professionals —

---

[1] George Milner, ed., *The Collected Writings of Samuel Laycock*, Oldham, London and Manchester, 2nd ed., 1908, p. 3.

[2] For sports generally see *The Badminton Library*, published in the eighties.

the wherry men of the Thames and the Tyne, but the gentlemen soon took over with improved boats and rowing styles.[1] Horse racing also excited a good deal of working-class attention. But the workers could make their own amusements. The Lancashire collier, down at least to the forties, had his full dress for holiday occasions, for though his body might be black beneath the finery he was a splendid sight in his white stockings, low shoes, very tall shirt-neck, very stiffly starched and ruffled. The equally colourful costume of the Northumberland miner had disappeared with the coming of Wesleyan sobriety in the early decades of the century. That of the Lancashire miner was not to last much longer; only his isolation preserved it so long. Distinctive dress in general disappeared, except for the Cockney 'pearly kings'. Yet emulation could still produce diversion, as miners, potters, and others, staged their cock-fight and bred their dogs for catching rats or coursing hares.

The music hall developed from the spontaneous display of talent in taverns. These saloon theatres, depending for their profits solely on refreshments, began in the thirties in London, and were soon popular with the lower classes and an element in the middle class. 'A Free and Easy Kept here every Tuesday' was a common notice outside public houses for the rest of the century. The theatre owners, supported by those who feared damage to the theatre, tried to restrict the repertoire of the halls, but after 1866 the music halls were freed from restrictions on what they might present. So could arise, with their admission charges and printed programmes instead of Chairmen, the Palaces of Varieties, the Palladiums, the Empires, the Hippodromes, the Pavilions. Men and women of humble origin found the secret of entertaining their own kind by offering characterizations of an extraordinary range from the aristocrat to the down-and-out, and by producing songs that were infectious. The contagion quickly spread as the halls appeared in all the cities and the stars moved in their orbits. Thomas Hardy in the seventies regretted that these urban songs and dances, brought to the villages by migratory labour, were replacing the immemoriably old ones of the countryside; thus could the new culture of agglomeration destroy that of the villages.[2] But there were gains to be placed against losses, for the music hall was the creation of individual performers, each of whom had to provoke the audience to response. For though they

---

[1] W. B. Woodgate, *Boating*, London, 1888, chapter XVI.
[2] G. W. Sherman, 'Thomas Hardy and the Agricultural Labourer', *Nineteenth Century Fiction*, 1952/3, p. 117.

were capable of 'crapulous buffooneries', they could only succeed if they were both interpretative and creative.[1] As they dealt with personal and domestic situations, and as they commented upon high politics and the great they became a very important means of self-interpretation to the working class.

The Working Mens' Club and Institute Union was founded in 1862, with middle-class support.[2] Because one of the principal objects was to get workmen to meet elsewhere than in public houses, the Clubs were at first on a temperance basis, but the members soon objected. They learned to regulate their affairs in a thoroughly democratic way, often providing for mutual improvement and enjoyment a hall for debate, a reading room, a refreshment buffet, billiards, bagatelle, and chess. By the eighties Working Mens' Clubs existed in all sizeable towns. They were both social and political, with a leaning toward the radical wing of the Liberal party, rallying points for 'penny-paper liberalism'.

In Lancashire and Yorkshire the cloth workers had found in the late eighteenth century a mode of self-expression that was not only to survive in their satanic towns, but to thrive. This was choral singing, especially of the great oratorios. By 1835 a very advanced development had taken place, with every town in the two counties with its choir rendering the sacred works, revelling especially in Handel, with fervent practice in homes and public halls. The brass band movement began in the north of England in the forties; bands appeared all over the industrial parts of Yorkshire and Lancashire, often attached to particular collieries or factories.[3] They attained an astonishing technique as they fought for supremacy within the strict rules of adjudication that governed the famous contests at Belle Vue, Manchester, beginning in 1853. In South Wales the chapel choir became a rallying point for Celtic musicality.

By the seventies two forms of entertainment had found a new popularity. Many of the workers discovered that the railway could carry them to the seaside, and in good times, they might well be able to spend a few days there. So began the rise of the seaside resort catering not for aristocrats but for the masses. At Blackpool, Sutherland, Lytham, Ramsgate, Margate, Brighton old and New, and elsewhere, a new

---

[1] Escott, op. cit., vol. II, p. 434.

[2] Rev. Henry Solly, *These Eighty Years*, London, 1893; B. T. Hall, *Our Sixty Years*, London, 1922.

[3] J. F. Russell and J. H. Elliott, *The Brass Band Movement*, London, 1926.

class of seaside landladies eagerly extended the lodger system to the annual taking in of whole families.[1]

Weekly diversion came in the seventies and eighties with the forming of the great cricket and football clubs.[2] Cricket produced that great drawer of crowds, W. G. Grace, as early as the sixties; the Australians visited England first in 1878.[3] In football, gate money produced professionalism, recognized by the Association in 1885.[4] So arose the vogue for non-participant sports, an opportunity for the focusing of loyalties and the working off of vociferous aggressiveness, especially by those who themselves enjoyed little esteem.

Reading, or being read to, offered further compensations to the urbanized.[5] The seriously minded artisan might read the provincial newspapers, those important organs for the creation of public opinion.[6] But the optimism of those who, from the later eighteenth century onward, had hoped for an enlargement of mind from the spread of literacy, received damaging blows from the skilful entrepreneurship of those who supplied the new market made possible by cheaper printing methods.[7] In the thirties and forties male youth and adulthood buried themselves in the penny dreadful, so efficiently supplied by Edward Lloyd and others, wherein they might sup upon the horrors of gothic romance with its deathsheads and spectres.[8] The liberal improvers, however, clung to their optimism, believing that good writing would drive out bad. But mystery and violence had universal appeal. Criminals had the special attraction of men who had stepped outside of society and, single handed, taken what they wanted. Even the *Boy's Own Paper*, founded in 1879 by the Religious Tract Society, was not for weak

---

[1] J. A. R. Pimlott, *The Englishman's Holiday, a Social History*, London, 1947; E. W. Gilbert, 'The Growth of Inland and Seaside Health Resorts in England', *Scottish Geographical Magazine*, 1939.

[2] Ensor, op. cit., p. 164; Bernard Darwin, 'Country Life and Sport', in G. M. Young, op. cit., chapter V.

[3] A. G. Steel and Hon. R. H. Lyttleton, eds., *Cricket*, London, 1888; E. V. Lucas, *The Hambledon Men*, London, 1907; G. Broadribb, *Felix on the Bat*, London, 1963.

[4] Montague Shearman, *Athletics and Football*, London, 1889, esp. part II, chapter IV.

[5] R. K. Webb, *The British Working-Class Reader, 1790–1848*, London, 1955; Richard D. Altick, *The English Common Reader: A Social History of the Mass Reading Public, 1800–1900*, Cambridge, 1957.

[6] Donald Read, *Press and People, 1790–1850, Opinion in Three English Cities*, London, 1961.

[7] E. S. Turner, *Boys will be Boys*, London, 1948.

[8] Thomas Frost, *Forty Years' Recollection*, London, 1880; W. L. G. James, *A Study of the Fiction Directed to the Urban Lower Classes in England, 1830–50*, Oxford D.Phil. thesis, I.T.A., 1960–61, no. 209.

stomachs, though it made some effort to canalize emotional energies into hobbies and sport.

But there was also a serious reading element among the workers, for the excitement of the new world opened by the printed page could be infectious. Durham miners went to school with their own children to learn to read; the Working Men's College in St Pancras, founded in 1854 by F. D. Maurice, succeeded because it was based upon the belief that adult education must be reciprocal between teachers and taught.[1] Efforts were made, especially by the Society for the Diffusion of Useful Knowledge, founded in 1827, the Working Men's Union, started in 1852, and the Pure Literature Society, established in 1855, to make good books available.

The only religious census ever taken, that of 1851, showed that, in general, the worker had very little to do with church or chapel.[2] The Established Church, so long accustomed to its place in a landowning countryside, was slow to respond to the new challenge, and was baffled when it did. The worker, especially if he had read Tom Paine and other secularists, had little interest in a church that seemed to have such little relevance to his situation.

Dissent had rather more effect than did the Church of England. There were notable religious revivals in 1859 and 1865. The Primitive Methodists converted substantial numbers of craftsmen, labourers, agricultural workers, and miners; the Evangelical revival also reached some of the workers through the Baptists, the Calvinistic Methodists, and the Congregationalists. Between the forties and the seventies the Church of England spent over £25 million on church buildings, with other denominations spending on the same scale.[3] Church and chapel could thus affect the lives of some very deeply.[4] The chapel especially could provide working men with opportunities for participation and self-expression within organized groups. But the nation as a whole could not be made much more than nominally Christian.[5] The Irish, the first and greatest of the immigrant groups, brought their own church, which nurtured them and launched a counter-reformation, beginning in the ports.

[1] J. F. C. Harrison, *A History of the Working Men's College, 1854–1954*, London, 1954.
[2] Briggs, *A. of I.*, p. 465; *Census of Great Britain, 1851. Religious Worship in England and Wales*, 1855.
[3] G. Kitson Clark, *The Making of Victorian England*, London, 1962.
[4] K. S. Inglis, *Churches and the Working Classes in Victorian England*, London, 1963; S. H. Mayor, *Organized Religion and English Working-Class Movements, 1850–1914*, Manchester Ph.D. thesis, I.T.A., 1960–61, no. 353.
[5] 'Why Skilled Workmen don't go to Church', *Fraser's Magazine*, 1869.

Out of the travail of industrialization came a new order of worthy men raised in the new urban setting. Because of them it became possible for English society to produce a new view of itself. From them came a love of freedom based upon the hope that the mass of men might cease to be under tutelage in the traditional way, however beneficent this might be, but should become capable both of intelligent thought about society and of respect for the general consensus of its members. These were the men in their stout broadcloths, who looked out from their lithoprint photographs with due *gravitas*, their mild vanity perhaps sustained, as photography spread, by the knowledge that their visages compared not unfavourably with those of their employers. Often these men had a loyalty and pride, confirmed by their regional speech, that could not be erased, making them unmistakably men of Yorkshire, Northumberland or Durham, South Wales, or London.

The effect of the new industrial society upon the countryman, though less dramatic, was profound. He had to come to terms with a countryside that was inexorably changing, in response to demands outside itself. The old sense of being part of nature, and so part of the basic processes of life, was beginning to be damaged. The countryman over countless generations had learned how to get the best results with animals and crops by long and slow adjustment, within a stable and conventionalized situation. Such knowledge, because it arose from direct experience of a particular context – was intensely personal, and consequently was difficult to express except to those who knew the same fields, pastures and woodlands. Against this kind of inarticulate knowledge came the challenge of new methods, the case for which ran in terms, not of a modification of some detail of application, but of a far-reaching change the implications of which would sweep away generations of lore of the old personal and particular kind. In this situation the countryman, so long accustomed to accept man's vulnerability to nature and the consequent need to conform to traditional wisdom, found himself confronted by the need to accept new parameters to life.

In the countryside, as in the towns, the old custodians of the traditional lore lost standing, for the new situation put the premium not on seniority but upon adaptability. New methods of using heavy soils with the help of deep drainage, the conversion from pastoral to arable, the use of new fertilizers, the controlling of the steam plough and other powered machines, were all adjustments of which only the young were capable. The village craftsmen and workers in local industries, approaching in numerical importance the farm workers themselves, were subject to the same eclipse as the factory system learned to supply the tools and furnishings of the countryside, lowering their cost. The young had the advantage over the old, not only in adjustment in agriculture, but in the towns also: the old found it hardest either to change their ways or to move.

But the speed of change must not be exaggerated.[1] Many farms remained small and short of capital so that revolutionary change was impossible. The Suffolk farm horseman continued to take immense pride in his charges, often growing special herbs in his garden for the preparation of highly secret remedies and conditioners. Whatever might be happening to the farm on which he worked, the labourer could continue to apply old skills and judgement to his own small piece of ground and to the management of his income. Poaching could add zest to life and variety to diet.[2] Often the village could produce men of a vision that was both wide and particular – aspiring, as Joseph Ashby did, that men might exercise responsible freedom within the village.[3]

10. THE RESIDUUM

The pauper and the criminal had been objects of experimentation in England at least since the sixteenth century. No really satisfactory treatment had emerged. Now industrializing society was producing unassimilated men and women on an immense scale.[4] They suffered greatly.[5] But a new humanitarianism had developed in the Britain of the nineteenth century, from which the greatest gainers were 'the poorest, the most dependent, and the most defenceless'.[6] This urge to help the casualties of society had to come to terms with human dilapidation in the urban, industrial, society.

[1] See Escott, op. cit., vol. I, chapter XI.

[2] James Hawker, *A Victorian Poacher [his] Journal*, ed. and introduced by Garth Christian, Oxford, 1961.

[3] M. K. Ashby, *Joseph Ashby of Tysoe, 1859–1919*, Cambridge, 1961. For other countrymen sensitive to their surroundings see Montague Weekley ed., *A Memoir of Thomas Bewick written by Himself*, London, 1961; Bernard Jones, ed., *The Poems of William Barnes*, London, 1962.

[4] Henry Mayhew, op. cit., vol. IV, 1862, 'Prostitutes, Thieves, Swindlers, Beggars'.

[5] See the illustrations by Doré in Gustave Doré and Blanchard Jerrold, *London. A Pilgrimage*, London, 1872, chapters XVII and XVIII.

[6] Macaulay, *Works*, vol. I, p. 332.

Men had to be coerced into changing their jobs if the national product was to rise rapidly enough to make possible the support of the mounting population. Some, especially the young and vigorous, might do so voluntarily, in response to the incentives of higher wages. But most had to be subjected to persuasion. This typically took the form of falling wages in obsolescent jobs, or absolute loss of employment. If the goad was to be effective there had to remain no alternative but to seek new employment. Any Poor Law that provided an income, however minimal, would arrest the essential process of adjustment, and, as the authors of the Report of 1834 argued, place the nation in a parlous state, if, while committed as it was to industrialization, it so acted as to forestall the operation of an essential mechanism.[1]

But to the poor the matter looked quite different. Most of them could not adjust to new tasks as the reformers wished – they did not know where to go geographically, they were offered no means of retraining, they had no reserves to cover the transition period, and if they left their houses it was very unlikely that they would find a dwelling place elsewhere. Finally, there was the new instability of the system: prosperity came and went; in bad times jobs were not to be had anywhere. The problem of the Poor Law thus became: how far should the workers be subjected to the discipline of starvation and exposure; how far should compassion be expressed through public policy?

In the countryside, before the enclosure movement culminating during the Napoleonic wars, there had been scope for the poor to scrub along with tiny parcels of land and casual jobs. In addition there was the Poor Law, steadily relaxed after Gilbert's Act of 1782 and especially after the general spread of the system adopted by the Justices of the Peace in the Parish of Speenhamland, in Berkshire, in 1795, whereby relief was related to the cost of living as measured by the price of the quartern loaf, and need took account of all children, legitimate or otherwise. In the same year, by the Settlement and Removal Act, a poor person, formerly removable if thought likely to become a public charge, could only be ejected from the parish if he had actually become chargeable.[2]

As a result all rigour went from the system. The municipal work-

---

[1] *Report of the R.C. for Inquiring into the Administration and Practical Operation of the Poor Laws*, 1834.
[2] Sidney and Beatrice Webb, *English Local Government*, 9 vols, London, 1906–29, vols. VII, VIII, IX, *English Poor Law History*.

houses were in a very bad state by the early thirties. The paupers within them were often dissolute, depraved, and promiscuous, and many of those outside showed an alarming tendency to accept their dependence upon the parish. The parish officers could no longer, as they had done since Tudor times, hurry new arrivals back onto the road as soon as they appeared.

But this easy permissiveness ended with the new Poor Law of 1834.[1] The decision was taken that the nation could no longer permit its poor to be in receipt of incomes without working for them, free of poorhouse discipline. Poverty was to be made painful; the state was to provide no incentives to idleness, but was rather to discipline those at the nether end of the income scale. The principle of 'less eligibility' was the keynote of the Report, whereby no one maintained by the parish was to be as well off as the poorest paid labourer in work. Outdoor relief was to be cut to a minimum, and segregation of the sexes, with consequent breaking up of families, was to be carried out wherever possible. Parishes were to be grouped in 'unions' so that a more viable unit of administration became available, more effectively answerable to the central government, though each parish still remained responsible for its own poor.

The three Commissioners for the enforcement of the Act, with Edwin Chadwick as their energetic and uncompromising Secretary, went about the task of implementing the policy of the Report. They did so with such vigour as to be quickly execrated as the 'three bashaws of Somerset House', their workhouses damned as bastilles.[2]

The new policy was understandable enough. The farming parishes, especially in the south, were indeed producing a frightening pauper population. But many such, if not most, had no alternative but to seek parish aid. The farmers, still suffering depression and in any case wishing, in the larger enclosed fields, to economize on labour, could not employ them. There was, moreover, a struggle between the landlords and farmers whose interest lay in the 'close' parishes, and those concerned with the 'open' ones.[3] In the former the land was owned by a few, making possible, through the control of the supply of cottages, the control of the number of poor likely to become chargeable. If cottages were allowed to fall into decay, and new building was not undertaken,

[1] Maurice Bruce, *The Coming of the Welfare State*, London, 1961, chapter IV.
[2] G. R. W. Baxter, *The Book of Bastilles*, London, 1841.
[3] Clapham, op. cit., vol. I, pp. 467–70.

the parish population diminished, families removed to open parishes, and walked to their work in the close ones, and thus were able to sustain their birth rate. Though some landlords of close parishes took great interest in their workers, the double system was liable to grave abuse, with the labourer the victim. The alternative for the farm labourer was movement to the towns. But many, especially those of mature years, could not face such an upheaval; moreover the generative power of the countryside produced, in many parishes, new lives in excess of the emigration. The result was great suffering.

In the new industrial towns of the north the Poor Law encountered not the relatively passive resentment of the farm labourer, but urban militancy. This disciplinary piece of legislation was bitterly resented by men and women who were without jobs, and who found their pleas to a reformed Parliament for an employment policy left unanswered.[1] Good harvests in the years immediately after the Act made possible an apparent reduction in pauperism, but in 1836 depression returned and with it bitter hostility.

In 1847 with famine and bad trade disturbing the nation, the duties of the Poor Law Commissioners of 1834 passed to the President of the new Poor Law Board, who was an M.P. and often a Cabinet member. A steady succession of inspectors produced invaluable Reports. By this time Poor Law Unions authorized by the 1834 Act covered most of the country, with their new and imposing workhouses. Strict segregation of the sexes had proved impossible because of building costs. The new Board was obliged in 1852 to recognize that resistance in London and in the north had made a flat refusal of outdoor relief impossible. Policy differed between regions; by the seventies refusal of relief was a reality only in some of the rural areas. Yet there was still enough rigour in the system to contribute to the conversion of labourers and their sons from agriculture to other tasks. In addition, the bad times caused a further relaxation in the Settlement Law; by the Act of 1846 those who had resided in a parish for five years without receiving relief acquired a 'settlement' there, that is to say became irremovable.

Though the country was gaining in prosperity in the sixties, the poor remained a serious problem. The cotton famine brought general unemployment in south Lancashire, causing the passage of the Public Works Loans Act of 1863 whereby parish unions and local authorities could raise loans for public works. The housing conditions of much of

[1] See below, Chapter 9, section 6.

the agricultural population, as described by Dr Hunter in his Report to the Lords of the Privy Council in 1864, were very bad. The Union Changeability Act of 1865 sought to improve the Poor Law by making the Unions more effective. The old distinction between the fund for common purposes and the parochial fund, used for relief, was done away with, so that the Union and not the parish administered the aid to paupers. Settlement was to pertain to the Union, so that the pauper was now entitled to relief in any constituent parish. This went some way to obviating the old distinction between open and close parishes, and was strongly resisted by a powerful landed group on this account.

By 1871 the case for bringing the Poor Law, Public Health, and local Government affairs under a single government department provoked an Act assigning to a Local Government Board this triple responsibility. The inspectorate of the Board, together with the Union Poor Law Guardians, launched a campaign against outdoor relief, and, in general, sought to return to the principles of 1834. The number of paupers fell, so that in 1884–86 they reached a new low as a proportion of population, constituting only 2·8 per cent. Though the strict meaning of this figure is difficult to determine, it would seem to suggest real diminution. On the other hand there was a concession to the poor under an Act of 1876, whereby the period of residence required to obtain a settlement was reduced to three years. Thereafter, with the accelerated mobility of persons, the idea of settlement gradually became a nullity, with the poor relieved in the parish in which they found themselves. Though the vagrancy laws still made it possible to award three months' hard labour for rogues and vagabonds, there was a formidable number of mendicants moving about the countryside, begging in the towns, and even mapping out the new suburbs for begging purposes. Thousands of doorways sheltered ''appy dossers', and the Embankment shortly after its completion in the seventies became a national dormitory for the down-and-outs. Prostitution was an ever-present problem.[1]

Yet the idea was still strong in the eighties that there was no way of separating the workers from the worthless, other than by test labour, that is, their reaction to arduous physical work in the stone yard, the parish saw pit, or even the treadmill. Men who showed themselves willing thus reassured the Guardians they were not parasitic, and so

[1] For a bibliography see O. R. McGregor, 'The Social Position of Women in England, 1850–1914', *British Journal of Sociology*, vol. IV., no. 1; Wm. Acton, *Prostitution considered in its Moral, Social and Sanitary Aspects*, 2nd ed., 1870; Wm. Logan, *The Great Social Evil*, London, 1871.

could make a relaxation of rigour possible. For as the desire to behave in a humane way toward the deserving poor grew, so grew the fear of making the nation vulnerable to the crime and social disturbance of the worthless. For the peace of mind of those responsible it was essential to have a test of worthiness; it is impossible not to feel sympathy for them in their bafflement. And though fear of excess population in a general sense had largely passed away, there was very real concern about increase among the least fit, for among the lowest class it was often the case that the more incompetent the parent for work, the larger his family, as if the one reason for his existence was to multiply incapables.[1] Finally, the Poor Law was very difficult to interpret, being in need of codification. It also contained many anomalies; those in receipt of relief in Battersea got 5s. 6d. per week while those in the adjacent union of Rotherhithe received only 2s. 4d.

Between pauperism and crime there was the army and the navy, for a high proportion of recruits, especially to the army, were men not otherwise employable. The effect on character was very bad: in the sixties nearly one-third of the soldiers stationed in the United Kingdom were admitted to hospital, each year, for venereal disease.[2] Army reform in the seventies had among its aims that the nation should no longer depend upon the waifs and strays of the population.

The prisons contained those who were total liabilities to society, men and women who had become wholly parasitic; with them were immured many who were more unfortunate than vicious but who, under such conditions, often changed for the worse. They were in the charge of a prison system that was sunk in promiscuity and squalor, jailers' tyranny and greed, and administrative confusion. Many of the worst were confined in the hulks, the decaying carcasses of ships moored in the Thames adopted as an emergency expedient in 1776.[3] Some were more fortunate in being placed in ships still capable of sea-going, which bore them as deportees to the colonies of Australia, a practice continued as late as 1853. It was ended by the Penal Servitude Act of that year.

The ending of deportation led to two developments: the provision of gaols and the reorganization of prison administration, and the ticket-of-leave system. The enormous new prisons of Millbank and

1 Francis Peek, 'The Workless, the Thriftless, and the Worthless', *C.R.*, 1888.
2 *Report of the Registrar General*, 1875, Supplement, p. lxix; *Report of R.C. on Contagious Diseases*, 1871.
3 W. Branch-Johnson, *The English Prison Hulks*, London, 1957.

Pentonville were built, and eventually a national responsibility for all prisons was accepted in 1877, under the Prison Commissioners. Though the new system brought a new administrative order, it brought also rigour and regulation, with the purging away of elements of tolerance formerly present at least in some places, and the adoption of close regulation, solitary confinement, and the rule of silence.[1] But in the direction of clemency and redemption the experiment was tried of allowing provisional remission of one-sixth to one-third of the sentence, with the prisoner released on ticket-of-leave. The garottings and violence which rendered the London pavements unsafe in quiet streets after or even before sunset were blamed by the frightened upon such men and their premature freedom.

Just as Victorian England, under the impetus of a few devoted men and women, was trying to amend its penal system, changes were going forward also in the criminal law, reducing brutality and severity. In 1800 there were 220 capital offences, in 1837 there were fifteen, in 1870 there were two only, murder and high treason.[2] The use of the pillory was restricted in 1816 and abolished in 1837, with the treadmill often a substitute. Dissection of murderers' bodies, formerly obligatory, was made optional for the judge in 1834, and was abolished in 1861. Hangings as public entertainments ended with the Act of 1868.

Greater compassion entered into the treatment of the insane.[3] But though the Lunatics Act of 1845 required all English counties to provide themselves with asylum accommodation, the lunatic often continued to share the poorhouse with the other inmates. The national police system was greatly strengthened. Peel's new force in London formed in 1830 was very effective; the result was a migration of criminals to other cities, with the consequent need to establish real forces over the country.

But in spite of prison reform, the development of new criminal law, and the institution of a national police system, crime, far from being purged from industrial society, became a confirmed and permanent problem. As general wealth increased, the element of decay at the lower end of society became more obvious. Crime for many had

---

[1] D. L. Howard, *The English Prisons*, London, 1960; Gordon Rose, *The Struggle for Penal Reform*, London, 1961; Henry Mayhew and John Binney, *The Criminal Prisons of London*, London, 1862.

[2] A. B. Dicey, op. cit., p. 29; Elizabeth Orman Tuttle, *The Crusade against Capital Punishment in Great Britain*, London, 1961.

[3] Kathleen Jones, *Lunacy, Law and Conscience, 1744–1845: The Social History of the Care of the Insane*, London, 1955; *Mental Health and Social Policy, 1845–1959*, London, 1960.

become a kind of hereditary occupation; something approaching a criminal caste had appeared in all the large cities by the eighties, involving society in a heavy burden, and creating a sense of insecurity. Retribution rather than the reform of the criminal remained the basis of policy.

CHAPTER EIGHT

The Other Orders

I. DIVISIONS AND DISTINCTIONS

THE potency of individualism in the life of nineteenth-century England was in no way inconsistent with a highly developed sense of class. Indeed the two tended to go together, for the radical changes of industrialization, stemming from individual initiatives, had the effect of altering the social structure, making men aware in a manner heretofore unknown of their relationship with their fellows. In agrarian society the gap between buyers and sellers of labour had been great indeed, with the result that class consciousness was much less developed. It was when social mobility increased, when questions of borderline status were raised, that definition and conventional tests became necessary. 'The truth is,' said *The Economist* in 1857, 'that the sense of social inequality cannot be completely felt without the amount of common culture which is needful to give a keen feeling of where the difference begins.'[1] It was the men and women of the industrializing class who most needed to discover an attitude toward themselves and others; only slightly less involved in this matter were those among the workers who sought to improve their position. Individuals needed a system of categories, first, whereby they might conceive their place in society, and secondly as a means of establishing a degree of solidarity sufficient for political action. But this did not necessarily mean a progressive confirming of mutually hostile classes; rather was there a good deal of assimilation, especially at the margins.[2]

The members of each of the principal groups composing society felt the need to form something of a corporate view of themselves, from which self-idealization was not always absent; at the same time within each group there was developed a concept of each of the other groups with which it stood in juxtaposition. All this meant that class

[1] 'Class Localities and Local Self-Government', *Econ.*, 20 June 1857, p. 669.
[2] For a general discussion, see G. D. H. Cole, *Studies in Class Structure*, London, 1955.

attitudes were a highly involved affair. Until very recently this complexity has been obscured by the emphasis placed upon the view held by the middle class, both of itself, and of the rest.[1] This situation arose because of the economic ascendancy of the middle orders of society, together with their greater articulateness.

The degree of stability present in society depended in no small measure upon the working of this process of mutual categorization. Class solidarity might produce a drive for change, or, quite the reverse, might act as a promoter of social quietism. Bagehot, in his fears for the stability of England, found comfort in contemplating the tyranny of the commonplace, whereby each man was kept docile by the urge to conform. But it was against this urge to seek security in conformity that Samuel Smiles and John Stuart Mill fought most bitterly, perceiving in it the death of personality and culture.

2. THE LANDED ELITE

The upper class in Britain, in spite of some complexity, was the most homogeneous and easily understood group. The essential test was the possession of a title or at least a coat of arms, accompanied by substantial land-holdings over several generations. The possessors of these were the heirs of the pre-industrial rent receivers, and their fate was linked with the prosperity of agriculture.

The landed men had their difficulties, along with other elements in the economy, down to the fifties.[2] They suffered heavily in the immediate post-war years when agricultural prices fell. The air was full of their complaints whenever the question of repealing the Corn Laws was raised. Many were still suffering from the spending excesses of their predecessors, for in Regency times the cult of heavy and ostentatious outlay had made serious inroads into many estates. Then for some twenty years rents were well sustained by the rising demand for foodstuffs, and efficient husbandry added to the gains. Indeed there are signs that improvements in agricultural efficiency were going forward briskly well before 1846. The landed upper class, though threatened in their dominance by new trading and commercial groups, enjoyed rising incomes along with everyone else, so that, though their proportionate share of the increasing national product was probably diminishing, its

---

[1] See Asa Briggs, 'Middle-Class Consciousness in English Politics, 1780–1846', *P. and P.*, 1956.
[2] See above, Chapter 5, pp. 180, 181.

absolute size was certainly large enough for them to maintain their way of life, and even to render it more splendid.

There were many unconvenanted benefits that fell to those owners of landed estates who had the initiative to reap them. Collieries and railways could yield large returns. Robert Stephenson told the Institution of Civil Engineers in 1856 that 'of the £286,000,000 of railway capital expended, it is believed that nearly one-fourth has been paid for land and conveyancing'.[1] In addition the value of agricultural land had improved rather than deteriorated. Coal, in particular, meant great additions to the incomes of some families, especially in Durham, Northumberland, south Yorkshire, Cumberland, Lancashire and south Staffordshire. The Marchioness of Londonderry, only child of Sir Henry Vane Tempest, inherited enormous property in County Durham, studded with collieries.[2] Her husband, after a military career of some note under Wellington, joined his wife in the development of her inheritance, opening new pits, building a railway, and founding a thriving seaport at Seaham. The Marquis was a true entrepreneur, but he was also thoroughly feudal in his dealings with other men. The Duke of Portland owned coal near Kilmarnock and built the port of Troon. The home of the Earl of Durham was surrounded and even undermined by his coal workings. The Duke of Northumberland, the Bishop of Durham (succeeded by the Ecclesiastical Commissioners), Earl Fitzwilliam, and the Duke of Norfolk, were other great magnates of the coal fields. On the whole such men did not exploit the coal directly, but leased the right to do so through the sale of royalties.

The increase in urban land values made further contributions to the wealth of the upper class.[3] In London the Dukes of Bedford, Portland, Norfolk, and Westminster, and others of the nobility, made great gains in this way. The Marquis of Bute owned the land upon which Cardiff Docks were built. In Liverpool the Marquis of Salisbury, and the Earls of Derby and Sefton, were large beneficiaries. In Sheffield it was the Duke of Norfolk; in Birmingham the Calthorpes, and in many other towns various branches of noble families. In general the system was to lease land for the construction of houses and buildings, for a period of years (100 years or less), in return for an annual ground rent. At the end of this term both land and buildings became the

[1] *Presidential Address, Minutes of Proceedings of the Institution of Civil Engineers, Session 1855–6*, London, 1856, p. 139. This is an excellent survey of railway problems.
[2] See above, Chapter 4, section 8, p. 162.
[3] David Spring, 'The English Landed Estate in the Age of Coal and Iron, 1830–1880', *J. Econ. H.*, 1951, p. 9.

property of the owner of the freehold, who could then charge a fine for the renewal of the lease.

But though the landed class could levy on the new wealth-creating process through their inherited properties, they were not wholly passive. A great nobleman might become the patron of considerable developments in building, harbour construction, mining, and industry at financial risk to himself.[1] The great gains that followed upon possession had to be placed somehow, so that a considerable addition to capital available for investment occurred.

Some members of the nobility could also gain by making their resplendent names available for company prospectuses. There was criticism of the men of business as early as the fifties for this sponsoring practice, both because it revealed a feeble worship of aristocracy, and because it gave an illicit air to the project concerned: 'The aristocratic puff is tried when no other puff would have much chance of success.'[2]

There are many signs that, from the accession of Victoria down to the eighties, the landed interest, in reaction to the loose spending of earlier generations, some perhaps affected by the Evangelical ethic, and by the example of royal respectability, sought to redeem the debts of the past, and to place their estates on a sound and self-perpetuating basis. In spite of the system of entail, not a few estates had reached the point at which large sales appeared to be necessary in order to avoid general collapse, including the great patrimonies of the Duke of Devonshire, the Duke of Buckingham, and Lord Sefton. Out of this situation came the realization that a change in the handling of great landed estates was imperative if the position of the class possessing them was to be maintained. By means of windfalls and retrenchment the aristocratic families redeemed a situation which by Victoria's accession in 1837 had become dangerous, and were thus to continue their hold upon government and the affairs of the nation.

The view the nobility held of themselves was not easy for others to determine, for they were not the kind to produce public apologies. The matter might arise when a sale of land for the settlement of debts was being considered, and one nobleman wrote to another for advice, or more generally, when the Radical opponents of the House of Lords raised their voices, provoking noble protests. A new exclusiveness had

---

[1] See Sidney Pollard, 'Barrow-in-Furness and the Seventh Duke of Devonshire', *Econ. H.R.*, 1955.

[2] 'The Want of Self-Respect in the Commercial Classes', *Econ.*, 19 December 1857, p. 1399.

been adopted during and after the French Revolution, and by the fifties it appeared as though this divorce from the rest of the population was becoming more pronounced.[1] There was a good deal of bewilderment among the lords as, with the progress of industrialization, the new kind of city poor, with its political volatility, increased.[2] The traditional mode of issuing charity at the door of the great houses to a docile peasantry, was now rapidly passing, to the puzzlement and concern of many a nobleman and landed gentleman. Some, like Sir Thomas Dyke Acland in the west country, sought not only to relieve immediate distress with soup dispensed from the manor-house boiler, but to promote betterment through rural education, cottage building, and a general defence of the countryside against uncontrolled urban development.[3]

Perhaps more important than the view held by the nobility of themselves was the light in which they were regarded by the other elements of society. There can be no doubt that the old mystique of aristocracy was very strong in Britain throughout the nineteenth century, in spite of outcries in the thirties over Parliamentary reform, and again in the fifties after the Crimean debacle. Many men, even among the groups who were most active in changing society, felt that stability, order, and status among the nations, were still only capable of preservation under a landed nobility.

With this belief often went a burning desire for recognition by the upper order of society. Thackeray in his *Book of Snobs* remarked: 'What Peerage worship there is all through this free country!'[4] Taine, in the sixties, quoted one of the greatest industrialists in England as saying: 'For we believe, we men of the middle class, that the conduct of national business calls for special men, men born and bred to the work for generations, and enjoy an independent and commanding situation.'[5] This attitude, about the same time, enraged the philosopher, T. H. Green. Corruption, by which he meant the holding out of the possibility of acceptance by the landed oligarchy, he wrote, 'is eating out the heart of the upper commercial classes ... flunkeyism ... pervades

[1] A. S. Turberville, *The House of Lords in the Age of Reform*, 1784–1837, London, 1958, p. 373.

[2] Ibid., p. 390.

[3] A. H. D. Acland, ed., *Memoir and Letters of the Rt. Hon. Sir Thomas Dyke Acland*, London, 1902.

[4] William Makepeace Thackeray, *The Book of Snobs; Sketches of Life and Character*, London, 1872; Turberville, op. cit., p. 395.

[5] Hippolyte Taine, op. cit., p. 155; see also 'The Advantage to a Commercial Country of a Non-Commercial Government', *Econ.*, 4 January 1862, p. 3.

English society from top to bottom'.[1] Though the radicals might fulminate against the lords, stressing their inequities and inadequacies, they could make little real impression upon these middle-class aspirants. Nor were those of the working class who bothered about the matter at all far behind the middle class in this. Indeed the aspirant to working-class respectability might well be susceptible to the idea of a patriarchal class, remoter and more exalted than the middle class whose failings he knew.

The roots of admiration for aristocracy ran very deep, drawing support from all manner of folk concepts of which the holders were only vaguely aware. There was the idea of permanence, in terms of the noble family; its members were thought to be motivated by a sense not only of their own honour, but that of their ancestors and descendants. The income and social position of the lord was so secure that he could and would judge the issues between other men solely upon their merits. Moreover the aristocrat and the substantial member of the gentry were believed to be in intimate contact with the countryside, a point of great importance in a society in which urbanization was not fully accepted emotionally.[2] The view was held, not without some justification, that other great European powers were ruled by their ancient landed aristocracy, and that Britain, in dealing with such societies, had better have representatives of a like kind. In the case of India the idea of the aristocratic pro-consul was very strong; some of these noble rulers of the sub-continent, like Dalhousie, played an important part in Indian economic development. There was also the idea that the aristocracy would continue as arbiters of taste. Many of these attitudes reached the workers through the large servant class, recruited from their number, and often returning to it with grand tales of the magnificoes of the great houses.

Right down to and perhaps beyond the Crimean War, in the fifties, to the days of Bismarck, war was an amateur game, a matter for lords and gentlemen, not very different, though on a grander scale, from the duel.[3] Men and their spirit meant more than arms and technology. This was the philosophy behind the Duke of Wellington's persistent support for the purchase of commissions – only under such an arrangement

[1] R. L. Nettleship, *Memoir of T. H. Green*, London, 1906, p. 168.
[2] For one squire's view of his role see Owen Chadwick, *Victorian Miniature*, London, 1960.
[3] The standard army history is J. W. Fortescue, *A History of the British Army*, London, 1899-1930; for an archetypal high Victorian soldier see Michael Alexander, *The True Blue: The Life and Adventures of Col. Fred Burnaby, 1842-85*, London, 1958.

would the right order of men officer the country's forces. When, in 1858, after the end of Company rule in India, a new army was established, Lord Ellenborough fiercely opposed the admission of sons of traders to cadetships.[1] Moreover, at the other end of the scale, the recruitment of the ranks from the waifs and strays of society, with discipline maintained (until 1868) by flogging, did not disturb men of feudal mind, for an impassable gulf between officers and men seemed the best arrangement for a fighting force, making it possible for officers to avoid emotional entanglement with the lot of the men they commanded. Superior diet and living conditions among the wealthy made for a striking physical contrast, causing officers to be conspicuous among their men. These sharp differences between classes too often meant that officers with little knowledge of war knew even less of their men. 'We call Lucan the cautious ass and Cardigan the dangerous ass' wrote an embittered subordinate of these military noblemen after service under them in the Crimea.

All this suffered a shock in the Crimea, provoking a wave of public anger against the aristocracy who had failed to meet this traditional responsibility, and condemning a profession of leisured command over an inferior social class.[2] For the new technology had ushered in the age of mass conscripted armies on the Continent. In England this reversal of the democratic trend was impossible, leaving efficiency, skill, and science as the alternative. Even so, the filth and near-starvation typical of Sandhurst continued until 1862; it required a mutiny in that year to bring improvement.[3] The old aristocratic amateur attitude to war appears in a more favourable light when, as in the fifties the production of a poison gas shell was discussed, the military opposed the cold logic of the scientists with the archaic analogy that such a weapon was like poisoning the wells of the enemy.[4]

Edward Cardwell instituted his army reforms in the late sixties and early seventies, in the face of the German threat, and in the teeth of the resistance of the traditionalists at home.[5] His Army Regulation Act of 1871 rested on the view that the pre-industrial attitude to warfare must be seriously altered. The navy could resist longer: as late as 1893 an

---

[1] 'Aristocratic Views of Trade', *Econ.*, 7 August 1858, p. 867.

[2] Cecil Woodham-Smith, *The Reason Why*, London 1953; Christopher Hibbert, *The Destruction of Lord Raglan*, London, 1961.

[3] Hugh Thomas, *The Story of Sandhurst*, London, 1961.

[4] Wemyss Reid, op. cit., p. 160.

[5] Brian Bond, 'Recruiting the Victorian Army', *V.S.*, 1962; Arvel B. Erickson, *Edward T. Cardwell: Peelite*, Philadelphia, 1959.

Admiral on manœuvres could insist upon an order the inevitable consequence of which was a collision costing 358 lives.[1] This could occur, in spite of the opinion of Lord Overstone, one of the greatest of City men, expressed as long ago as 1860, that Britain was now in so sophisticated a condition in her financial, industrial, and trading life, that a sustained defence against invaders of British soil, once they had achieved a bridgehead, was almost impossible.[2]

There were, of course, plenty of other grounds for misgiving by the middle and lower orders of society about their aristocratic superiors. During the American Civil War there was a good deal of upper class support for the slave-owning South; the workers, in spite of the sufferings of the cotton famine, largely supported the North. The penny-dreadful novelettes of mid-Victorian times were full of profligate peers and debauched and seducing scions against whom there was no redress.[3] Drunkenness and dissipation in the upper reaches of society had been very bad before the Queen's accession, and even the austerity of the Royal couple could not put it down altogether. But the courage and good sense of Prince Albert made it possible for the monarchy to assert itself against the aristocracy in matters of morals. There was, moreover, a very powerful tendency to cruelty among many of the upper class: brutality in public schools was commonplace, and the power urges of judges, senior sailors, soldiers, and even highly placed churchmen, were apparent in their approval of violent corporal punishment. Rudeness, and the many refinements of the snub, were developed to an art, and permeated downward through society. Among concrete grievances held against the nobility were the game laws, for the coverts and deer forests of England and Scotland were a monopoly of the landed. Game, gamekeepers, and game hunters had the run of a high proportion of farms, to the grave annoyance of the tenants. By the sixties poaching assaults and murderous affrays were frequent as the countryfolk conducted their guerilla warfare against the hated gamekeepers.

On the other hand, there were circumstances by which these aspects of nobility could be mitigated, especially in the eyes of the working class. Drunkenness was by no means confined to the upper orders, and many a working man who sought relief from boredom might have

---

[1] Richard Hough, *Admirals in Collision*, London, 1959.
[2] *Report of the Commissioners Appointed to Consider the Defences of the United Kingdom*, London, 1860, pp. 90–2.
[3] Margaret Dalziel, *Popular Fiction 100 Years Ago*, London, 1957.

sympathy with fellow escapists. Nor was brutality a matter which greatly horrified even its victims. It was part of the aristocratic ideal that men with authority should not be stultified by qualms about using it. Even the effect of the game laws could be compensated, at least in part, by the bond of sport, bringing aristocracy and commonalty together: race meetings, greyhound racing, cricket matches, cock-fights, and prize fighting.[1] The barbarous practice of releasing pigeons to be shot at provided both sport and spectacle for ladies and gentlemen, and for humbler watchers. For well past the mid-century most men and women at all levels of society knew horses as aids in work or play, and though the application of steam to work and transport meant the slow dissolving of this social bond, it was still strong at the end of the century. There was, in fact, in spite of aristocratic misgivings about the urbanized proletariat, something of an affinity between the lower portion of society and the upper.[2]

The aristocratic ignorance of the business world was, by the late sixties, being noted, for though noble fortunes might have substantial industrial ingredients, their possessors had, with a few exceptions, little real acquaintance with the way in which commerce and industry operated. Vaguely remembered classical education was no longer enough in the midst of problems calling for 'close figures and modern detail'.[3] The lower orders in the seventies applauded Arthur Orton, the butcher from Wapping, who challenged aristocracy and law in his extraordinary imposture as the heir to the Tichborne fortune.[4] Yet Lord Salisbury, with his reputation as an absolutely honest and trustworthy man, ruled the newly enfranchised England for twelve of the last fifteen years of the century.

Signs of change in the economic position of the nobility became increasingly apparent by the eighties. The new cheap cereals from Canada and the United States had seriously damaged rents. Coal incomes lagged as industrial recovery was delayed. The decline of economic predominance, hidden during the golden age of agriculture, was now obvious, though the possession of land was more concentrated that ever.[5] There were signs also of a revival of aristocratic extravagance. The old tactic of marriage to commercial and industrial heiresses

---

[1] 'Thormanby', *Boxers and their Battles*, London, 1900.

[2] T. H. S. Escott, *Personal Forces of the Period*, London, 1898, p. 173.

[3] 'The Legislation of the Next Parliament', *Econ.*, 15 August 1868, p. 926.

[4] Douglas Woodruff, *The Tichborne Claimant*, London, 1957; Michael Gilbert, *The Claimant*, London, 1957.

[5] See above, Chapter 5, p. 184.

provided some relief; to the native supply there were added by the eighties young ladies from America, anxious to become titled so that the basis of social exclusiveness in the United States, by then seriously impaired, might be renewed.[1]

The long resisted intrusion by the middle class in their own right into the nobility was rapidly accelerating.[2] Alexander Baring, financier, became Baron Ashburton in 1835. Edward Strutt, millowner, the first of the industrial peers, had entered the Lords in 1856.[3] In 1885 came the City peerage for Nathaniel de Rothschild, after several refusals by Victoria; at the same time the current head of the rival house of Baring received the same honour. The son of the great Brassey was ennobled in 1886. The 'Beer Barons' made their entry in the eighties and early nineties: the Guinness brothers, Arthur and Edward, respectively in 1880 and 1891, Sir Henry Allsopp and Sir Michael Bass, the titans of Burton-on-Trent, both in 1886. The doors were thus opened for a new concept of nobility.[4] 'For in England', wrote Bryce 'great wealth, skilfully employed, can, by using appropriate methods, practically buy rank from those who bestow it.'[5]

But a title did not mean full social acceptance. There existed, in fact, two levels of peerage. The older aristocrats were often in closer touch with the countryside and its people, and usually took the land with its crops and animals as the fundamental background of their lives. For the new peers, their landed estates represented their surplus assets, the bulk remaining, together with the real focus of their lives, in industry and the towns. The new men might indeed become very efficient owners, but neither this nor intermarriage could dissolve the differences quickly.[6]

In the House of Commons the landed upper class had encountered the rivalry of the new plutocracy a good deal earlier. The Reform of 1832 did not open the floodgates, as had been feared, but certainly the dominance of the landed members began to be challenged. After 1867 the process was accelerated, though even then the landed element in the House remained powerful. The real break-up came in the eighties,

---

[1] Lady Elizabeth Eliot, *They All Married Well*, London, 1960.

[2] This and many other matters are illuminated by family articles and appendices in Hon. Vicary Gibbs, *et al.*, eds., *The Complete Peerage*, 12 volumes, London, 1910 to 1959.

[3] Ralph E. Pumphrey, 'The Introduction of Industrialists into the British Peerage; A Study in Adaptation of a Social Institution', *A.H.R.*, 1959.

[4] See H. J. Hanham, 'The Sale of Honours in Late Victorian England', *V.S.*, 1960.

[5] James Bryce, *The American Commonwealth*, London, 1889, p. 604.

[6] L. G. Pine, *The Story of the Peerage*, Edinburgh, 1956, p. 224.

signalled by the Third Reform Act of 1884, enfranchising the farm labourer; and the County Councils Act of 1888, putting an end to the Quarter Sessions, the traditional means of the landed interest for governing the countryside, and instituting elected councils.[1]

There were further grounds upon which the role of aristocracy was questioned in the eighties. The ideas of Henry George about the taxation of land values were making a strong impact in the United Kingdom, and radicals could make much play with the heavy concentration of land ownership in noble hands.[2] With the advance of industry and the passing of social and political initiative to other groups, the case for so heavily endowed an aristocratic class was rapidly growing weaker. Yet it would seem that not even among the last brilliant flowering of the patricians in the eighties, with Salisbury Prime Minister, and Balfour, Randolph Churchill, and Rosebery the coming men, was there any real anticipation of the approaching eclipse of their class.[3]

3. THE MIDDLE GROUND

The difference between the English and the continental aristocracies was to many people in England a point of self-conscious pride – how that the former had always had connections with other orders of society whereas the latter was a closed caste, a true *noblesse*. Just as the upper reaches of the middle class in England could lap over into aristocracy, so too within the middle class itself there was no rigid distinction. Indeed the heir to a dukedom essaying a marriage with a lady high in the middle class might discover that her relations ran a frightening way down the scale. But there could be plenty of bad feeling when the pace of social preferment appeared to be hurried; this was discovered by both the spouses of the railway giants, Mrs Hudson and Mrs Brassey, as they were conveyed by their husbands without preparation or intermediate halts to the upper reaches of society.[4] For such rapid ascents could do damage to the extraordinarily refined distinctions that had come to operate at the margin of each social group. Precedence at an ordinary London dinner-table had become very complex and

---

[1] For the cost side see W. B. Gwyn, *Financial Aspects of a Parliamentary Career in Aristocratic and Democratic Britain: a Study of the Expense of Membership in the House of Commons, with Special Reference to the Nineteenth Century*, London, Ph.D. thesis, I.T.A., 1955–56, no. 661.

[2] See below, Chapter 10, section 10, p. 430.

[3] Ensor, op. cit., p. 71.

[4] A Foreign Resident, *Society in London*, London, 1885, p. 238.

very rigid by the eighties, so that ludicrous and sometimes insoluble problems might arise in getting the company into the posture for eating. Indeed one contemporary observer found the principal explanation of the reserve of Englishmen, proverbial by the eighties but new in the twenties, in the social pitfalls thus created.[1]

This curious system whereby society was open, yet divided by a great number of delicate though fiercely guarded barriers, had roots running back to the later Middle Ages when the new entrepreneurship in trade and commerce began to appear, and the charms of the alderman's daughter were enhanced by her dowry. In Britain, in contrast to the Continent, the system of primogeniture had long operated to confine the family inheritance to the eldest son, obliging all the others to seek their own fortunes, thus pushing them not only into the professions, but into commerce as well.[2] This had many curious side effects. Such a son kept for long periods in strict subordination to his father's will might well look forward without excessive grief to being orphaned.[3] He would then suddenly find himself in charge of the entire expendable portion of the family fortune, and so with power over all its indigent members. If the heir was sensible all might be well, but if not, like the frantic Duke of Buckingham, he might involve even the wealthiest family in great difficulties. The mania for establishing a line through an eldest son proved contagious, as newly arrived men of business sought to perpetuate their names.

The nineteenth century saw the elaboration of a system of raising children that went back at least to Tudor times. It was now to serve not merely as a means of unifying the élite under a common outlook, but also of allowing new recruitment and the spread of the system downward. The upper classes, in fact, largely delegated the upbringing of their children, putting them in the care of others, rather like Plato's guardians. The nursery was the centre of a system approaching foster-parentage, under which the children were raised not by the wood-cutter and his wife in the heart of the forest as in the story books, but by the family servants in another part of the house. It is hardly surprising that the nursery could be a place where weaklings fared ill, with swaddling, the professional wet-nurse, heavy diet, inappropriate clothes, insensitive discipline and purgative medicaments.[4] Parents were on

---

[1] Escott, *England*, vol. II, pp. 28, 29.
[2] See above, Chapter 5, p. 178.
[3] George Ramsay, *An Essay on the Distribution of Wealth*, Edinburgh, 1836, p. 429.
[4] Magdalen King-Hall, *The Story of the Nursery*, London, 1958; Frederic Gordon Roe, *The Victorian Child*, London, 1959; Marion Lochhead, *Young Victorians*, London, 1959.

visiting terms with their children, a relationship that could produce uncritical idealization as the respect conveyed by loyal servants went uncorrected by more intimate contact. The heirs of wealth and power were thus in a state of semi-isolation, a condition that continued at the subsequent stages of prep school, public school, and university, punctuated by visits home that could be almost idyllic, with parents seldom caught off guard or challenged to come to terms with their children under conditions of stress or *ennui*.

So close could such early contacts with gardeners, stable boys, and the rest of the outdoor servants become that the young scion throughout life could enjoy the comfortable feeling of being intimate with the lower orders, and understanding them. Governesses aimed at preparation for school. There a code of attitudes, with strong institutional custodians, was available.[1] The public schools, like the nurseries, dated from earlier generations, but with Thomas Arnold's great work a new synthesis was available, authoritatively dispensed. Through Arnold, in a sense, the middle class took the educational initiative, for he, at the town Grammar School of Rugby from 1828 to 1842, provided boarding education for the sons of the higher professionals and of the larger business men, resting his system upon a concept of the Christian gentleman, of reasonable cultural sensitivity, with a high moral code and a developed sense of social responsibility.[2] This last was induced by entrusting much of the school discipline to prefects. Gradually the old barbarities of Winchester, Harrow, and Eton were diminished as teaching ceased to be a mixture of indifference and arbitrary action, and teachers sought to aid their pupils positively rather than by relying upon the formation of strong character as the only means of survival. Aristocrats and plutocrats could thus mix and merge in the post-Arnoldian public schools, and so in after life.[3] 'Look at the bottle-merchant's son and the Plantagenet being brought up side by side', said Matthew Arnold in an imaginary conversation with a Prussian visitor to Eton in the seventies. 'None of your absurd separations and seventy-two quarterings here. Very likely young Bottles will end up by being a lord himself.'[4] But the 'godliness and good learning' of

[1] See Vivian Ogilvie, *The English Public School*, London, 1957. For provocative discussions see Brian Simon, *Studies in the History of Education, 1780–1870*, London, 1960; G. F. Lamb, *The Happiest Years*, London, 1959.

[2] T. W. Bamford, *Thomas Arnold*, London, 1960; David Newsome, *Godliness and Good Learning*, London, 1961.

[3] See 'Public Schools', *C.R.*, 1867, p. 160, for an indication of cost.

[4] Matthew Arnold, *Friendship's Garland*, London, 1871, p. 28.

Arnold's ideal were challenged in the seventies and eighties by the new and more extroverted combination of 'godliness and manliness', with Charles Kingsley as the leading exponent of 'muscular Christianity'. Attempts to use public inquiries or legislation to bring in general reform of the public schools dissolved in the innocuous Act of 1868.[1]

Earlier in the century the wealthier Nonconformists had maintained their separate academies, or had sent their children to the Scottish universities, but from the mid-century there was an increasing tendency for the two elements to come to terms within the same system. Lesser Etons multiplied to meet this need as the middle class grew.

University life continued the process, but in the direction of relaxing supervision, leaving the young man to be tested by his own devices, rather than subjecting him to a final course of training that capped in vigour and comprehension his former education. The new cult of sport was an obsession at the universities by the sixties, with elaborate rites and conventions for the men of brawn; the debates at the Unions allowed prematurely portly youngsters to anticipate the graces of the Chambers of Westminster.[2] The main function of the universities, in short, was social, not intellectual, permitting graduation by the pass degree, an almost derisory requirement.

On the whole those to whom the upper classes entrusted the forming of their children served them well, both in the immediate task and, almost as a by-product, in providing for a somewhat extroverted and inarticulate aristocracy a satisfying concept of itself. There could be further mingling and merging in later life in the privacy of clubs where the loyalties and ethos of school could be renewed.[3]

But there were signs that as the universities began to rediscover a positive identity they were rather less appropriate as the crown of the aristocratic and upper middle-class child-raising process.[4] The University of London, founded in 1826, provided alternative criteria of higher education, unfettered by the dogmatic principles and assumptions that had been so closely woven into the older universities as to have become imperceptible to their products.[5] Pusey, Newman, and others of the Oxford Movement in the thirties, began the long process of disturbing the easy consciences upon which the system rested;

[1] See above, Chapter 7, section 8, p. 258, n. 3.

[2] Percy Cradock, *Recollections of the Cambridge Union, 1815–1939*, Cambridge, 1953.

[3] Charles Petrie, *The Carlton Club*, London, 1955.

[4] J. P. C. Roach, 'Victorian Universities and the National Intelligentsia', *V.S.*, 1959.

[5] F. G. Brook, *The University of London, 1820–1860, with Special Reference to its Influence on the Development of Higher Education*, London, Ph.D. thesis, I.T.A., 1957–58, no. 524.

Matthew Arnold continued to do so in the sixties, and Pater was leading a movement for aesthetic appreciation in the seventies. The thoughts of all of these men were likely to cause disturbance of the erstwhile healthily extroverted states of mind of young undergraduates. From the great headmasters of the public schools, Henry Montagu Butler (1833–1918), John Percival (1834–1918), E. W. Benson (1829–96), and Edward Thring (1821–87), came new ideas and new pressure upon the universities.[1] The ending of religious tests in 1871, the admission of women to degrees in 1878 by the University of London, followed by Oxford and Cambridge in the eighties, and the removal of the celibacy requirement upon fellows of colleges in 1882, meant that the universities were altering themselves to meet new needs. Mark Pattison and others called for university leadership in research in the new disciplines and in the renovation of the old; Jowett on the other hand sought to reinvigorate the teaching of the universities.[2]

Yet in spite of incipient elements of disruption, the system was still capable in the eighties of producing men of immense self-assurance, able to confront anyone in the world, however high or low, with the steady gaze to which all English boys were to aspire. The type of the upper-class English gentleman was firmly established – a man robust, conservative, somewhat opaque in intellect, closely bound by a code that took no account of those who were afflicted by internal uncertainties or flights of emotion. Those who could gain entry to the system, however dubious the status of their forebears, could acquire the same outlook, at least to some degree, with all its power and limitations. Not least in importance came the assurance to deal with the continuously elaborating *punctilio* of society. There were, however, dangers. The aristocratic proclivity to use others and then discard them may have been strengthened by such conditioning. The highly peculiar background of boyhood could hardly fail to induce a superbly unconscious arrogance and insensitivity toward servants and other inferiors.[3]

Relatively few who had come from trade or industry reverted to it for their adult career, except perhaps among the merchant bankers. Such middle-class men were now ready to enter the esteemed pro-

[1] G. R. Parkin, *Life of Edward Thring*, London, 1898.

[2] Mark Pattison, *Suggestions on Academical Organization*, Edinburgh, 1868. Sir Geoffrey Faber, *Jowett, A Portrait with Background*, London, 1957.

[3] See Alice Fairfax-Lucy, *Charlecote and the Lucys*, Oxford, 1958, where the 'fun' of domestic service is described.

fessions: the church, the law, the universities, the army or navy, or the higher civil service.[1] Where there was an insufficiency of such posts at home, they might be found in the Empire, especially in the Indian Civil Service or judiciary, providing a professional class to alien societies. Where a profession might seem too binding, such young men might join the ranks of leisure, either as ornaments to grace the London season, or as cultured amateurs willing to play their part in politics, local government, and the magistracy, or in the various attempts at improving the lives of the less fortunate.

But though aristocrats and plutocrats could find much common ground in the upper reaches of education, it must not be thought that all differences were smoothed away or that all members of the higher bourgeoisie thus lost that distinctive identity that had been so conspicuous in the early years of northern industrialism. John Cobden and Bright called, not without effect, upon their class to resist the urge to aristocratic pretensions.[2] 'The insatiable love of caste that in England, as in Hindostan, devours all hearts', wrote Cobden with some exaggeration, 'is confined to no walks of society, but pervades every degree, from the highest to the lowest.'[3] For many it was religion that stopped assimilation. The old Nonconformity that had given identity and security to so many rising families was not so easily dissolved. Very often it had provided the framework of family life, with the chapel, the Christian Sunday, the Bible, and family prayers. Evangelicalism and Methodism had breathed new life into dissent in the early years of the century. Some of the more respectable Methodists came to terms with the Established Church, but the Oxford Movement, in seeking to renew the soul of the Church of England, was vehement against the Methodists, pushing most of them into a confirmed dissent. Soon they were the largest of such groups. Many of the great industrial families of the north clung to Nonconformity throughout the century. It was from it that the erstwhile screw manufacturer, Joseph Chamberlain, came (retired from business with a comfortable fortune, aged thirty-eight) to demonstrate a new radicalism in civic politics, and to assert in

[1] For the consequent shortage of employment in such careers see F. Musgrove, 'Middle Class Education and Employment in the Nineteenth Century', *Econ. H.R.*, 1959; for debate see H. J. Perkins, 'A Critical Note', *Econ. H.R.*, 1961; Musgrove, 'A Rejoinder', *Econ. H.R.*, 1961. For the mingling of such families see N. G. Annan, 'The Intellectual Aristocracy', in J. H. Plumb, ed., *Studies in Social History*, London, 1955, p. 243.

[2] J. L. Sanford and Meredith Townsend, *The Great Governing Families of England*, Edinburgh and London, 1865, p. 6.

[3] *The Political Writings of Richard Cobden*, London, 1867, vol. I, p. 131.

pointed fashion radical hostility to the landed men and their institutions. His ally John Morley, son of cotton manufacturers on one side and of shipowners on the other, had much the same background. It was probably among the lesser families that radical zeal was greatest, for the prospect of assimilation to the culture forming in higher society was sufficiently remote as to have no great effect upon their minds. So was strengthened the famous Nonconformist conscience, formed in earlier generations and still nurtured by its great ministers, standing for private and public integrity, a court of opinion separate and distinct from the Established Church, to which the great issues of the day must be referred, the sensibilities of which no politician could afford to ignore, and upon which many learned to play.[1]

4. THE MIDDLE ORDERS

In retrospect and to many contemporaries the emergent middle orders were the hero-villains of the nineteenth-century scene. They were heroic in renovating society, in sponsoring new wealth, in increasing man's capacity to manipulate his physical surroundings, in filling the ancient gap between workers and owners of land, in sponsoring a political system with an ever-broadening franchise, and in constituting themselves an element of society sensitive to new possibilities, yet remaining robustly pragmatic. As villains they were responsible for embarking themselves and everyone else on an irreversible course leading no one knew where, the unperceiving liberators of forces that could be neither understood nor controlled, creating a society maturing toward an unprecedented vulnerability to manipulation, consolidating in their own hands the elements of wealth and power, seemingly indifferent to the effects of their actions in bringing pain and even destruction, and too often philistines or imitators, unable to develop their own cultural expression to fill the gap left by the destruction of agrarian society.

That each of these often mutually inconsistent views should express a truth, is indicative of the complexity of the groups concerned and the changes they manifested over the decades. When Taine visited England in the sixties bewilderment overtook him as he sought order in this confusion. Some observers could sense a kind of pattern and feeling among the middle class: Disraeli, with his taste for paradox and

---

[1] E. P. Hennock, *The Role of Religious Dissent in the Reform of Municipal Government in Birmingham, 1856–1876*, Cambridge Ph.D. thesis, I.T.A., 1955–56, no. 370.

practicality went to the Queen herself for perceptive judgements about them.[1] It is possible, however, to over-estimate what the Queen may really have known of the new business baronage. It was composed of families that were complex in themselves, and were involved in a continuously elaborating relationship with other families. The middle class included the new regional oligarchs whose position was consolidating by the later sixties – the small group of men and families in control of the great industrial complexes, constituting, with the great merchant bankers the *haute bourgeoisie*: the families of Peases, Backhouses, Lambtons and Bells on the north-east coast, the Guests and Crawshays in South Wales, Tennants, Dixons, Bairds in Scotland, the merchant princes and the shipowners of the outports – Rathbones, Forwards, Booths, Holts, the textile families of Manchester, Illingworths of Bradford and Halifax, Forsters, Baineses, and Kitsons of Leeds.

These provincial élites and their London counterparts often contained a considerable foreign element attracted to Britain by greater opportunities. These contributed not only to business expansion, but to the pressure for better technical education and to the mitigation of cultural philistinism in the northern cities.[2] Among these men of alien origins the Jewish community played a very important part, especially in finance and trade. There were really two such Jewish groups: the old, established, and wealthy families of Anglo-Jewry headed by the Rothschilds and recruited by newcomers like the Sassoons, and the submerged new Jewry expelled from Europe and consumed by the need for acceptance.[3] This complicating element of the English middle class was no less anxious than the denizens to acquire landed status.

In some cases the vitality of the first generation might be carried to the second and beyond, but very often the sons and especially the grandsons of pioneers showed a decline of energy or interest. This might not be serious, but it might occur when new difficulties were challenging the family enterprises; there were not a few cases in which the business difficulties of the seventies coincided with just such a family

---

[1] E. F. Benson, *As We Were*, London, 1930, p. 27; for a middle-class view of the middle class see Alison Adburgham, *A Punch History of Manners and Modes, 1841–1940*, London, 1961.

[2] See above, Chapter 3, section 6, p. 96.

[3] See above, Chapter 4, section 2, p. 111; Lloyd P. Gartner, *The Jewish Immigrant in England, 1870–1914*, London, 1960; Robert Henriques, *Marcus Samuel, First Viscount Bearsted*, London, 1960; S. N. Behrman, *Duveen*, London, 1952, chapter 2. For English difficulties in coming to terms with the Jews see Harry Stone, 'Dickens and the Jews', *V.S.*, 1959.

decline. There were cases too where older members of the family persisted in the control of policy well after their vitality had faded, sometimes because they wanted to, and sometimes because they had no confidence in the new generation. Industrial dynasties were beginning to encounter the problem that had so long haunted the landed aristocracy — how to ensure continuity both of blood and assets.

Not all of those who continued in business were innovators nor sought to be titans. By the fifties if not earlier the industrial sector of the economy was producing its 'worthies' — men conducting enterprises that were in the main repetitive or imitative. As an industry approached its phase of maturity, with the potential for new inventiveness declining, as in textiles, an increasing number of those in charge of firms found that though they had occasional minor challenges to meet, the mode in which business was conducted was largely settled. These were not necessarily small men, but were often in charge of large concerns which, once launched, could continue to grow with the general expansion of the economy.

As older lines of output encountered these limitations, new elements of commerce and industry were coming forward, producing new men and families. The great age of the Victorian steamship owner was just dawning in the sixties. Metallurgy was making immense strides with steel replacing iron. So too with engineering. The chemical industry was assuming new proportions and a new versatility producing Muspratts, Albrights, Brunners, Monds. And all the while new men of parts were seeking new opportunities.[1] From these came new draughts of vitality, by which the position of the great business families in the life of society was continuously renewed. But in spite of these new injections the world of business was probably producing an increasing proportion of less aggressive men.

In the heroic phase of the family, while its founders were establishing their great enterprises, there was no great tendency for incoherence of outlook to appear. For the members of the new race of tycoons had much in common. It has been argued that in the early phase of industrialization there were profound forces at work, of a psychological kind, inducing an aggressive spirit of commercial and industrial enterprise and generating an ethos that inspired and even obsessed men with the urge to launch new ventures and to extend old ones. There can be little doubt that the new individualist approach to religion and social life gathering momentum after the Reformation was far more

[1] See above, Chapter 4, section 12.

favourable to the emergence of a class of entrepreneurs than the continued acceptance of ancient authority and the relegation of the pursuit of wealth to an inferior and even contemptible role in society. The entrepreneurs of the nineteenth century, great and small, had managed to find a formula that justified the vigorous pursuit of market gain.

In the case of the first half of the nineteenth century the idea of a British business class dominated by a cult of working and saving must be treated with caution. Indeed, entrepreneurs seem to have fallen into two main categories. There were those who, once having placed their capital in an enterprise, took very little interest in its workings beyond the receipt of profits, delegating responsibility to managers and foremen, and who were the targets of bitter working-class criticism for their failure to do more. There was also the other and more typical kind – the men who did, indeed, take a close interest in their undertakings, but did so, not as a form of self-discipline, but in eager enjoyment. Nor was there much morbid addiction to saving. Foreign observers were amazed and sometimes disgusted at the level of spending they saw among the new tycoons of Britain. The notorious contrast between the French and the British was frequently drawn in these terms: the French got rich by saving, the British by heavy toil and lavish spending.[1] 'Extravagance', said Smiles, addressing himself to all classes of society in the seventies 'is the pervading sin of modern society'.[2] The fact that there might be drastic cutting of the family budget in times of depression was indeed a form of thrift, but this could as well be construed as an act of propitiation for a fall from grace. Nor must we believe too easily that the urging of thrift upon the impecunious working class implied a belief that it should be practised by all.

The second half of the century may well have seen in Britain a greater emphasis upon the virtues of disciplined work and thrift. But it remains doubtful how far this was due to a deeply rooted set of 'puritanical' ideas. The later Victorian business man may have produced or at least induced a view of himself that was something of a retrospective self-justification. As, in the second half of the century, he began to perceive that his role in society must eventually become dominant, and that it would become subject to scrutiny, he might well, in a kind of instinctive collaboration with his fellows, begin to elaborate the lines of a group apology. What better than to stress the work-ethic, of

---

[1] Taine, op. cit., pp. 26, 173.
[2] Samuel Smiles, *Thrift*, London, 1875, p. 233.

such great antiquity, and of such immense justificatory power? From the fifties onward the concept of men who during most of their waking hours were wholly preoccupied with business gained much greater acceptance. With the migration out of the cities by the business class, the old connection with the life of the community within which they worked was much weakened; in its place came the distinct and regularized habit of attendance at the office. In this way it was made more obvious that the business man was 'working', and often for very long periods.

For the most part, the vigorous men of business were notoriously unintellectual. They had little use for the idea of sitting and reflecting, or of the elaboration of abstract concepts. Their mental processes were the business analogues of the engineering thinking of Stephenson; just as his mind was organized like a workshop, so theirs were organized in terms of business relationships.[1] In both cases, reflective leisure seemed to have no place in the important business of life, and certainly could offer no basis for self-justification. Indeed, it was necessary to place the vigour of things done and decisions taken against the passivity of savouring and assessing.

But this limitation of outlook had its dangers, both economically and politically. Men of business by the sixties had discovered that to hold a seat in Parliament was to increase one's eligibility for directorships. Their unwillingness to face the growing importance of science and its fundamental significance from the point of view of their own activities, was part of their businessmen's blindness. As they assumed their places in Parliament the new men were a source of worry to observers: 'They have excellent ideas as to their own business to which they have been accustomed', wrote *The Economist*, 'but they have very few, and those bad, ideas as to the subtle and new political business to which they have not been accustomed.'[2]

It would seem that these business executants were possessed of a very high level of physical energy. The new kind of opportunity situation had allowed a new kind of self-selection within society to operate; out of it came the new captains of industry, free, in their phase of vigour, from pseudo-aristocratic conventions, motivated by drives that were highly personal, intensely obscure, enormously powerful, and sustained by the will to dominate. But the mellowing process among the families of the more successful men of business, like recruitment,

---

[1] See above, Chapter 3, section 2, p. 81.
[2] 'The Probable Result of the Elections', *Econ.*, 8 July 1865, p. 813.

was continuous. Many, perhaps most of those who gave the great lead in matters of conscience and social amelioration in the first half of the century, and who then and subsequently constituted a kind of intellectual aristocracy, lived upon wealth gained in the later eighteenth and very early nineteenth centuries.

5. THE LOWER MIDDLE CLASS

But not all men of business could operate on the grand scale. A new order of lesser bourgeoisie developed rapidly from the mid-century or earlier, composed of those men whose character and opportunities precluded them from major ventures. Among these were many smaller manufacturers, struggling against the trend toward greater scale, together with shopkeepers, dealers, milliners, tailors, local brewers, millers, coal merchants, in short the 'third rate men of business', forming a growing lower adjunct to the middle class.

They were tainted with the stain of trade, involved in a triple opprobrium. Dealing in small quantities, being in intimate contact with, and even touching, the goods themselves, and directly receiving money payments from hand to hand, constituted the trinity of shameful acts. These criteria, derived from an agrarian aristocracy, however devoid of ethical content, were a sufficient basis of judgement for those whose opinions determined who was on the right side of the prestige gap that separated tradesmen from others. To those engaged in minor trade and industry there should be added the rapidly expanding army of office workers in both business and government, school-teachers, railway officials, the emergent managerial class, and the 'lesser' professions of accountancy, pharmacy, and the various branches of engineering.

But even those of the lower middle class found compensations. They too maintained a conceptual distinction, no less arbitrary, separating themselves from their inferiors – the workers by hand. Moreover, as pressure to extend the franchise became irresistible after the mid-century, the upper middle class and some of the nobility sought, through the principle of property qualification, to limit the extension of the franchise in such a way as to embrace mainly men unlikely to be infected by revolutionary ideas or to respond to inflammatory appeals.[1] Indeed the importance for the operation of the political system of these men was much stressed by the eighties.[2]

[1] See below, Chapter 9, section 8, pp. 367, 368.
[2] G. J. Goschen, *Essays and Addresses on Economic Questions, 1865–1893*, London, 1905, p. 224.

This judgement contained much truth. Many members of the lower middle class, as they sought income and identity in their task of distributing the growing flood of goods and services, were without strategic advantage of any kind. Yet they were exposed to chance and to engineered pressures, doing a large part of the stock-holding of the community, though in small parcels. The natural consequence was a conservative disposition. This attitude might well extend beyond business to an obsessive conventionality, necessary both to reassure their customers as to their business soundness and themselves as to their own station. So indeed the lesser men of business might become custodians of stability, a kind of drag-chain on hasty political action.

But by the same token such attitudes could be destructive of imagination and of zest. It was among this order of society, rather than among the greater traders and manufacturers, that English puritanism cast its most powerful blight over self-expression and the exploration of temporal pleasures. The Nonconformist tradesman who became a pillar of his chapel and a custodian of morality among fellow members of the congregation, could join with other like-minded men to become censors over a large part of society. Moreover it was very easy for these attitudes to induce a sense of loneliness and isolation in those who held and preached them, so that an Englishman's home, though his castle, might also be, at least to some degree, his prison.

According to Marx this lower middle class could not survive. The *grande bourgeoisie* would simplify the national scene in two directions, ousting the landed class from its long-held dominance as the owning element in society, and thrusting the *petite bourgeoisie* down into the wage-earning proletariat. But, though anxious to make a recognized place for themselves in the nation, the greater men of business did not wish to destroy either aristocracy or the lesser traders. For the first could only be disposed of in measurable time by penal taxation, which by the third quarter of the century would hit profits as hard as rents; the destruction of the second would mean the eclipse of an order of men who, for many reasons, including the stability of the nation, there was no desire to destroy.

It was true enough that many groups of smaller men could not maintain themselves indefinitely. Their chief enemy was the improvement in technology, causing the economies of scale to oust the small producer from one industry after another. In textiles the trend had become clear as early as the forties; by the seventies and eighties it was spreading to other industries, including the metal trades, and was about to overtake even traditional local industries – milling, brewing, tailoring, and retailing. But there were considerable time lags in many industries, new products came forward to provide new opportunities for small beginners, and the economy as a whole was growing, so that the small producer and petty trader was still very important in the eighties.

Some elements of the lesser bourgeoisie were not to languish but to thrive. The growth in the scale of enterprises, and the spread of the limited liability company brought new status and new incomes to the men who had served as managers in concerns formerly run by partners.[1] So too with professional men providing services, especially accountants, surveyors, notaries, and engineers. As such elements began to perceive where their future lay they aspired to provide better education and professional training for their sons. To master their continuously elaborating professions, these men had to develop a steadiness and self-discipline that would carry them past severe academic and professional examinations; this they were prepared to do for there was no other way to advancement.[2] Such close application to closely construed problems could not but leave its mark. Out of the lower middle class there was coming by the eighties an order of society that, far from being thrust down among the wage-earners, was eventually to offer its own challenge for supremacy.

The lesser bourgeoisie had, however, no real access to the system of education shared by those above them. Some might enter the older public schools as Scholars or Town Boys, and thus find themselves embarked upon a rapid ascent once they had purged their stigma of social inferiority. But these fortunates were few. It was to the local grammar school that the son of the tradesman looked for his education. By the mid-century there was strong feeling at Oxford that the university should take more men of modest income, but there was considerable shock in 1858 when Oxford examiners in seeking for talent from provincial schools had to reject a high proportion of the candidates; the middle classes were rebuked for failure to produce a better result, and for their lack of interest in Owen's College, Manchester. But the grammar schools, some 1,000 in all, were not really aided by their endowments, for these were not applied with vigour and

[1] See above, Chapter 7, section 2, p. 221.
[2] A. M. Carr-Saunders and P. A. Wilson, *The Professions*, Oxford, 1933; Barrington Kaye, *The Development of the Architectural Professions in Britain: a Sociological Study*, London, 1960, M. A. Dalvi, *Commercial Education in England during 1851–1902; an Institutional Study*, London Ph.D. thesis, I.T.A., 1956–7, no. 531.

imagination, but were largely monopolized by the Church of England and were rather a cause of complacency. The revelations of the inquiry of 1868 prompted the belief that education would positively benefit from the loss of these monies, for then constructive thinking could take place.[1] Improvement was on the way, especially after the Endowed Schools Acts of 1869 and 1874. Yet the immense disparity of educational opportunity, with the consequent differences in outlook, was only slightly modified by the eighties.

The changing condition of the middle class affected their behaviour at a most important point – the size of their families; it was among them that the tendency to smaller households was strongest. Lengthy periods of education, the desire to achieve a standard of living that was satisfactory both in real and in prestige terms, reduced the number of children. As the death rate fell, middle-class couples, as well as those of the skilled artisan class, finding their homes filling more rapidly than had been the case with their parents, might well consider whether, in the interests of a higher standard of living for the family and better prospects for the children, numbers might be curtailed.[2]

6. THE PROBLEM OF EMOTIONAL ACCEPTANCE

As society industrialized it became necessary for the aristocracy and the greater and lesser bourgeoisie to adjust themselves emotionally and intellectually to the bewildering discontinuity that assailed them. This need was felt especially by the mellowed element of the middle class who could free themselves of the preoccupations of business and the blindness of conventionality, by becoming publicists, university dons, teachers in the better schools, members of the minor professions or traders whose businesses permitted of leisure. Beyond the struggle of attitudes and reactions lay the need to locate and understand particular evils and abuses, and to become articulate about them. The complaints of the victims themselves were often loud and sustained. But the spokesmen of the working class were ineffectual in arousing sympathy or attention, partly because they were suspect as irresponsible agitators.[3] The middle class had to produce largely from within itself the means of coming to terms emotionally with the new kind of society,

---

[1] See above, Chapter 7, section 8, pp. 258, 259.

[2] J. A. Banks, *Prosperity and Parenthood*, London, 1954.

[3] D. C. Morris, *The History of the Labour Movement in England, 1825–1852: the Problem of Leadership and the Articulation of Demands*, London Ph.D. thesis, I.T.A., 1952–53, no. 664.

and the means of reaching the consciences of its own members deaf to the elemental complaints of the lower orders.

Striking the balance between good and evil was a difficult matter. Some contemporaries were aware of the relentless forces at work in society, especially those associated with population growth, and of the race being run between these and the nation's ability to produce both for home consumption and for sale abroad. Some too were aware of the crushing tasks or tedious routines of earlier society, many of which had been lessened by machinery. Though few were as rhapsodic as Ure, many knew what he meant when he wrote of the factories: 'In those spacious halls the benignant power of steam summons around him his myriads of willing menials, and assigns to each the regulated task, substituting for painful muscular effect on their part, the energies of his own gigantic arm, and demanding in return only attention and dexterity.'[1] But Ure's own idiom of the machine summoning its menials and ruling them had its own element of horror, to which his technological enthusiasm had made him insensitive.

Men and women of the necessary intensity of feeling and acuteness of perception to attempt the task of understanding, or at least of sensing the new direction of man's fate were in fact forthcoming; persons for whom systematic analysis was impossible, but whose task, performed in personal isolation, was to help the middle orders to discover their *feelings* about the course of society. These were the poets and novelists.[2] They came from a wide variety of economic backgrounds, but all were unable simply to accept life as they found it; each was a pent-up force of inner tensions erupting in the form of a vision of distortions and wrongs. Such writers became, if not the conscience of the age, at least the source of manifold assaults upon a conscience that had more or less insulated itself against the calamities that were afflicting so many. Improvements in printing (new types, the application of steam power), new binding methods, cheaper paper, the rise of the lending libraries as thriving businesses, the new initiatives among publishers, all provided the physical means of access to the middle-class mind.[3]

The romantics of the old and new centuries, Blake, Shelley, Coleridge, and Southey, protested in high literary style against the passing of the older agrarian world, and though they disliked and criticized the

---

[1] Ure, *Philosophy of Manufactures*, p. 18.
[2] Jeremy Warburg, ed., *The Industrial Muse: The Industrial Revolution in Poetry*, Oxford, 1958.
[3] See above, Chapter 4, section 12, p. 175.

new, they were at least as much concerned with what was being lost, as with the incipient evils that were to replace former good.[1] A strain of anti-intellectualism operated in such men, a hostility to science and calculation as alien to the true feelings and instinctive responses of Rousseau-ite man, now in greater danger than in any former time of being imprisoned by artificialities.[2] But just as such writers were weak on analysis, so too they were seldom very exact in social observation.

A new precision came with the first serious novels dealing with the new social situation, appearing in the forties: Disraeli's *Sybil*, Kingsley's *Yeast*, and Mrs Gaskell's *Mary Barton*.[3] Disraeli with his striking statement of the existence of two nations within the one, the rich and the poor, dramatized the fact that the benefits of the new productivity had gone to a minority and the losses to the majority.[4] Charles Kingsley showed how vulnerable were the weaker members of society to the stronger.[5] Mrs Gaskell, in intensely human terms, described how even men of high mind and courage, like John Barton, could be ground down to criminality by the demands and constrictions of a workman's life. The fifties saw further novels of grim social protest.

Though the novelists were revealing something of the degradation that was perverting or destroying human potential beneath the middle class, these demonstrations of sufferings did not of themselves lead men of feeling to clarity of view. They in fact were added to the elements of bewilderment at work within the middle classes themselves. Might it not be that the entire trend toward the mechanization of society was wrong? The railway brought a new crisis of feeling. Wordsworth inveighed against it for its violence to nature. Dickens, like Goethe a good deal earlier, sometimes saw in it the great exemplar of a new blind force, irresistible and universal, carrying humanity into the void toward Death itself.[6] Browning protested in symbolic fashion against the dehumanization of society by industry.[7] Charles Reade took major social wrongs as the main themes of his novels.[8] George Eliot carefully

---

[1] See below, Chapter 10, section 3, p. 401.

[2] Walter E. Houghton, 'Victorian Anti-Intellectualism', *J.H.I.*, 1952, p. 291; Peter Viereck, 'The Poet in the Machine Age', *J.H.I.*, 1949, p. 94.

[3] See Kathleen Tillotson, *Novels of the Eighteen-Forties*, Oxford, 1954.

[4] *Sybil, or the Two Nations*, London, 1845.

[5] R. B. Martin, *The Dust of Combat: A Life of Charles Kingsley*, London, 1959.

[6] For Dickens's reactions to industrial society see P. A. W. Collins, 'Queen Mab's Chariot Among the Steam Engines: Dickens and "Fancy" ', *English Studies*, 1961.

[7] David V. Erdman, 'Browning's Industrial Nightmare', *Philological Quarterly*, 1957.

[8] E. G. Sutcliffe, 'Charles Reade and his Heroes', *The Trollopian*, 1946, p. 7.

studied the economic literature of the day so that her fiction might reflect reality.[1]

Perhaps the greatest of those who sought to deal with these impractical problems upon which all else depended was Thomas Carlyle.[2] He was the major sage of Victorian times for those who sought guidance on what they ought to think and feel as they participated in the great renovation: 'England', he said, 'is full of wealth; yet England is dying of inanition.' His remedies were universal education, emigration, and especially, work. With Ruskin, though he carried out his own self-supervised schemes, there was the same air of being more effective in protest than in solution.[3] Matthew Arnold in the late sixties raised a powerful voice for a new humanism to combat the new faith in material power. 'Faith in machinery', he wrote, 'is our besetting danger.' He envisaged a highly developed class structure, the Barbarians, the Philistines, and the Populace, each element with its uncritical view of itself, each prepared, for political reasons, to play on the others by specious appeals, with consequent danger of cultural sterility, social and political barrenness, and possibly disruption and anarchy. The remedy was for all orders of society to seek 'sweetness and light' through the pursuit of 'right reason' and 'the best self'.[4] By the seventies a new aestheticism was making its appearance with Walter Pater as its prophet — calling those of sensibility to renew their capacity to enjoy the world of beauty, an appeal which, though it vigorously condemned ugliness, could serve to divert attention away from the need to remedy it.[5]

The universities too made their contribution to the struggle to adjust to the new society. For all their passivity in the teaching of undergraduates, and for all their slowness to place due emphasis upon science and technology, they nevertheless produced men who struggled hard and not without effect to find answers to the ancient problems of values and beliefs. This was particularly true of Oxford. The Oxford Movement, the muscular Christians, the sharers in Matthew Arnold's plea for more sweetness and light, and the followers of Pater, all owed

[1] Walter Francis Wright, 'George Eliot as Industrial Reformer', *Pubs. Modern Language Assoc. of America*, 1941.

[2] See below, Chapter 10, section 3, p. 402.

[3] John Tyree Fain, *Ruskin and the Economists*, Nashville, 1957.

[4] Matthew Arnold, *Culture and Anarchy*, London, 1869; Raymond Williams, *Culture and Society*, London, 1958, chapter 6.

[5] Walter Pater, *The Renaissance*, 1st ed., London, 1873, new ed., Sir Kenneth Clark, ed., 1961.

much to the fact that in the great phase of economic expansion the scientists and technicians remained, in the English universities, the underdogs.

The romantics were not alone in their distrust or hostility toward social analysis. The churchmen too were in difficulties. Should they elaborate their religions at a high level of generality, with emphasis upon the strengthening of the soul and the preparation for future felicity, or should they descend to commitment on the issues of the day? Though there were many who entered directly into the political struggle, there were also many others who, as the conditions out of which the Christian Church arose, producing a general synthesis between ideas of God, man, society, and nature, became ever more remote, sought to gain continued acceptance for ancient beliefs that now stood high above the conditions of life.[1] Not a few men of intellect and character, of evangelical upbringing, found that new ways of thinking made the theology on which they had been raised inadequate, and so rejected it.[2] But the moral earnestness they had also inherited was less easily shed, so they turned to rationalism.[3] But others were satisfied that their religion should continue to have no compelling implications that they should strive to amend, much less revolutionize, society. A religion involving a good deal of formalism, but no less satisfying for that, was the focal point of most family routines. From the pulpits of church and chapel, especially the latter, the middle class received what was for most of them their only contact with attempts to understand the questions man has always asked about himself and his place in the universe.

Even the liberal view, for all its dependence upon political theorists and economists, was anti-intellectual in a sense, for insistence upon the self-help of the individual could stand in the way of positive thinking about social improvement. A long line of publicists including Mrs Marcet, Harriet Martineau, Maria Edgeworth, and Samuel Smiles arose among the middle class to impress upon the workers the limitations of their situation, calling for acceptance and discipline in social behaviour, especially restraint on family size and upon collective

[1] John Clifford Gill, *The Ten Hours Parson: Christian Social Action in the Eighteen Thirties*, London, 1959.
[2] Olive Brose, 'E. D. Maurice and the Victorian Crisis of Belief', *V.S.*, 1960, pp. 227–48.
[3] A. C. Pigou, ed., *Memorials of Alfred Marshall*, London, 1925, J. M. Keynes on Marshall, pp. 7–9; Royden Harrison, 'Professor Beesley and the Working-Class Movement', in Asa Briggs and John Saville, eds., *Essays in Labour History*, London, 1960, p. 205.

pressure for higher wages. Yet they also sought to spur the worker to realize himself as an individual.[1]

7. THE PROBLEM OF SOCIAL ACTION

Amid all this emotional and intellectual confusion the conscientious members of the middle classes, and, for that matter, of the landed, found it difficult indeed to become effective in making their contribution to betterment.[2] Some, like the Christian Socialists, sought a means of effective alliance with the workers so that, together, a solution might be found.[3] Some were driven to work themselves to exhaustion, compelled as much by their own need for expiation as by the need to ameliorate. Some acted in a spirit of uninquiring philanthropy, with a minimum of intellectual concern. Many of the earliest attempts at relief were of this kind. The medieval idea of charity was still strong, but, as in early times, unthinking giving could do grave damage. The substantial gifts from West End churches to East End parishes often increased the degradation of the London poor. Casual wards, night refuges, free dinners, seemed to make vagrancy worse.

More positive good could be done by choosing a defined project. The organization of Sunday Schools by Robert Raikes and the Reverend Thomas Stock in Gloucester in 1780 served as a model for a general development that commanded the services of devoted men and women who felt that to meet the great urge of so many of the working class to have access to books and newspapers, using a combination of secular and religious instruction, could only be for good.[4] Support was given to Mechanics' Institutes for the same reason. Effort was expended on the Savings Banks movement.[5] So too with housing: Peabody and others brought their energies to bear at a point at which the need for action seemed self-evident. Young ladies exposed themselves to indignity

[1] R. K. Webb, *Harriet Martineau: A Radical Victorian*, London, 1960; Vera Wheatley, *The Life and Work of Harriet Martineau*, London, 1957; Elizabeth Inglis-Jones, *The Great Maria: A Portrait of Maria Edgeworth*, London, 1959; Samuel Smiles, *Self Help, with Illustrations of Conduct and Perseverance*, with a centenary introduction by Asa Briggs, London, 1958; Aileen Smiles, *Samuel Smiles and His Surroundings*, London, 1956; J. F. C. Harrison, 'The Victorian Gospel of Success', *V.S.*, 1957.

[2] See Herman Ausubel, *In Hard Times: Reformers Among the Late Victorians*, New York, 1960.

[3] See below, Chapter 9, section 8, p. 364.

[4] G. Kendall, *Robert Raikes*, London, 1939; J. K. Meir, *The Origin and Development of the Sunday School Movement in England from 1780–1880, in Relation to the State Provision of Education*, Edinburgh Ph.D. thesis, I.T.A., 1953–54, no. 24.

[5] See above, Chapter 5, p. 209.

and even danger when helping in the Ragged Schools. Barnardo was carried on a wave of revulsion and conscience to create Homes for the street-arabs of London. The Charity Organization Society offered a means whereby relief could be given to those of the deserving whose need was dire.[1] Young men went from Oxford to the East End of London to make contact with the poor.[2]

Much good was achieved by such efforts.[3] But the effect was less than it might have been. Two elements were lacking. The middle-class improvers, for all their philosophy of individualism, had no real programme for releasing the initiatives of the workers, and perhaps did not in their hearts believe this to be desirable, preferring a kind of tutelage. The atmosphere surrounding most Mechanics' Institutes was such that the manual workers soon declined to have anything to do with them, leaving the middle-class sponsors with classes composed of clerks and lesser tradesmen.[4] In spite of this they made a real contribution to public education. The early public libraries, following Ewart's Act of 1850, often suffered from the same taint of patronage.[5] The proposal by the philanthropic Miss Coutts to clothe the female indigents in her Home for Homeless Women in dull uniformity provoked Dickens to protest.[6] Indeed he campaigned against the drabness that seemed to infect middle-class aid to the poor. The whole effort at relief was haunted by the fear that the problem might become greater if enjoyment by the recipients should creep in. It is hard indeed to see how *ad hoc* amelioration from above could do other than promote frustration in its sponsors and resentment in its recipients.

Secondly, to make a scientific attack on the major ills of society called for the adoption of new ideas about diagnosis and therapy, and the acceptance of a new kind of ameliorator, the trained professional. This too was difficult, for the Benthamite investigator and administrator was a slur upon the liberal creed, standing for the need for men who

[1] Charles Loch Mowat, *The Charity Organization Society, 1869–1913, its Ideas and Work*, London, 1961; Helen Bosanquet, *Social Work in London, 1869–1912: a History of the Charity Organization Society*, London, 1914; C. Woodard, *The Charity Organization Society and the Rise of the Welfare State*, Cambridge, Ph.D. thesis, I.T.A., 1960–1, no. 626.

[2] J. A. R. Pimlott, *Toynbee Hall: Fifty Years of Social Progress, 1884–1934*, London, 1935.

[3] A. F. Young and E. T. Ashton, *British Social Work in the Nineteenth Century*, London, 1956.

[4] Thomas Kelly, *George Birkbeck, Pioneer of Adult Education*, Liverpool, 1957; Mabel Tylecote, *The Mechanics' Institutes of Lancashire and Yorkshire before 1851*, Manchester, 1957.

[5] W. T. Murison, *The Public Library*, London, 1955.

[6] K. J. Fielding, 'Miss Burdett-Coutts; Some Misconceptions', *Nineteenth Century Fiction*, 1953–4.

acted not by emotion or instinct, or in conformity with a generalized system, but on a reasoned and demonstrable basis. Edwin Chadwick was perhaps the great prototype of such men, unfortunately combining a general affront to the liberal minded with extraordinary ineptness when personal and political tact was required.[1] Chadwick made his enemies among all classes, but it was his own order of society that found it hardest to accept this man whose sense of the rightness of his policies was so great as to make any service to susceptibilities unnecessary. Others however, like Sir James Stephen, could employ more discretion, keeping near the centre of power but out of the public eye as a senior civil servant, and so reconciling the middle class to the need for the professional improver.[2] Even he received a classic rebuke for 'Pharisaism of intellect', and 'the pedantic spirit and oligarchic pride of official accuracy'.[3] Also of great assistance was the elaboration of the Royal Commission, investigating the problems of the nation under the command of the sovereign.[4]

Even when there was acceptance both of the need to liberate working-class initiative and to adopt scientific procedures, there could still be much impotence and frustration. The failure to renovate the nation's system of education, involved as it was in a tangle of attitudes and obstructions, was one of the most striking examples of the difficulties involved in remaking a traditional society.[5]

It must not be thought that the middle class were always ineffectual in their attempts at social improvement. Though *laissez-faire* could serve to justify much urban negligence between the Municipal Reform Act of 1835 and the later sixties, as civic initiative finally broke away from the old inertia, some men of business made a number of curious discoveries. There was an exhilaration in handling the affairs of a great municipality akin to that of business, but with certain extra elements added. The scale of things was so great. Thinking and gathering data in civic terms broadened the outlook and, though in some ways a distraction from business, could nevertheless add something to it. To make a city viable in terms of water supply, gas, sanitation, roads, and bridges demanded a widening of horizons and involved the challenge of thinking about the productive process in quite a different sense. It became clear that whatever was said in the name of political economy

---

[1] S. E. Finer, *The Life and Times of Sir Edwin Chadwick*, London, 1952.
[2] *D.N.B.*, LIV, p. 613.
[3] 'Sir James Stephen on the Sense of the Nation', *Econ.*, 5 February 1859, p. 141.
[4] See H. M. Clokie and J. W. Robinson, *Royal Commissions of Inquiry*, Stanford, 1937.
[5] See above, Chapter 7, section 8.

about the impropriety of public action and ownership the simple fact was that there were some jobs that private enterprise could not handle.

Not only did the improvement of the cities depend upon civic initiative, so too did the efficiency of industry. Such services as the gas and water supply were so intimately involved with civic government through the laying of mains and the maintenance of streets that it began to appear ridiculous to leave such jobs to private companies. Indeed, there were profits being made; why should these not go to relieve rates? This was work that might well come to public notice and bring public reward. The fountain of honour found it easier to distinguish among northern worthies if to business eminence there was added civic achievement. Thus did men who would have been indignant at the name of socialist begin to modify their ideas though not their vocabulary. When new towns were brought into being as at Middlesbrough and Barrow-in-Furness the sponsors themselves had to undertake civic planning. But sponsors of industrial villages, like Sir Titus Salt with his Saltaire, built some twenty years after 1851, had few effective imitators until very late in the century.[1]

On the whole it would seem that the business men, with some exceptions, showed a tendency to leave the messy problems of health and sanitation to the doctors and bureaucrats, concerning themselves mainly with the great civic enterprises.[2] The greatest exponent of civic improvement was Joseph Chamberlain who in a few hectic years in the seventies in Birmingham showed what could be done under the impetus of a powerful will.[3] But his efforts also revealed the power of hostile forces. On the whole, business men's government of the great cities in the second half of the nineteenth century, though it could carry through great works called for by pressing need, could do little directly for the betterment of the social life of the mass of the people.

But civic government was but a small part of the social improvement to which members of the middle class contributed. The children of the first generation of the Clapham Sect, the evangelicals who had done so much to bring humanitarian reform at the turn of the century, continued to contribute to the rise of new sensitivity, fighting against the ferocities of the criminal law, hanging, and flogging.[4] There were

[1] Robert K. Dewhirst, 'Saltaire', *Town Planning Review*, 1960; see below, Chapter 9, section 7.

[2] W. Ashworth, 'British Industrial Villages in the Nineteenth Century', *Econ. H.R.*, 1951.

[3] Briggs, *Victorian Cities*, chapter V.

[4] E. M. F. Howse, *Saints in Politics: the Clapham Sect and the Growth of Freedom*, London, 1952; E. M. Forster, *Marianne Thornton: 1797–1887, A Domestic Biography*, London, 1956.

also the great Unitarian families, who maintained a duality between private business and public service.[1] William Rathbone V was a partner in a great trading house; his brother and nephew accepted a division of labour within the family whereby William was largely relieved of business concerns so that he might devote himself to politics and philanthropy.[2] Sir Benjamin Heywood, Manchester banker, responded to the challenge of bad housing and lack of education with a programme of local improvement, aided by David Winstanley, schoolmaster.[3] The experience of awakening social awareness could be painful indeed — it almost destroyed the prosperous young Charles Booth in the sixties.[4] There were others, including William Ewart with his public libraries and his fight against capital punishment, and James Silk Buckingham, pioneer of town planning ideas.[5]

The throwing open of the Civil Service to entry by competitive examination was recommended in 1854.[6] It came, in part, in the following year with the establishment of the Civil Service Commission; after 1870 the competitive principle was in full operation except in the Foreign Office. These changes came with the recognition that the tasks of government were becoming such that however dear the old enchantments of patronage, the nation had to take at least elementary safeguards against incompetence. But the examination retained a strong classical literary flavour; some members of the middle class could not help wondering whether, if the business to be conducted by the various arms of government was to continue to grow, it would not be better if it were entrusted to the character and methods of business men. Such men would be particularly effective in serving that great canon of Victorian finance — economy. The demand for efficiency was particularly strong during the Crimean War and thereafter: much would be gained, wrote *The Economist*, 'if we could place efficient men of business in all the principal posts of our public departments'.[7]

---

[1] Raymond V. Holt, *The Unitarian Contribution to Social Progress in England*, London, 1938.

[2] Mary Stocks, *Eleanor Rathbone*, London, 1949, chapter 2.

[3] Edith and Thomas Kelly, eds., *A Schoolmaster's Notebook*, Manchester, 1958.

[4] T. S. and M. B. Simey, *Charles Booth: Social Scientist*, Oxford, 1961.

[5] W. A. Munford, *William Ewart, M.P., 1798–1869*, London, 1960; James Silk Buckingham, *National Evils and Practical Remedies, with the Plan of a Model Town*, London, 1849.

[6] Briggs, *A. of I.*, pp. 442–4.

[7] See above, Chapter 8, section 2, p. 288; 'Economy False and True', *Econ.*, 19 January 1861, p. 60.

8. HOMES AND HABITS

However disturbing the struggle to come to terms with the new forms of society may have been for the sensitive and thoughtful members of the middle class, for many of them there was an uninhibited acceptance of the new situation within which they lived their lives. Misgivings could for most of the time be relegated to parts of the mind that did not operate directly upon conduct. For the first time in any society a very considerable proportion of its members could turn their attention toward the problem of disposing of income. In so doing they encountered, just as did those below them in the social scale, the problem of spending.[1] For the new middle class, like the workers, were without experience of the new kind of budgeting demanded of them; it consisted, not in knowing how to allocate scarcity, but what to do with new plenty.

There was a natural tendency to look to the aristocracy for guidance, and to reproduce in middle-class homes, suitably scaled down according to income, the kind of house and furnishings preferred by the élite.[2] But imitation seldom produces a graceful outcome; in any case there might well be nothing that could be borrowed to meet new needs and feelings. The older wealthy families were in possession of eighteenth-century furniture and ceramics before Waterloo.[3] A good deal of the former fell out of fashion as being too frail and spindly, and was discarded or relegated to the servants' hall or the maids' attics. Newer families, in any case, had to buy. As they did so the supply ran out and the industrialist undertook to increase it. But quantity began to make its own demands. About 1830 the furniture shop, perhaps the last stronghold of craftsmanship for the home, was ceasing to be the workplace of joiners and carpenters, but was becoming a showroom for articles manufactured on a less expensive site; the invention of the coil spring in 1828 caused upholstery, the last of the crafts done on the business premises, to pass to the factories.

As factory production developed, articles were designed to conform to the needs of machines and their standardized operations. Once this was achieved, shape often became static because of the costs of re-tooling. The use of new materials like rolled metal plate conditioned

---

[1] See above, Chapter 7, section 5.

[2] For guides to furnishings see R. Edwards and L. G. G. Ramsey, *The Regency Period, 1810–1830*, London, 1958; *The Early Victorian Period, 1830–1860*, London, 1958; Ralph Dutton, *The Victorian Home*, London, 1954; John Gloag, *Victorian Comfort*, London, 1961.

[3] See Winslow Ames, 'Inside Victorian Walls', *V.S.*, 1961, p. 159.

the shape of many things. So the challenge of taste assumed a new form as society sought to assimilate quality and quantity.[1] There were references to 'Brummagem art', 'the result of mechanographic or mechanoplastic means, in paper, silk, cotton, clay or metal'.[2] These contrivances, it was said, would kill all life and freedom in design and execution, and paralyse the imaginative faculties. Even where mechanization was relatively slight down to 1860, as in cabinet making, the demands of higher output affected both organization and product: designs agreed among manufacturers were kept in production as long as possible in order to avoid labour disputes, as is apparent from the London Cabinet Makers' Union Book of Rules.[3] Parliament did what it could to help the nation in its dilemma by founding a School of Design in 1837: in 1841 there were government grants for schools of design in Manchester, Birmingham, Glasgow, Paisley, and Leeds. Prince Albert in 1853 promoted the Department of Science and Art.[4] For it was becoming apparent that, for good or ill, output for both home and export consumption had to be extended, for the nation could not go back to simpler things: 'The dinner may not be what an ascetic moralist can consistently approve', said *The Economist*, 'yet still the cook lives by it.'[5] But Ruskin and William Morris did not accept this. Their contemplation of man's new artefacts caused them to turn to the society in which they were produced. Corrupt products, including the most obvious of all, architecture and decoration, were a challenge to social renovation and not to an acceleration of machine production.[6]

Many Victorians, vigorously ascending the income scale, largely unconscious of any necessity for canons of good taste, showed their self-assurance by their purchases, producing a lavish and hectic eclecticism. This appeared in its most striking form in painting and sculpture, where the craftsman was still in command of his materials. It produced Landseer with his stags, pheasants, and rabbits, with copious effusions of blood, gathered in by highly bred hounds for the delectation of demure children, and William Etty, fluently recording in various improving allegories a suppressed eroticism. These were the things that

[1] John Steegman, *Consort of Taste 1830–1870*, London, 1950; Charles L. Eastlake, *Hints on Household Taste in Furniture, Upholstery*, and other details, London, 1868.

[2] Sir Francis Palgrave, 'The Fine Arts in Florence', Q.R., 1840, pp. 324–6.

[3] Elizabeth Aslin, op. cit., pp. 22, 27.

[4] Quintin Bell, *The Schools of Design*, London, 1963.

[5] 'The Exhibition of 1862', Econ., 26 April 1862, p. 453.

[6] Sir Kenneth Clark, *The Gothic Revival*, 3rd ed., London, 1962, p. 218.

sold, that produced the academicism that was to divert Millais to fashionable portraits and picture stories, and to earn for Landseer over a quarter of a million pounds. The Royal Academy remained strongly representational, even when photography had helped to push French painters into radical experiments with light and form; no doubt the British attachment to the pictorial and the literal was related to the fact that the principal source of income for rising young British artists was book illustration.[1] To produce anecdotes in the name of art might seem to one subsequent generation like apostasy to the standards of a monied middle class; to another, perhaps further removed, such works may have their charm, especially when painted by craftsmen with a reliable realism.[2] They have recorded the Victorian house, the family, the dramatic occasion, the nemesis for breaches of a rigid code.

The Queen and the Prince Consort stood high among the arbiters of middle-class taste; their enthusiasm for brilliance and colour at Osborne and Balmoral brought vigour to interior decoration, and their addiction to Winterhalter's paintings reflected the same literalism beloved by so many of the Queen's subjects. Philistine all this may have been, but at least it was vigorous. So too was the vogue of the naturalist style, stemming from Paxton, of the conservatory or winter garden, bringing nature into the house. Ferns, pampas grass, and aspidistras spread to the drawing-room, and thence down the social scale.

To both imitators and those who knew what they liked the mellowed generation could be of great assistance. Having learned through school and university something of the attitudes of which upper-class homes were the reflection, they could help the members of their family to pass from imitation to at least partial participation in the setting of tone. As for those who backed their own taste, some at least, in accepting the pre-Raphaelites (first exhibition 1849) were beginning to respond to less obvious treatment.[3]

Sensitivity to correctness of dress became obligatory on a much higher proportion of society than ever before.[4] This caused swings of fashion on an altogether new scale, especially among women. When the murderess, Mrs Manning, died on the public scaffold in 1849, the vogue for black satin gowns expired with her, though only temporarily.[5] Though it may be possible to relate trends in dress to trends in

[1] R. C. K. Ensor, *England 1870–1914*, Oxford, 1936, p. 156.
[2] Graham Reynolds, *Painters of the Victorian Scene*, London, 1953.
[3] A. Paul Oppé, in Young, op. cit., Vol. II, pp. 159–76.
[4] Charles H. Gibbs-Smith, *The Fashionable Lady in the 19th Century*, London, 1960.
[5] John W. Dodds, *The Age of Paradox*, London, 1953, p. 390.

national experience, the suppliers in the second quarter of the nine-teenth century had no such knowledge; in any case it was the subtle season-to-season variations that left clothes on the shelves.

Aesthetics did not exhaust home life; there were the pressing matters of home management. Here too the members of the middle class had to learn how to adjust continually to their situation. The fact that Mrs Beeton's *Household Management* had in the sixties run through five editions within four years and rapidly ran through many more showed how complicated the management of a considerable household had become, requiring 'information for the mistress, housekeeper, cook, kitchen-maid, butler, footman, coachman, valet, etc.'[1] The fact that Mayhew could describe a distinct trade in London of collecting the fats and other goods that were either waste or plunder from well-to-do-kitchens, demonstrated the need for close oversight, or for a housekeeper to whom all these troublesome matters might be re-ferred.[2] By the seventies it was beginning to be realized by the upper and middle classes how much a comfortable home life depended upon servants. Suggestions, seldom acted upon, were made to give them greater freedom, to provide schools of cookery, and to improve the operation of the house by a better water supply, waste disposal, speak-ing tubes, and lifts.[3]

Among practical concerns was the question of personal cleanliness and sensitivity to smells.[4] In the early nineteenth century neither was highly developed. The daily bath, that symbol of Englishness, was, like cotton textiles, an import from India, brought home by the Duke of Wellington.[5] It was adopted, though slowly, by the more fastidious of the wealthy members of society; as late as the thirties all classes of Englishmen showed no great concern over the state of their persons; bathing was rare, typically the hands and face were washed and the rest of the area cared for only by brushing dirt from the clothes.[6] The principle of 'active exhalation' was still highly regarded in the fifties whereby a healthy person exhaled from the skin, not amounting

1 See Nancy Spain, *The Beeton Story*, London, 1956; H. Montgomery Hyde, *Mr and Mrs Beaton*, London, 1951.

2 Henry Mayhew, op. cit., vol. II, p. iii, 'Of the Buyers of Kitchen-Stuff, Grease and Dripping'.

3 Mrs E. M. Kean, 'Confederated Homes and Cooperative Housekeeping', *R.B.A.*, 1873, Section F, p. 195.

4 For a history of domestic sanitation see Lawrence Wright, *Clean and Decent: The Fascinating History of the Bathroom and the Water Closet*, London, 1960.

5 Young, op. cit., I, p. 87.

6 U.N.O., *The Determinants and Consequences of Population Trends*, 1953, p. 52.

to perspiration, thus actually repelling dirt.[1] Coachmen and grooms waiting at dinner were not always clear of the odours that accompanied their service to the lower animals. By the mid-sixties the wealthy bathed daily, with the middle class taking its weekly bath and daily wash. The cult of cleanliness was thus slowly seeping down in society. Those anxious to improve their standing had increasing recourse to soap and water; to the other gaps in society was added the gulf of personal noisomeness. But though it was possible to wash the person, and personal linen, woollen clothes defied cleaning except by brushing and spot removal.[2] Moreover ironing was confined to linen, except when a suit of clothes was about to leave the hands of the tailor.[3]

9. THE INTEGRITY OF THE FAMILY

The homes created by the upper classes became the matrices that formed their children, either positively, or by revulsion.[4] The family was the citadel of Victorian upper and middle-class society, in many cases in fact, and in almost all in convention.[5] For the family to be stable it was necessary that man and wife should find terms upon which a continuously harmonious and fruitful partnership might rest. The relationship between the sexes was, of course, radically affected by economic and cultural revolution.

A strong male supremacy was asserted. The men who formed and ran the new industry were not only able to bring to bear over their women folk the ascendancy that their position created, but were also disposed to continue to enjoy advantages that came from an earlier form of society. For though single women kept their own earnings and held their own property, if they had any, not so their married sisters, whose husbands had the absolute right to take their wives' property as their own. The records of the Savings Banks are full of pathetic instances of humble women trying to keep secret a tiny nest-egg. A Bill to change the law in 1868 roused great alarm as an attack on the unity of the family under the sovereignty of the husband. The idea of female

---

[1] Abraham Hayward, *The Art of Dining*, London, 1853.

[2] For lighting, both private and public, see William T. O'Dea, *The Social History of Lighting*, London, 1958.

[3] Ames, op. cit., p. 161.

[4] Muriel Jaeger, *Before Victoria*, London, 1956; M. J. Quinlan, *Victorian Prelude: a History of English Manners*, New York, 1941; Janet Dunbar, *Early Victorian Woman: some Aspects of her Life, 1837–57*, London, 1953.

[5] H. L. Beales, 'The Victorian Family', in H. Grisewood, ed., *Ideas and Beliefs of the Victorians*, London, 1949.

passivity was much encouraged; a woman was much more feminine if she waited to be told both what to think and what to do. This idealization appeared in art, showing women as of a more delicate but purer clay than objective experience might suggest.

But it is clear that such a set of concepts was not wholly satisfactory. It was strongly inhibitive upon women of initiative, who could not fail to be excited by the new opportunities that had emerged in male society. Though aristocratic women had always had a high degree of freedom, those of the middle class were closely bound. With a diminishing amount of menial work to be done the prospect was one of useless leisure.[1] To boredom was added humiliation in the case of those who did not marry yet were expected to accept a virtually functionless dependence. Such women were increased in number by large-scale emigration of males. The position of the aunt was well established by the sixties, but taking care of the family aged or the children was not an adequate role, not was the job of governess, where humiliation and under-payment were too frequently combined.

Difficulties among middle-class males showed in a different form. The aristocrat, with his long tradition of arranged marriages, necessary for economic reasons, had become used to the idea of romantic love outside wedlock. The courtesan flourished where there was wealth, leisure, freedom, and acceptance, rather than censoriousness. In the early part of the century this tradition was still strong; it was to weaken though not to disappear.[2] But the middle-class male found it more difficult to enter into such a custom. He was inhibited by a stricter morality; moreover he had, ostensibly, married for love and not convenience. But he was subject to new strains, for with the struggle for business success, often came changes in character and outlook that, added to the ravages of the years, made his early partner less able to hold his affections. Taine, uninhibited by English conventions, pondered this problem in the sixties: English wives bore too many children, 'which deform them'; youthful good looks too often passed into broody vacuity: 'I have in mind', he wrote, 'two or three such matrons, broad, stiff, and without an idea in their heads.' Moreover, for reasons that are only now beginning to be explored, respectable Victorian wives were raised to regard the act of procreation as a necessary but

---

[1] Emily Shirreff, *Intellectual Education, and its Influence on the Character and Happiness of Women*, London, 1858, chapter VIII.

[2] Harriette Wilson, *The Game of Hearts: Harriette Wilson and her Memoirs*, Lesley Blanch, ed., London, 1957.

rather repulsive duty.[1] Perhaps some part of the sympathy for victims of the industrial revolution should go to these middle-class women whose problems of adjustment were so appalling.

There was a thriving trade in pornographic literature that eventually provoked the Obscene Publications Act of 1857, but was not stopped by it. The cheaper newspapers relied a good deal upon sex. Prostitution reached alarming proportions.[2] By 1841 there were well over 3,000 brothels in the Metropolitan district of London alone. Many critics, including Josephine Butler, put much of the blame for this on the exclusion of women from so many occupations.[3] Venereal disease increased, not only in garrison towns, but generally. Advertisements for 'restoratives', curing 'debility resulting from the early errors of youth' embellished the pages of *The Economist* in the sixties. White slavery provoked public protest.[4]

Over the discussion of such matters there was cast the most binding taboo, and reference to them in novels could only be of the most guarded or the most skilful kind. The male did not permit discussion; the ostensible reason for the ban was the harm this might do to tender womanly susceptibilities. But it was made quite clear that women who fell into error could expect nothing but dire retribution.

The number of cases in which stresses within the family could no longer be contained increased, for the double standard, affording to men freedoms denied to women, was no real relief. Previous to 1857 divorce was only available through a most costly private Act of Parliament; in that year came the Matrimonial Causes Act setting up a special Divorce Court.[5] Even the new facility was discriminatory. Mrs Henry Wood with her novel *East Lynne* of 1861 demonstrated the agonizing fate of the female divorcee. Adultery alone was ground for divorce when sought by a husband; when the wife was the applicant there had to be additional grounds of cruelty, desertion, or bigamy. The less responsible newspapers made great play with scandalous news from the Divorce Court.

The development of the divorce system did little to impair the male

---

[1] Keith Thomas, 'The Double Standard', *J.H.I.*, 1959.

[2] See above, Chapter 7, section 10, p. 276.

[3] A. S. G. Butler, *Portrait of Josephine Butler*, London, 1954; Josephine Butler, *Personal Reminiscences of a Great Crusade*, London, 1896.

[4] Charles Terrot, *The Maiden Tribute*, London, 1959.

[5] O. R. McGregor, *Divorce in England*, London, 1957; Griselda Rowntree and Norman H. Carrier, 'The Resort to Divorce in England and Wales, 1858–1957', *P.S.*, 1958; *Royal Commission on Law of Divorce, First Report*, 1853.

position. But there were stirrings that were to disturb it. John Stuart Mill took up the feminist case, arguing that a fuller life was necessary for women, and that the only way to provide this was to find functions for them beyond child-bearing, and to recognize them as having status and authority so that, among other freedoms they might not be press-ganged into childbirth as the navy had press-ganged sailors.[1]

There were thus continuous and mounting stresses in the marriage relationship. Moralists and churchmen condemned the conduct that resulted, and many husbands and wives, under the influence of such teachings, simply bore the strains to which they were subjected. Public opinion expressed itself when in 1871 the Prince of Wales, having narrowly missed being named in divorce proceedings, was hissed at Epsom.[2] The vigour of the debate on divorce made many wonder how long the idea of the sanctity of the family could continue. In quantitative terms, there were, between 1876 and 1885, some 460 divorce petitions per year.[3]

There were signs in the sixties of single women seeking to escape from the prison of domesticity to wider functions.[4] They had both to fight free of 'the feelings and prejudices of those nearest and dearest to them', and to penetrate new fields of action.[5] The rise of the nursing profession was one way out; so too was school teaching.[6] For the most determined there were experiments in social welfare. Middle-class women, in fact, in strong contrast to their working-class sisters, had to fight for the right to work.

Only in the eighties was real momentum gained for the emancipation of women domestically and politically. By then the long pioneering work in women's education by Miss Beale and Miss Buss, and many others, together with the advocacy of men like Mill, had converged with mounting frustration to make possible the beginning of a new era.[7] Wives too were gaining ground; by the Act of 1870 married

---

[1] John Stuart Mill, *The Subjection of Women*, London, 1869.

[2] Ensor, op. cit., p. 142.

[3] McGregor, op. cit., p. 36.

[4] For the fifties see John Duguid Milne, *Industrial and Social Position of Women in the Middle and Lower Ranks*, London, 1857.

[5] Margaret Simey, *Charitable Effort in Liverpool in the Nineteenth Century*, Liverpool, 1951, p. 68.

[6] Brian Abel-Smith, *A History of the Nursing Profession*, London, 1960; A. Tropp, *Elementary Schoolteachers as a Professional Group, 1800 to the Present Day*, London Ph.D thesis, I.T.A., 1953–54, no. 603.

[7] Josephine Kamm, *How Different from Us: A Biography of Miss Beale and Miss Buss*, London, 1958.

women acquired some rights to their own income; the Married Womens' Property Act of 1882 extended these, constituting a charter of freedom under which the law at last conceived it possible for the family to contain two income receivers and two owners of property.

PART FOUR

Conflict, Control and Comprehension

ALONG with the personal and family adjustments of the industrializing society came the need to remodel the political system. For effective action to be possible, the aggrieved, with their allies, had to make a challenge for power. In order that the increasingly precarious economy should not break down in consequence, the political struggle had to be so conducted that the new ideas of equity and responsibility were reconciled with the conditions of continuous economic growth. This involved the attempt to understand and make explicit the principles by which the behaviour of the economy was regulated.

CHAPTER NINE

The Politics of an Industrializing Society

---

1. THE DISTRIBUTION OF POWER IN 1815

BETWEEN the second and the eighth decade of the nineteenth century the old distribution of power became increasingly archaic, for the new masters of trade and industry would not remain in indefinite tutelage to the traditional landed aristocracy. Nor would the workers continue without question to accept the widening authority of the men of business. These successive challenges were made in a highly complex way.

The victor of Waterloo, the Duke of Wellington, was, and remained for another thirty-six years, the great exemplar of older society. Government at the centre and in the shires was in landed hands. The House of Lords was almost wholly so; the House of Commons, on an archaic franchise and a ludicrous pattern of constituencies, was also a landed preserve. At Quarter Sessions the same men governed in the countryside. There had been those who, before the French Revolution, dissatisfied with this arrangement, had agitated for a more representative system. Such claims had come both from disgruntled members of the professional and trading classes, and from protagonists of the commonalty following Tom Paine and others. But the war against revolutionary France involved putting down such disturbing claims. In 1815 the old political system stood intact.[1] But, much less in the public eye, the new industrialists had begun, well before 1815, to find a common interest, learning how to organize and wield effective lobbies. The realization was being forced upon even the most traditionally minded legislators that the sinews of war were coming increasingly to depend upon the new industrial products.

---

[1] See below, Chapter 10, section 3, p. 440.

The struggle for power was to have two great foci.[1] There was the question of government in the traditional sense at Westminster, where the general rules of economic and political behaviour were set. Here the middle classes were slowly preparing to make their challenge to the landed, seeking a legislature that would provide conditions of growth for industry and trade even when these conflicted with agriculture, and which, at least as important, recognized the status of the new groups in the national society. Secondly, there was the struggle within industry over the distribution of the product – between capital and labour as to the share proper for profits and wages.[2] But this second level had no true independent existence, for the right of the worker to organize was regulated by Parliament. Both under the Common Law doctrine of restraint of trade, and under the Statutes of 1799 and 1800 the right of the worker to combine was closely regulated.[3] It therefore followed that the question of the distribution of the national income, affecting the conditions governing capital formation, and the further question of the workers' incentives to produce, were all in the hands of an archaic legislature.

In general the new industrialists were in favour of the powerful restrictive measures enforced against labour by the courts. It thus became necessary for the workers to seek means of altering this situation. They could do so either by a political challenge, aiming at power in Parliament, or by an industrial one, seeking to organize in unions to such effect as to alter the behaviour of the legislature, and perhaps, indeed, to cause its renovation or eclipse. Both courses involved the immensely difficult problem of finding effective leadership.[4]

2. THE PRESSURES CONTAINED

The war had produced grave discontents, with occasional eruptions, but on the whole the plea of patriotism quietened the disgruntled and the nation accepted the need to fight for the liberties of Englishmen against the continental threat.[5] But with the collapse of trade and

---

[1] G. D. H. Cole and A. W. Filson, *British Working Class Movements: Select Documents, 1789–1875*, London, 1951.

[2] See above, Chapter 7, section 4.

[3] M. D. George, 'The Combination Laws Reconsidered', *E.H.*, 1929; A. Aspinall, *The Early English Trade Unions: Documents from the Home Office Papers*, London, 1949.

[4] D. C. Morris, *The History of the Labour Movement in England, 1825–1852: the Problem of Leadership and the Articulation of Demands*, London Ph.D. thesis, I.T.A., 1952–55, no. 664.

[5] For the political scene in general see Briggs, *A. of I.*, chapter 4.

industry in 1819 fears and grievances, briefly deferred during the short post-war restocking boom, were expressed in mounting clamour and disturbance. Unemployment and reduced wages made the plight of many workers deplorable indeed. Their pleas to their aristocratic, landowning rulers produced no policy for their aid, and indeed little sympathy. The Corn Law of 1815 seemed an affront to a hungry people.[1] Protests were organized in the attempt to make vocal their sufferings, apparently in the pathetic hope that if a sufficient demonstration was made it would be followed by effective action.[2]

There were two possibilities. The landed Parliament might itself take steps to relieve the situation. There was available a great variety of suggestions. Official pensions and sinecures, so lucrative to their holders and so grave an offence to the suffering, should be ended. Taxation should be levied according to ability to pay, rather than lie with such disproportionate weight upon the poorer members of the community. The receivers of interest upon the national debt, now worth so much more in consequence of falling prices, should accept an equitable adjustment. The deflationary pressure due to the return to the gold standard should be eased. The government should create employment by direct action.

But the government found such a programme unacceptable. Except for a feeble effort in the form of a Poor Employment Act of 1817, intended to promote public investment, none of these solutions was tried.[3] Nepotism and official rewards had been part of the system for so long that they had acquired the justification of usage; in any case, how could such a measure greatly aid a situation characterized by a collapse of markets? A revolution in taxation would be resisted by all but the workers, for all had agreed with enthusiasm to the ending of the income tax in 1816; moreover, the landowner was already more heavily taxed than he wished to be. There was little sympathy for tampering with the National Debt, for fear of the confusion that might result, and the possible damage to the government as a borrower in the future. Nor was there any willingness to reverse the deflation, now that a staple and automatic currency was once more possible. The economic programme of the complainants thus seemed wholly lacking both in appeal and efficacity.

[1] See above, Chapter 2, section 2, p. 9.
[2] R. J. White, *Waterloo to Peterloo*, London, 1957; F. O. Darvall, *Popular Disturbance and Public Order in Regency England*, Oxford, 1934; Elie Halévy, *History of the English People in the Nineteenth Century*, vols. I and II, London, 1924–26; J. L. and B. Hammond, *The Skilled Labourer, 1760–1832*, London, 1919.
[3] M. W. Flinn, 'The Poor Employment Act of 1817', *Econ. H.R.*, 1961.

The second alternative was to press for a change in the constitution itself so that under a new system of election, giving the nation the power to choose its governors through a new franchise, a new order of men might come to power more able to govern effectively. This was an even more difficult proposal to carry through a legislature to which such potentates as the Duke of Northumberland, the Duke of Norfolk, Lord Darlington, and Lord Lonsdale each sent nominees who ran into double figures. Such a proposal quite naturally seemed to be exposing the nation to mere anarchy.

Thus the government had no solution other than to wait in the hope that after a period of adjustment the difficulties would pass.[1] Complaint became more general and more vocal. The thoughts of responsible men were simplified by demagogues and agitators, increasing the concern of the government for civil peace. But the mass of the workers, though prepared to protest, could not be persuaded from their almost pathetic legality. For they had no real wish to bring in a revolution. Their responsible leaders were deeply imbued with the need for consent to accompany change. They wished to spur their custodians to positive thought about the present, and as a last resort to persuade them to abdicate.

Yet nerves were stretched on both sides. The government knew its police in London were incapable of disposing by force of workmen's agitation, as could be done in Paris: the situation was even more dangerous in the provinces. The frustrated workers found that they could make no effective contact with their rulers. Crisis came with the mass meeting at St Peter's Fields, Manchester, in August 1819.[2] Henry Hunt, a leading agitator, was addressing a vast gathering when yeomanry and soldiers attacked it. There was a stampede, causing eleven deaths and many injuries. The Six Acts of 1819 followed, effectively putting an end to Civil liberties.

The government was at last clear about its policy. Earlier repressive measures were to be extended to impose a period of waiting on the workers. Within its limits the policy was successful.[3] But there was a feeling that it was a near thing: Lord Liverpool as Prime Minister looked from his window on London and pondered, 'What can be stable with these enormous cities? One insurrection in London and all is lost.'[4]

[1] W. R. Brock, *Lord Liverpool and Liberal Toryism, 1820 to 1827*, Cambridge, 1941,
[2] Donald Read, *Peterloo: the Massacre and its Background*, Manchester, 1958.
[3] See above, Chapter 2, section 2, p. 11.
[4] Val. R. Lorwin, 'Working Class Politics and Economic Development in Western Europe', *A.H.R.*, 1957, p. 341.

3. THE ECONOMIC ROLE OF THE STATE

Popular agitation could take little precise account of the state of the economy and its needs, or of the many almost imperceptible changes that legislators had accepted under the pressure of trends of which they were only vaguely conscious.

By the end of the Napoleonic Wars England had moved much further from authoritarian government than had her continental rivals. Her traders were uninhibited by regional clogs upon movement like divided jurisdictions, tolls, and dues, and had in fact achieved a unified national economy. The state had been steadily withdrawing from interference or surveillance in particular markets. The food trades, for so long the legislators' responsibility, were now virtually free. The old laws against forestalling, regrating, and engrossing in the Corn Market, so long intended to protect the consumer, had been abandoned. The operation of public granaries, an emergency device on many early occasions, had also gone. By 1815 the only relic of a once elaborate system of domestic price control was the Assize of Bread.[1] But though it remained in the provinces until the thirties it was really moribund. In London it was ended in 1815 by Act of Parliament on the petition of some 800 master bakers.[2] This freeing of the internal food market reflected the end of the kind of shortage that could occur in the past; greater relative stability of supply, due in part to improved transport, meant the withdrawal of the state from the market.

The state, by the repeal of the Elizabethan labour statute in 1813 and 1814, had also withdrawn from the wages bargain.[3] For 250 years the attempt had been made, with intermittent vigour and much regional variation, to set the wages to be paid in each trade. The Justices of the Peace charged with this difficult duty had sometimes prescribed in terms of maximum wage rates and sometimes in terms of minima. But from 1814 onward the free market was to be the determinant of the price of labour. Moreover, the law was to be used to insist that bargaining must take place on both the demand and supply sides of the market. The worker if he wished to raise his wages had to do so by individual bargaining with his employer, and by implication had to trust in Adam Smith's view that competition between capitalists for

---

[1] S. and B. Webb, 'The Assize of Bread', *E.J.*, 1904.
[2] Clapham, op. cit., vol. I, p. 345.
[3] Smart, op. cit., vol. I, p. 368.

the services of labour will ensure that the benefits of increasing productivity are passed to the workers.[1] The legislation of 1799 and 1800 also denied to employers the right to combine to force wages down, but this was a much more difficult provision to enforce, and there is ground for doubting that there existed any vigorous will to do so. Whereas the old Common Law prohibition of combination had been intended to stop the workers from challenging the verdicts of state regulation, the Combination Laws were to ensure that they did not interfere with those of the labour market.

One of the traditional ways of doing the latter had always been through the control of apprenticeship. Journeymen, as well as masters, had frequently had a real interest in ensuring that the recruitment to their trade was kept under control. In 1813 petitions were presented by a very considerable number of masters and journeymen, asking that the Elizabethan apprenticeship period of seven years should be made more effective. But other masters petitioned in the opposite sense, and the verdict of Parliament in 1814 went in their favour. The Apprenticeship Laws (on land, though not at sea) were repealed so that so far as the state was concerned the recruitment of labour, like the settlement of wages, was a matter lying between masters and men.[2] Those who had invested much time in their own training were now deprived of state protection, but those who were anxious to obtain and train labour for new projects were wholly free to do so. The abolition of wage control, the enforcement of market bargaining, and the ending of the old apprenticeship system were all moves in the direction of a free labour market.

It looked briefly as though the right to combine had been achieved by the repeal in 1824 of the Combination Acts. This was largely the work of Francis Place, the intellectual breeches maker of London, who showed great skill in lobbying, and Joseph Hume, the radical politician, who employed a high degree of political skill in mustering evidence and in putting the repealing Act before a Parliament dozing in the relative industrial peace then prevailing.[3] The subsequent behaviour of the workers, who rushed to form new unions, awakened the legislators with a start: angry deputations of employers descended on the government and a new Act of 1825 was passed. Combination remained a

---

[1] See below, Chapter 10, section 1, p. 384.

[2] Clapham, op. cit., vol. I, p. 207; T. K. Derry, 'The Repeal of the Apprenticeship Clauses of the Statute of Apprentices', *Econ. H.R.*, 1931.

[3] Graham Wallas, *The Life of Francis Place, 1778–1854*, rev. ed. London, 1918.

legal right of the workers, but they were encumbered with risks in its application, including that of prosecution for conspiracy. Much of the old system of control remained in operation affecting trade, production, the conditions of labour, and the monetary system. The Statute Book, in fact, was full of enactments that did patent violence to the concept of individualism; each such provision had its protagonists and its apology.

The Navigation Laws of the times of Cromwell and Charles II, largely aimed at the maritime power of the Dutch, still required that Britain's trade be carried mainly in British bottoms.[1] This created monopoly conditions for British shipowners and builders, allowing inefficiency to flourish, exempting British operators from the penalties that competition might enforce. The costs of sea carriage were kept up, to the bitter resentment of the traders. There was still great power in the ancient idea that the state must protect merchant shipping as a nursery for seamen, upon whom the safety of the people depended. Yet, though restrictive, the Navigation Laws were liberal in the sense that all British citizens had equal right to their protection.

Not so where the old principal of chartered monopoly still survived, with the Hudson's Bay Company, the Levant Company, and the East India Company. The Levant Company, with its monopoly of trade with Egypt and most of the Middle East, was almost moribund. The East India Company, though it had lost its monopoly to India in 1813, was still immensely powerful. The plea, long maintained by the Company, that to expose India to free trade would do great damage to her traditional economy and society had been rejected. But in China the Company managed to preserve its position so that, as before, the supply of British goods to the Far East was funnelled through the Company and the corresponding monopoly of Chinese merchants, the Hong.

In one area the state had imposed, in the interests of freedom, a great new restriction. In 1807 the legal supplying of slaves to the West Indies, or anywhere else, was stopped throughout the Empire. This was done on ethical grounds, but it outraged those concerned with the trade, who in their resistance to this impairment of their right to freedom of action apparently thought themselves entitled to derive support from the idea of natural liberty.

The state maintained a wide range of controls on exports, especially

---

[1] *Report of S.C. on Navigation Laws*, 1847; *S.C., House of Lords*, 1847–48; W. S. Lindsay, *A History of Merchant Shipping*, London, 4 vols., 1874–76, vol. III.

upon those goods thought basic to Britain's industrial pre-eminence. To send machinery abroad was forbidden; so too was the emigration of skilled artisans capable of installing and operating it. But emigration of skilled artisans was freed in 1824. Indeed the government was prepared to give mild encouragement, especially from trouble spots like Glasgow.[1] The export of coal, so important as a source of power, was effectively prohibited by a duty of 70 per cent, a provision aimed especially at France. The export of wool was also forbidden. This prohibition had long historic origins, going back to medieval times when England was the great wool supplier to Europe, and could thus hope to sponsor the home cloth industry by inhibiting foreign manufacturers through curtailment of their supply of material. The home woollen manufacturers were certainly anxious to continue the prohibitory statutes, and were still hopeful of a monopoly of the world's woollen manufacture. But the home producers of raw wool wanted the right to sell freely abroad, as their predecessors of medieval times had done.

Down to 1815 there had been an import duty on raw cotton for revenue purposes, equal to something like one-fifth of the value of the raw material. But this was no serious restraint upon trade. There was also a tax on the import of foreign wool; this was raised in 1819 from one penny per pound to sixpence in the interests of protection for the home grower. But the wool growers were bound to encounter the hostility of the cotton manufacturers, whose future expansion depended upon the principle of free entry for imported raw materials and freedom from retaliatory action by foreign countries.

The elaborate range of import controls, each with its peculiar and hard-fought history, reflected the conflicts of interests involved. Britain had long depended upon the Baltic for timber imports; the long war showed how vulnerable this could make her to any power capable of closing the Baltic. Accordingly, in order to encourage colonial suppliers, timber duties were imposed such that foreign imports in 1815 were paying a tax of some 65s. per fifty cubic feet, while colonial timber entered free.[2] This arrangement was especially cherished by Canada. But it meant that those who wished to build, both for industrial and domestic purposes, and those who bought timber for the enormous range of other uses to which it was still applied, had to pay more than the world price. Moreover, it served to protect the ship-

[1] Smart, op. cit., vol. I, p. 440.
[2] G. R. Porter, *The Progress of the Nation*, new ed. London, 1847, p. 380.

owners, for those with old and sometimes almost worn-out ships were often saved from the competition of new ones because of the high cost of construction. The duties on hemp had a similar effect. Impediments to iron imports further affected trade with Baltic countries.

The preferential duty on sugar served to protect the West Indian producers against competition from the East Indies under Company monopoly. This reflected the past political strength of the West India interest, but was bound to be challenged now that the Company's grip on India had been broken. There was also a growing consumers' demand for sugar, causing impatience with the West Indian case for protection. Even more formidable was the duty on non-Empire sugar. Tobacco had borne very high taxes during the war; it continued to do so.

The restrictive policy of the East India Company operated in another way; it served also as a control on imports. The Company monopolized the supplying of tea to Britain, maintaining a price much higher than the world level. The home silk manufacturers were given absolute protection against the foreigner, through a prohibition of imports of silk articles. There was also a formidable duty on raw silk; this meant that the industry was treated as a luxury one, with high prices and small output, and little technical progress, accompanied by a very great deal of smuggling. The Navigation Laws served as a general restriction on foreign trade, especially colonial produce.

But it was home agriculture that constituted the principle element in the protective scheme; the most important control on imports was the duty on corn. Here again tradition played an important part, for it had always been a matter of government concern to protect the incomes of landlords and farmers through the control of the entry of foreign grain, though this object had always been balanced against the needs of the populace for food. But by 1815 the situation was far more difficult than it had been in the eighteenth century. The margin of cultivation had been pushed far out, the farmers were often involved in long leases at wartime rents, there was a general fall in prices in which corn shared disproportionately. It seemed as though the case, so long accepted, for stability in this sector of the economy, was now stronger than ever. The upshot was the celebrated Corn Law of 1815, under which no foreign corn was admissible until the price at home had risen to 80s. per quarter. The importation of malt was very heavily taxed. The pastoral farmers were not slow to demand 'equal favour' in the form of an increase in the duty on imported butter and cheese, obtained in 1816.

Favour to home industry was still in many cases carried beyond protection into bounties or subsidies. The whale fishery was still so aided in 1815. So too were the herring fishery, linen manufacture and the exporting of manufactured silk. In the case of sugar there was a kind of inadvertent subsidy due to the inept operation of the system of drawbacks through which those who had paid duty on imported sugar were refunded this sum when the sugar was re-exported, very often recovering more than the duty paid.

Excise duties were a further important element in the fiscal panoply, having important but arbitrary effects. These were taxes on home production, and were thus additions to domestic costs that were within the control of the state. Most were for revenue purposes, and had been increased during the stringent years of war. Yet it seemed that they could hardly be discontinued to any great extent, for the government, after 1815, though spending less, had lost the Property Tax, repealed in 1816. The tax on leather was onerous, for the manufacturers of footgear and other leather goods, having lost their wartime contracts, were beginning to think in terms of export markets, and felt that the tax they paid on the hides they bought was a serious disadvantage. Salt bore a tax of 15s. per bushel, a vexatious addition to the cost of living, and a serious element in the cost of production in agriculture. Soap was also heavily laden with taxation; so much so that soap-making hidden from the eye of the excise men was an important industry. The woollen manufacturers, great users of soap, complained bitterly of the tax. Paper too bore a heavy impost; those on glass and bricks served, like the timber import duties, to raise the cost of building. Nor do these items by any means exhaust the long and troublesome list.

Most of these taxes involved the government in choosing between the claims of contending industries, or between the need for revenue and the danger of doing social and economic damage. Moreover they were exceedingly difficult to collect, prompting an astonishing display of ingenuity on the part of evaders that might have been better applied to improving the efficiency of production. They also involved the costly maintenance of a large quasi-police force. The excise officers themselves could be obstacles to the improvement of methods; the procedure used in glass manufacture was not the most efficient, but that of which the excise officers could best keep track; a tanner was strictly forbidden to combine the trade of currier, for it was as the leather passed from the one to the other that the excise man made his levy. Such a situation encouraged the continuance of the petty production methods of medieval times. The excise was an incoherent set of provisions, without real principle, and certainly not an effective means whereby the state could improve the efficiency of the economy.

The state was still involved in the regulation of production. Yorkshire cloth was still required by Act of 1765 to be inspected as to quality and stamped by an elaborate set of searchers and supervisors. Scottish and Irish linens were subject to the same system. There were Statutory Worsted Committees in all the chief areas of production with extraordinarily extensive powers.

The geographical mobility of labour, too, was regulated by the state; the Elizabethan Settlement Laws were still in operation. A man or woman who became chargeable on the parish rates could be forced back to the parish of birth or to that parish where he or she had acquired a 'settlement'. This was a serious impediment to movement about the country. But there was no other way of making a provision for the poor without abandoning the principle upon which the system had rested since Elizabethan times, namely parish taxation. Moreover, there was great danger that the expanding towns would find themselves confronted with a disproportionate burden in bad times, with the further result that they might be damaged in their ability to develop new enterprises. The system was not effective, however, with respect to Irishmen and Highlanders, who enjoyed greater mobility in England than did Englishmen.

The Poor Law continued on long-standing principles. In England every person had a right to a basic minimum standard of living in the parish of his birth or settlement. But the rigours with which the Tudors had enforced this combination of immobility and relief had gone. From 1795 it had been the general practice to afford assistance on the Speenhamland system.[1]

Though the managers of the monetary system were provided with a governing rule of conduct, derived from the gold standard system, serving to relate the domestic economy to those abroad, the market for borrowing and lending continued on the archaic arrangements adopted in Tudor times.[2] A ceiling had then been placed on interest rates. The level set in 1713, namely 5 per cent, was still in force. This meant that when a shortage of credit occurred and interest rates rose, once they reached this level no further competitive bidding among borrowers

[1] See above, Chapter 7, section 10, pp. 273, 275.
[2] See above, Chapter 6, section 2.

might occur, so that the allocation of credit between them could not be, legally, on the basis of price. But so great had the need for such a mechanism become that a wide and involved system of evasion had grown up. The trader and financier could usually find a way round the law, but the position of the landowner seeking to raise money by mortgages was much more difficult. Many men, including Bentham, had long held the view that the prohibition of usury was out of date.[1] But parliamentary proposals to change the law were steadily rejected, even in the face of the strong opinion of the Select Committee of 1818. Successive bills down to 1825 intended to do away with control of the rate of interest failed to pass; the commercial crisis of 1825 showed how great a nuisance it could be. But for some ten years after 1825 the market rate of interest was usually less than 5 per cent, a circumstance that made it possible to shelve discussion. Legislators seemed to have immense difficulty in facing this sanctified relic of the later Middle Ages. No one was prepared to argue positively in its favour, yet the government could always reject change on the ground that the public mind was not prepared, and embrace the question-begging argument that the law was not in fact an effective barrier to credit transactions. Even though Parliament consisted mainly of landowners, who often found great difficulty in borrowing, it could not determine to end the system.

There was another important respect in which older ideas maintained their out-moded grip on the minds of men. During the long war, Pitt and Addington had developed the property or income tax so effectively that at long last it was possible to assign fiscal burdens more or less in proportion to net income.[2] But this had involved individuals in the necessity of reporting to the public authority annually as their incomes accrued. Against so doing there was the strongest hostility. The property tax provoked a formidable list of adjectives of execration. So unanimous was the public hostility that the tax would only be borne in the face of an enemy. When a brief respite of peace came in 1802 it was immediately repealed and again in 1816 as soon as was practicable, after disposing of Napoleon, it was again withdrawn, this time in the teeth of the Chancellor's appeal that it should be

[1] J. Bentham, *Defence of Usury*, London, 1787, reprinted in W. Stark, *Jeremy Bentham's Economic Writings*, London, 1952, vol. I; Elie Halévy, *The Growth of Philosophical Radicalism*, London, 1934, pp. 110–13.

[2] A. Farnsworth, *Addington, Author of the Modern Income Tax*, London, 1951; A. Hope-Jones, *Income Tax in the Napoleonic Wars*, Cambridge, 1939. For the general fiscal history of the nineteenth century see Sydney Buxton, *Finance and Politics; an Historical Study, 1783–1885*, London, 1888.

The Politics of an Industrializing Society

continued for at least a little while longer.[1] Here was a dilemma. The only way to remove the state from the industrial and trading sectors of the economy, upon which the extensive and involved system of tariffs and excise weighed so heavily, and at the same time provide adequate means for the services of the state, was to continue the tax on incomes. But opinion of an almost hysterical intensity, aggravated by governmental mishandling, was against it. There remained, however, the 'assessed taxes' — on men servants, dogs, guns, carriages, armorial bearings, together with legacy and stamp duties.

Though the legislature could apparently dispose of the problem of the property tax, and with it the principle of payment according to income, by ending it, no such action was possible with respect to the National Debt. This had been increased enormously by the war from some £248 million in 1793 to £839 million in 1817.[2]

In 1827 interest on the debt (£29 million) accounted for over one-half of the total of the public expenditure of the United Kingdom (£56 million).[3] This greatly aggravated the taxation problem. So large a public debt meant that what was virtually a new order of society had come into being, namely the fundholders. These had made loans to the government in a time of high prices; now as the price level tumbled, the real value of the securities thus acquired rose, as did the real burden of the interest payments. Not surprisingly the fund-holders, gaining as the economy contracted, were the object of much criticism. The implications of so vast a debt and so large an interest bill were most disturbing to the liberal thinkers of the day. Ricardo, to the consternation and bewilderment of many of his friends, advocated a kind of surgical operation: a levy upon capital in order that once for all the debt with all its troublesome implications could be done away with.[4] But Ricardo's suggestion found no more favour than did that of Cobbett's proposal for an 'equitable adjustment'.

It is hardly surprising that the post-war Parliaments, beset by clamours for strong action to relieve distress and restore conditions of expansion, yet involved in an immensely complicated system derived from the miscellaneous and often ancient contents of the Statute Book, found themselves baffled, and, unwilling to accept the *a priori* answer to their

[1] F. Shehab, *Progressive Taxation: a Study in the Development of the Progressive Principle in the British Income Tax*, Oxford, 1953, pp. 60–9.
[2] E. L. Hargreaves, *The National Debt*, London, 1930, pp. 108, 134.
[3] Clapham, op. cit., vol. I, p. 318; also W. Page, op. cit., vol. II, p. 38.
[4] P. Sraffa and M. H. Dobb, ed., *The Works and Correspondence of David Ricardo*, Cambridge, 1951–55, vol. V, pp. 41, 51, vol. I, pp. 247–9.

difficulties offered by the political economists, continued the traditional course of piece-meal adjustment at those points where action was most needed, where the course was clearest, and where political pressure was greatest.

4. THE ATTACK ON OLIGARCHY 1832

The gradual lessening of official control of the economy was far from meeting the political aspirations of the day. Though economic recovery came in 1821, insecurity was still general, the gains of expansion seemed to go to millowners and landlords, and the exclusion of the mass of men from the process of government was an increasing affront as the concept of 'the will of the people' entered more and more heads. The agitation for the reform of Parliament, so vigorously put down after 1819, had never died and was ready now to break out with renewed vigour. Though the high Toryism of Eldon, Sidmouth and Castlereagh had given place to the liberal Toryism of Canning and his followers, this was not enough to meet the new aspirations of the workers. In 1830 the French revolted again, overthrowing the rulers imposed upon them after Waterloo and disturbing the peoples of Europe anew with their aspirations toward full democracy. Trouble began in Kent, following the example of the French peasants, led by the mysterious 'Captain Swing', and soon spread to sixteen counties. Strikes broke out in the cotton districts. An alarmed government was obliged to pacify the countryside by encouraging the magistrates to impose vigorous penalties; some 457 men were transported.[1]

A flood of reform literature covered the country, with Cobbett a leading contributor. Following the Birmingham example Political Unions appeared in the great cities, capable of mustering mass demonstrations on an unprecedented scale.[2] In spite of the close discipline practised by the unions, fears for the security of property were inevitable. Indeed the orderly pursuit of a political end by such great gatherings suggested a frightening growth of conviction. The demand was plainly made for the enfranchisement of male adults, secret voting through the ballot, and shorter Parliaments.

The new Whig government taking office in 1830 found itself

[1] E. Halévy, *A History of the English People in the Nineteenth Century*, vol. III, 2nd ed., London, 1950, pp. 7, 15.

[2] Asa Briggs, 'Social Structure and Politics in Birmingham & Lyons, 1825–1848', *B.J.S.*, 1950; 'Thomas Attwood and the Economic Background of the Birmingham Political Union', *Cambridge Hist. Jr.*, 1948; see also above, Chapter 4, section 6.

holding the dog by the ears. Its members, to a considerable degree, owed their electoral success to their flirtations with the cause of reform. Other pleas and demands were now being made, as was inevitable in the atmosphere of intense political excitement. Retrenchment and the reduction of taxes now seemed a pale programme indeed. The government was now confronted with the problem of dissipating the pressure and excitement without producing either a revolutionary explosion or an unworkable system. For though the Whig leaders wished to diminish the influence of Court and Cabinet, it was no part of their intention to destroy the power of the great landlords among whom they themselves were numbered. Middle-class Benthamite radicals were busy sharpening this dilemma, providing arguments and support for an extension of the franchise.

The government sought an escape in the form of a Bill abolishing the nomination boroughs (some 120) and the tiny constituencies (some 46), in all about one-quarter of the membership of the Commons, redistributing these seats among the great centres of population. In addition the franchise was to admit adult males in the boroughs who inhabited houses with an annual rental of £10 upward thus doubling the existing electorate of half a million. The right to a voice in the nation's affairs was to depend upon having a considerable stake in them: property was to be secured by giving the vote to men of property. The duration of parliaments was to be unchanged and the ballot was not mentioned. But to many Tory ears the proposals sounded so radical they were greeted with derision. King William IV, however, was grateful for the limitations of the programme, and accepted it. The nation too was happy with such terms and showed its mind in general rejoicing. But it was not to be; the Bill foundered in the Commons, and Parliament was dissolved.

A hectic election of one week followed, to choose the men who would not merely govern, but who would decide upon the future mode of government. The Whigs were returned with a great majority Three months' intensive parliamentary debate followed, for principle and implementation were inextricably mixed in defining a borough, in the grouping of towns, in the balance between the industrial north and the agrarian south, in the representation of London, and in the validity of the £10 franchise as a test of voting status. The Bill, slightly modified, was thrown out by the Lords during the night of 7–8 October 1831. So great was the reaction that there were fears of revolution. The rioters were out in many areas, the anti-Bill bishops were the object of public execration; the palace of His Grace of Bristol was destroyed. Moreover economic difficulties were returning with bad harvests, unemployment, and falling wages.

As the stresses of the struggle mounted, so the fundamental divergence in outlook between the elements of the reform movement became explicit. The middle class, through the National Political Union, strictly limited its case to the need to circumvent the noble obstacles to the Bill. Among the workers, in the excitement of so long and bitter an argument, there had developed the 'National Union of the Working Classes'; it reverted to earlier, much more radical, demands. Basing its claims upon natural right, it called for the abolition of all hereditary privilege, and for manhood suffrage, the ballot and annual parliaments. But working-class concern must not be exaggerated; in Manchester and Leeds journalists seeking to rouse working-class support for reform found that indifference only gave way to interest in the rather paradoxical hope that a wider franchise would be used against its middle-class protagonists to bring in effective Factory Acts.[1]

The Bill was presented to Parliament for the third time. Various formulae had been changed in the construction of constituencies and the allocation of seats, but the proposals were substantially the same. Two more months of wrangling followed. As the Lords further procrastinated, the revolutionary danger mounted. The King was forced to give an undertaking to dilute the peerage by sufficient new creations to carry the Bill. At last it passed the Lords.

The new House of Commons contained an overwhelming Whig majority. The Whigs were to rule, with a brief interlude, from 1832 to 1841, with Melbourne succeeding Grey as Prime Minister in 1834. It was not a business man's House; rather, it was indistinguishable in social composition from its predecessor. As late as 1857 there were scarcely three Members of Parliament who had held office who did not belong to the same classes that shared the government of the state before 1832; of these none held high office in Tory administrations.[2] But the way was now open for a very gradual change in composition as more middle-class men of business origins turned to political pursuits and learned the arts of persuasion.

The Whigs had now to crown their only partly intended constitutional revolution with a programme of legislation mainly aimed at economic change. Just as they had been the heirs to political programmes

[1] Donald Read, *Press and the People, 1790–1850*, London, 1961, chapter IV.
[2] 'The Election, Its Issues and Its Opportunities', *Econ.*, 21 March 1857, p. 305.

mooted at all levels of persuasion from the mild to the revolutionary, so too in nearly all the leading aspects of economic and social life they found proposals of varying degrees of radicalism and at various stages of gestation.

The administration tackled the problems of economic stability and more effective town government, the first with the Bank Act of 1833, the second with the great Municipal Corporations Act of 1835, destroying the often amiable oligarchies of the older cities, instituting new machinery in the new, putting civic affairs onto a new basis of democratic government.[1] A grievance of long standing was dealt with: the Tithe Commutation Act of 1836 settled a vexed question of ecclesiastical taxation.[2]

A striking series of humanitarian measures were passed. The owning of slaves in the colonies had involved arguments going back for at least a generation; the new Ministry in 1833 freed the slaves and forbade the practice of owning, paying no less than £20 million to those deprived of human property.[3] The Factory Act of 1833, setting minimum working conditions for children and teenagers in the factories, was the first real infringement of millowners' liberties in the interest of a group incapable of self-defence.[4] It was followed by the Chimney-sweep Acts of 1834 and 1840, to protect a group of children in whom vulnerability had encountered feckless rapacity to the most horrifying degree. (Two further Acts were required, 1864 and 1875.) Women and children were no longer to be employed underground after Lord Shaftesbury's Coal Mines Act of 1842. So was begun the trend toward the elaboration of a new code of control necessary to replace the long archaic remnants of the old gild system.

Tentative steps were taken to deal with the question of national illiteracy.[5] The Stamp Duty on newspapers, so inhibitive on the creation of effective public opinion, was reduced to one penny in 1836. The Penny Postage, an inestimable boon, came in 1839.

The Reform Ministry thus had a real claim to the gratitude of the workers. But their treatment of the poor more than countered all their other actions; it gave rise to the bitter charge that the Whigs, far from honouring the obligations arising out of the Reform agitation, had betrayed the workers. The Poor Law Act of 1834, with its refusal of

---

[1] See above, Chapter 6.
[2] See above, Chapter 5, p. 181.
[3] W. L. Mathieson, *British Slavery and its Abolition, 1823–1838*, London, 1926.
[4] See above, Chapter 7, section 7.
[5] See above, Chapter 7, section 8.

outdoor relief, was bitterly resented, especially in the north. Yet the government dared not let the old system of more or less indiscriminate relief of the indigent continue unchanged.[1] Moreover, for all the benefits of Whig paternalism, the politically conscious workers, having shown that their power to debate and to organize was matched by their self-discipline, were affronted that the old aristocratic indifference to their right to participate in government could so soon reassert itself, and that so few middle-class voices were raised in protest.

5. FURTHER LIBERATION OF THE MARKET MECHANISM

At a less exciting level the dismantlement of the old techniques of economic control had been proceeding steadily since 1815. Here the clue to action was the consideration of highly specific problems, with the discussion of general rights or principle kept to a minimum.

In 1822 a great tangle of obsolete trading restrictions was removed. In 1825 Huskisson carried his Navigation Act, codifying the law.[2] Conditions were made easier for those engaged in the re-export trade. Yet the Navigation system remained seriously restrictive, continuing, as it did, regulations intended to retain the long sea voyages for British ships.

Other nations were becoming increasingly restless under the British system of discrimination against foreigners, and began to retaliate. In order to arrest the competitive growth of these annoyances, Britain was forced to come to terms first with the U.S.A. in 1815, and subsequently with other countries. So considerable were the benefits of so doing that in 1823 another of Huskisson's Bills became the Reciprocity Duties Act.[3] Under it, over the next seven years, treaties of reciprocity were concluded with most European countries, the U.S.A. and the new South American Republics. The trend continued and gained a new character in the negotiations with Prussia. The latter having no colonies to open to reciprocal trade agreed instead to accord to Britain most favoured nation treatment, a principle that was to be rapidly extended. The scale and intricacy of transactions was now making the administrative task of tariff collection demanded by the old system impossible.

The long-standing problem of the relative positions of the West and East Indies in the national economy underwent important changes.

---

[1] See above, Chapter 7, section 10.
[2] Alexander Brady, *William Huskisson and Liberal Reform*, Oxford, 1928, pp. 122-9.
[3] Clapham, op. cit., vol. I, pp. 331, 333; Brady, op. cit., pp. 92-4.

By an arrangement of 1819 British goods could be carried to the Company's ports in China in American ships.[1] This had the effect of preserving the Company's prohibition of British free traders in China, but allowed American free traders to push British goods in Chinese markets. This was bound to appear highly anomalous to British traders, though it meant increased sales for British manufacturers. In 1833 China, the last element in the East India Company's trading monopoly, was thrown open by the Reform ministry. The Levant Company gave up its charter as early as 1821.

It had long been thought right to inhibit the East India Company in competing with the West Indies in supplying the home market with primary goods. But with private trade with India open after the Act of 1813, and with Indian agriculture responding to the new European markets, it was inevitable that the claim for parity between East and West Indies should become much stronger. The West had been favoured both with preferential sugar and coffee duties, and the West Indian planters were not without potent arguments that their preference should be continued. In particular, during the agitation for the legal abolition of slave owning in the Empire, the West Indian planters had convinced themselves that they were entitled not only to the compensation prescribed under the Act of 1833, but that in addition, because of the new labour problems that would arise, they should enjoy continued protection against the produce of the East Indies. But the system involved so great a limitation on the growth of consumption of the articles concerned that the government could resist no longer. In 1835 both elements of the Empire were placed upon the same footing. Foreign sugars were still effectively excluded by an additional tax of 5¼d. per pound. The government was able to continue to resist the admission of Cuban and other sugars partly because of the argument that they were still slave produced and ought not to compete with the free sugar of the British colonies.

Other controls on imports were weakening. The timber duty was reduced in 1821. Three years later silk manufacture was placed on a new basis; the duty on raw silk was lowered and it now became possible for foreign manufactures to enter though they were required to pay a duty. The reduction in the price of raw material caused a burst of prosperity and activity in the industry, though this was checked by the crisis of 1825.[2] Tobacco remained the subject of severe taxation, con-

[1] Clapham, op. cit., vol. I, p. 332.
[2] Porter, op. cit., p. 217.

343

tributing a good deal toward protectionist opinion in the United States. The import duty on raw cotton was halved in 1833 and ended in 1845.

But it was with respect to home industry that the movement toward freedom of trade was most pronounced. The tax on salt, reduced in 1823, ended in 1825. The duty on soap was substantially lowered in 1833 and that on paper in 1839. The old taxes on coal carried coastwise or inland were removed in 1831. The ending of the beer tax in 1830 was worth about £1 per year to the average labouring household, a relief celebrated with a great burst of drunkenness.[1] Almost all the products thus unburdened showed an increase in consumption that compensated for the reduction in the rate of tax.

The attitude to exports too was changing. The attempt to deprive foreigners of British goods was becoming increasingly unrealistic. In 1825 the export of wool was permitted. In the same year, though the export of machinery was still subjected to a general prohibition, a system of Board of Trade licensing was adopted, which pushed the regulation further toward nullity.

The system of bounties was in its last phase. Those available to the whale fishery and to the exporters of silks went in 1824. The herring fishery lost its support in 1830 and linen in England in 1832. In industry the remaining elements of regulation were going the same way. In the twenties the sealing and stamping Acts ceased to operate in the Yorkshire cloth trades. The regulation of Scots linens was abolished in 1823. The worsted committees, though continuing in existence, were passing into effective abeyance as the domestic industry with which they were concerned was gathered into the factories. In 1836, Parliament formally ended the duty of the Justices of the Peace to supervise the price of bread and to regulate bakers' profits.

The old controls on transactions in money and in shares were rapidly disappearing. The Bubble Act of 1720 was repealed in 1825, making possible joint-stock companies with transferable shares. Another legislative restriction, Sir John Barnard's Act of 1733, intended to limit speculative transactions, though it remained on the Statute Book until 1860, was dead by the early thirties. It was clear, too, that the principle of flexibility should be extended to borrowing and lending: in the Bank Act of 1833, the old usury law, so long evaded, was made inapplicable to bills of less than three months' term when discounted by the Bank of England: men might now bid as high as they liked for short-term funds.

[1] Clapham, op. cit., vol. I, p. 560.

With the nation increasingly committed to industry, the legislature, short of attempting a reversal of this trend, had no alternative but to simplify trading procedures, to seek by example to encourage other nations to diminish their economic controls, and to make available to its own citizens a wide range of primary products at the world price, free of duties.

6. THE WORKERS' SEARCH FOR A FORMULA: SOCIALISM AND CHARTISM

Though bitterly frustrated by the Reform Act, the workers' leaders could not generate another great agitation until the fatigues of the early thirties had passed. But there could be no lasting rest, for the aspirations for a new system were now so strong that, in spite of setbacks, they were bound to find expression.[1]

Whereas the working-class reformers had fought for the franchise without any great precision of mind about the programme of legislation that was to follow victory, Robert Owen, basing himself on views developed during a highly successful manufacturing career, was contemptuous of mere political reformers, holding that no tampering with the constitution would meet the needs of the new industrial society.[2] It was dangerous to seek to place power in the hands of an ignorant and debased commonalty. Regeneration must come first; the constitution could easily be adjusted afterwards. Productive communities organized on a voluntary cooperative basis would demonstrate how, instead of seeking fulfilment through mutually destructive competition, men might combine to produce both efficiency and brotherhood.[3] No one would be threatened and no one coerced, for as the success of the communities became manifest the nation would follow the demonstrated path. This programme attracted both aristocratic and middle-class attention and support, for the idea of obviating conflict between classes by a means that combined high output with social reconciliation was most attractive.

But the Owenite communities were failures. Owen had hoped that they would regenerate their members so that the early troubles of

---

[1] For a moving case of agrarian unrest focused upon messianic delusion see P. G. Rogers, *Battle in Bossenden Wood, the Strange Story of Sir William Courtenay*, Oxford, 1961.

[2] See above, Chapter 7, section 7, p. 246. For an anthology of opinions by and about Owen, see A. L. Morton, *The Life and Ideas of Robert Owen*, London, 1962; below Chapter 10, section 4, pp. 407-9.

[3] Ralph Miliband, 'The Politics of Robert Owen', *J.H.I.*, 1954.

adjusting from the competitive society to the cooperative would be overcome and a continuous growth of both individual and communal character would follow. Instead the regenerative process was defeated by the assertion of attitudes that were older than industrial society.[1] Owen returned in 1829 from America where he had established New Harmony in Indiana. He had gone to a new society in the hope that it was less confirmed in mistaken ways, but was disappointed.[2] The early support from the upper orders of society at home had been destroyed by the gratuitous parading of Owen's heretical views on money and religion, for these seemed to reveal him not as innocuous experimenter but as doctrinaire revolutionary.

The idea of cooperation both for production and selling was by no means wholly barren. The county of Lanark had produced a Victualling and Baking Society as far back as 1800, long before Owen was to give the county's name a more resounding celebrity. Nineteen such societies in England and Scotland, founded before Victoria's accession in 1837, were still trading in 1898.[3] Finding capital for the Villages of Cooperation had always been difficult; retail cooperation was seen by some members of the movement as a way of providing it.[4] The idea that men should form their own enterprises to cut out the great proliferation of middlemen, many of whom were guilty of adulteration and short weight, was natural enough. But special conditions were required for success, including a soundly organized scheme designed to do a job realistically conceived, carried out by men not debased by industrial subservience, but of a high order of intelligence and dedication.

The cooperative principle, moreover, had to come to terms with trade unionism.[5] The great outburst of militancy that had followed the repeal of the Combination Acts in 1824 had by no means spent itself. But hard times postponed the formation of large unions until 1829 when the Spinners were organized; the Potters followed in 1830, with the Builders soon after. The moving spirit in the new unionism was John

[1] W. H. G. Armytage, *Heavens Below: Utopian Experiments in England, 1560–1960*, London, 1961.

[2] A. E. Bestor, *Backwoods Utopias; the Sectarian and Owenite Phases of Communitarian Socialism in America: 1663–1829*, Oxford, 1950.

[3] E. W. Brabrook, *Provident Societies and Industrial Welfare*, London, 1898, p. 135.

[4] Sidney Pollard, 'Nineteenth Century Cooperation: from Community Building to Shopkeeping', in Asa Briggs and John Saville, eds., *Essays in Labour History*, London, 1960, p. 74.

[5] For the unions generally see S. and B. Webb, *The History of Trade Unionism, 1666–1920*, London, 1920.

Doherty; its primary objects were the raising of wages and the shortening of the working week.

There were thus two ideas that might offer the basis for the revival of a general attempt at change: cooperation and the unions. Owen was the biggest figure on the scene, with a number of notable followers capable of skilled demonstration of cooperative ideas.[1] The Owenites stood for non-coercive change through the cooperative principles; the more militant unionists stood for a mass demonstration of workers' solidarity, the preliminary to a direct political challenge, with the final form of society obscure. Neither side could carry the other in argument, and neither could be successful alone. Was it then possible to unite the two approaches?

Before the main struggle was joined, the Owenites embarked upon an ancillary movement. It was to combine the principle of cooperative marketing, intended to obviate the parasitic middleman, with a system of labour value, that would by-pass a defective monetary and credit system no less replete with exploiters. In 1832 the first Labour Exchange was opened in London, followed by another in Birmingham. To them workers brought their products, receiving vouchers or Labour Notes stating the estimated cost of the raw materials and the amount of labour time embodied in the goods. The estimation of the latter seems to have been left to the workers themselves. When purchases were made, the Notes were tendered at the Exchanges, under the same principle of labour equivalence. But the Exchanges were little more than a side-show; Owen seemed hardly aware of their lack of success as, resting upon a fine disdain for consumers' demand, they became congested with unwanted articles. His mind was now working upon a breath-taking scheme.

The trade union movement was to seek membership and unity on a national basis.[2] The power thus generated would make possible a peaceful application of the principle of cooperation over the entire face of industry. For each enterprise, including workers and management, would cease to follow the rules of the competitive game, and would become a brotherly association or syndicate. So would bitter conflicts be ended and new energies released for the productive task. Though there was a sound element in the industrial psychology of this, seeking

[1] See below, Chapter 10, section 4.
[2] G. D. H. Cole, *Attempts at General Union: A Study in British Trade Union History, 1818-1834*, 2nd ed., 1953.

to diminish the sharp dichotomy between masters and men, the plan took no real account of the way in which the syndicates would be related to one another, the role discharged, however faultily, by market competition. The proposal for a general congress of such cooperatives or syndicates, sitting instead of Parliament in London, regulating the production of the nation, with goods exchanged on the basis of labour equivalents, seemed a disconcerting prospect for all those who were immune to the Owenite spell.

But immense numbers had yielded to it. The trade unions in 1833 grew at a rate unknown in all their experience either earlier or later. They were united in the Grand National Consolidated Trades Union. With so hectic an increase in membership (with perhaps 800,000 unionists by November) the very scale of success began to work its alterations. Excitement was inevitable; so too was a new emphasis upon militancy. Even Owen himself was briefly intoxicated, forgetting his misgivings about putting political power into uneducated hands. If Owen could succumb, it is not surprising that those whose approach was more embittered by direct experience should go much further. A general strike was called for. Soon Owen found that militancy had become dominant. He and the moderates tried to divert the movement away from strikes toward cooperative production and limitation of the working week. The fighting unionists preferred direct action.

Owen's judgement of the behaviour of the employers was as unrealistic as was his view of the workers. It was now the turn of the masters to show their strength. They did so by a most efficient set of lock-outs, demonstrating that though even more loosely organized than the workers, they could move effectively to assert their solidarity. The closing of the works was accompanied by the enforcement of the 'document' – a signed undertaking by individual workers not to threaten strike action again. The climax came in March 1834 when the Dorchester Labourers were sentenced to be deported. By this time Owen was pleading for restraint and industrial peace; the national executive of the Grand National shared his reaction. The militant leaders, James Morrison, editor of *The Pioneer*, and J. E. Smith, editor of *The Crisis*, were both expelled from the movement and their journals suppressed. But the constituent unions scorned conciliation, going down to piecemeal defeat. The Grand National was all but destroyed by the late summer of 1834. Its collapse meant the end, also, of the Labour Exchanges. Owen was then sixty-three; he had a further twenty-four years of life.

It might be thought that such a failure would convey lessons of caution and conservatism, suggesting that close and sound organization with a precise definition of objectives were essential. But the thoughts and emotions that had been stirred were too powerful to subside. The Reform agitation, so closely followed by Owenite socialism, had brought a great many men to such a degree of political consciousness that they could not revert to passivity. Moreover the brief respite of prosperity ended with the return of depression in 1836. The economy had entered a phase of critical depression that was to last for at least six years.[1]

The Chartist movement was the response to this situation. The Peoples' Charter was issued in 1838; its rejection by Parliament for the third time in 1848 was the effective end of the movement. These ten years saw the last direct threat from the English working class to the social and political framework upon which the industrializing process rested.

Chartism was something of a political paradox. The movement itself contained an extraordinary diversity of men who were never within sight of an agreed programme of economic and social change. Yet their political demands, though not new, were utterly precise. They called for six specific alterations in the constitution – adult male suffrage, the abolition of the property qualification for M.P.s, the payment to them of a salary, the secret ballot, equal electoral areas, and annual parliaments. Such an exactness of objectives was attained by the device of leaving questions of policy unmentioned, and concentrating on altering the principles of selection upon which the choice of governors rested. Sound choosing of men would produce sound policies, with the choosing done secretly and often.

But this obviation of differences was only maintained with great difficulty. There were disagreements that sprang from the way in which hard times and a changing way of life affected different localities: the Chartism of Birmingham, Sheffield, Leeds, Bristol, all had their distinctive elements.[2] There were differences that stemmed from the divergent characters of the leaders. The moral force element of the movement, led by William Lovett, the drafter of the Charter, and his London Working Men's Association, sought to work through argument and persuasion expressed through a reasoned programme, in

[1] See above, Chapter 2, section 4.
[2] Trygve R. Tholfsen, 'The Chartist Crisis in Birmingham', *I.R.S.H.*, 1958; Asa Briggs ed., *Chartist Studies*, London, 1959.

alliance with the middle-class philosophical radicals who followed Bentham.[1] But others, like their predecessors in earlier agitations, could not think and act in this desiccated way. Such men found force, at least as a threat, irresistible. Thomas Cooper was a bitter man who could not purge his soul of emotion and even violence.[2] Bronterre O'Brien, unlike Lovett, was contemptuous of Owen's efforts to gain support from the powerful and wealthy for their own eclipse; he called upon the workers to strive for their own salvation. They were not to be side-tracked into the Corn Law agitation. They were not to become bemused by the middle-class concept of individualism, but were to seek effective solidarity. This was perhaps the hardest injunction of all, for the Chartist leaders were men of powerful character consumed with a sense of mission, and therefore liable to collisions with their fellows who, though sharing their ends, were committed, in the unassailable recesses of their beings, to other means. George Julian Harney believed that only revolution could improve the workers' conditions; he called himself the English Marat and invoked the spirits of Danton and Robespierre, especially in the hope of international revolution.[3] Feargus O'Connor expressed a Cobbett-like hatred of millowner and factory, together with a wish to return to older values through land reform.[4] Though there may have been some vestige in the minds of some of these men of the old respect and affection for the landed class, they had none whatever for the new men of business, perceiving with almost obsessional clarity the power they would eventually wield if not challenged. But though a few, like Harney, realized the need for both a working-class party and a developed Socialist programme, the physical force Chartists, like earlier attackers of the constitution, had little real idea what they would do if the recourse to violence did produce a rupture of society. This turmoil of ideas was presented from platforms all over the country, to a population largely illiterate, yet capable of responding to O'Connor's insistence that their local needs could only be met by a national programme.

The years of greatest support for the movement were between 1838

[1] William Lovett, *Life and Struggles of William Lovett in his pursuit of bread, knowledge and freedom*, London, 1876.

[2] Thomas Cooper, *Life, Written by Himself*, London, 1872; *Eight Letters to the Young Men of the Working Classes*, London, 1851.

[3] A. R. Schoyen, *The Chartist Challenge: A Portrait of George Julian Harney*, London, 1958.

[4] D. Read and E. Glasgow, *Feargus O'Connor, Irishman and Chartist*, London, 1961; see also Rachel O'Higgins, 'The Irish Influence in the Chartist Movement', *P. and P.*, 1961.

and 1842. Discussion proceeded, organization advanced, conventions were held. Rioting in Birmingham and South Wales did something to demonstrate the value and risks of such action, landing Chartist leaders in prison.[1] The first National Petition was presented to Parliament in 1839 and was rejected. A second London Convention was held and a second petition shared the fate of the first in 1842.[2] Strikes and rioting were renewed, with Cooper and O'Connor, refusing the constitutional verdict, among the principal instigators. The most notable of the disturbances was the apparently spontaneous 'Plug Plot' in south Lancashire, not directly connected with Chartism, based upon the idea that the workers should coerce their masters by emptying the boilers of the steam-engines, arousing once more the futile hope of a general strike. But trade was bad; indeed such a stoppage of work was favourable rather than otherwise to employers embarrassed by excessive output. There were plenty of men, especially in the industrial north, whose patience was being exhausted; more than two-thirds of the employable workmen in Sheffield in 1842 were either on short time or wholly destitute; the handloom weavers had even less to lose. A third Convention and a third Petition in 1848 suffered parliamentary rejection, though not without provoking considerable official trepidation, with the Duke of Wellington assuming his last command, the defence of London.[3]

So ended the agitation. The economy was now ready to begin its long ascent, aided by repeal of the Corn Laws, the railway, and the expansion of other economies.[4] The Ten Hours Act of 1847 seemed some assurance that the governing classes were at last taking some cognisance of the vulnerability of the industrial worker to exploitation. O'Connor had already begun another attack: in 1845 he had formed his O'Connorville National Land Company intended to promote peasant proprietorships through cooperative land purchase; the project found many who wished to participate, but was wound up in 1849.[5]

Though Chartism expired in the year of European revolutions its active members did not cease to have influence upon the English political scene. Old Chartists, like Ernest Jones, continued to propagandize and popularize for many years; indeed there was a personal bridge between

[1] David Williams, *The Rebecca Riots: A Study in Agrarian Discontent*, Cardiff, 1955.
[2] G. Kitson Clark, 'Hunger and Politics in 1842', *J.M.H.*, 1953.
[3] F. C. Mather, *Public Order in the Age of the Chartists*, Manchester, 1959.
[4] See above, Chapter 2, section 5.
[5] Joy MacAskill, 'The Chartist Land Plan', in Asa Briggs, ed., *Chartist Studies*, London, 1959.

the old valiants of Chartist days and the new militants of the later eighties.[1] Many such men and their wives held to the ideas of Owenite tocialism through cooperation, and to even earlier aspirations toward a society dedicated to equity and fellowship. At the local level in parsicular those who had served after 1838, with their energetic and steadfast work, continued to exercise great influence through trade unions, the cooperative movement, temperance, education, local preaching, civic government and in many kinds of voluntary association for mutual improvement, all this long after the excited multitudes had been pacified by prosperity. Most of such men were craftsmen of one kind or another, miners, and small traders, who by their own efforts at self-improvement had to seek their own answers to the great questions affecting society. Among them were discussed the problems of the practical implementation of the better life, so conspicuously omitted from the public programme of the Chartist movement; among them was most strongly felt the urge, culminating in a demand, that they should have some part in making the laws that all must obey.

There were many who opted out of England, removing in increasing numbers to the colonies and especially the United States.[2] Emigration societies flourished, and trade unions helped their members to go. With them went a great deal of frustrated radicalism, to express itself in a new situation, but their departure weakened the Chartist proters, especially in the later forties.[3]

7. THE OPEN ECONOMY: THE FORTIES AND AFTER

The career of Sir Robert Peel included two dates of the greatest importance in the history of British fiscal policy. In 1842 he brought back the income tax; in 1846 he repealed the Corn Laws.[4] He thus both made possible the placing of the fiscal system upon a new footing, with all that was implied for the economic role of the state, and demolished the great central citadel of the protective system, the tax on foreign corn.

---

[1] John Saville, ed., *Ernest Jones: Chartist, Selections from the Writings and Speeches*, London, 1952.

[2] Wilbur S. Shepperson, *British Emigration to North America. Projects and Opinions in the Early Victorian Period*, Oxford, 1957, 'Industrial Migration in Early Victorian Britain', *J. Econ. H.*, 1953; M. L. Hansen, *The Atlantic Migration, 1607–1860, A History of the Continuing Settlement of the United States*, Oxford, 1940.

[3] Clifton K. Yearly, *Britons in American Labor: A History of the Influence of U.K. Immigrants on American Labor, 1820–1914*, Oxford, 1958.

[4] See above, Chapter 2, section 5.

But it is by no means clear that the changes going on in his mind, or in those of his supporters, were as dramatic as the changes in policy.[1] The officials of the Board of Trade in 1840 made their views felt through the Report of the Select Committee on Import Duties; the Report convincingly demonstrated how wasteful and inhibitive the indirect tax structure was.[2] But direct taxation of incomes was still abhorred by many. There was an apologetic air about the revival of the income tax; Peel presented it merely as a means of filling the gap in the revenue that would occur in the short run in consequence of his various tariff reductions. It was his hope that the income tax could be dropped once the customs and excise revenue had recovered through enlarged consumption; he had no wish based upon principle to rest fiscal burdens on incomes.

In 1842 Peel revived the efforts of Wallace and Huskisson to clear up the tariff confusion. The following year he made the export of machinery, once so fiercely insisted upon, free. But he was not in any advanced sense a free trader. He could appreciate the arguments of liberation, but those pointing in the opposite direction were also powerful in his mind. The greatest tariff and trade issues had to do with food: the Corn Laws and the preference on sugar. Then came the question of the merchant marine as protected by the Navigation Laws, followed by the timber duties. The sugar problem was simplifying itself. The West Indies were now far gone in dilapidation; more important for British expansion were the South American countries, especially Brazil and Cuba. But the West Indian interests were still powerful enough to continue to ensure that such sugars were kept subject to a higher duty than that on British colonial produce.

The greatest public debate on economic policy in the Britain of the nineteenth century centred upon the case for repealing the Corn Law of 1815. The latter had been modified in 1828 and in 1842 by the adoption of a sliding scale, making the import duty dependent upon the home price; as the latter rose the duty fell. But farmers and landlords were still effectively protected against foreign corn so that to a very considerable degree their incomes were insulated from world markets as supplied by Poland, Prussia, Russia, and the United States. In effect the Corn Laws were a mechanism not only for the control of food imports

---

[1] For the attempt to provide the Conservative party with a set of general ideas, see R. B. McDowell, *British Conservatism, 1832–1914*, London, 1959; Arvel B. Erickson, *The Public Career of Sir James Graham*, Oxford, 1952.

[2] Lucy Brown, *The Board of Trade and the Free Trade Movement, 1830–42*, Oxford, 1958.

but also for the regulation of relative incomes between the agricultural and industrial sectors of the economy. The effect went further than this, for the ability of the economy both to maintain stability and to grow could be much affected by the law.[1] The issue was: should Britain create the conditions for her economic integration with Europe and the world, accepting the full implications of industrialism?

The full range of arguments on the respective sides was elaborated.[2] The opponents of the Corn Laws used the Smith-Ricardo model of the economy as the general framework of their case.[3] Their simplest argument was that of cheaper food, which must, other things being equal, produce higher real wages. This could become a highly emotive plea, for bread had a kind of Biblical sanctity, so that interference with purchases from the cheapest source could be regarded as an affront to divine providence. Then, too, with cheap bread, manufacturing costs could be lowered or their rise dampened, for the maintenance of the labour force required less in terms of real effort. If Britain bought food from abroad, offering in return the goods in which she had a great manufacturing advantage, this would help to keep costs down all round. In this way it was possible to reconcile the argument of higher real wages with that of lower manufacturing costs. But benefits on the side of imports were only part of the picture. Exports too would be larger, for (as the problem appeared to business men), one of the greatest difficulties encountered by them in practice was how to obtain returns, how to find some range of goods that could be brought into Britain in return for manufactured exports. Foodstuffs and raw materials were ideal from this point of view, for they did not challenge British manufactures in the home market.

Even more important, if foreigners could pay their international debts in such goods, their incentives to develop industry at home would be lessened; or, to give the argument in more direct terms, keeping out foreign corn meant encouraging foreign manufacturers. This view could be even further extended to mean that the most effective way to discourage foreign industry was to let foreign primary products in freely. In addition there would be greater stability of prices, for corn dealers would no longer find the various ports opened and shut in their

---

[1] For the operation of the Laws see J. D. Chambers, op. cit., p. 79.

[2] G. Kitson Clark, 'The Repeal of the Corn Laws and the Politics of the Forties', *Econ. H.R.*, 1951. For a Whig magnate, enemy of the Corn Laws, see David Spring, 'Earl Fitzwilliam and the Corn Laws', *A.H.R.*, 1954. For 'the Bonaparte of Free Trade', see L. G. Johnson, *General T. Perronet Thompson, 1783–1869*, London, 1957.

[3] See below, Chapter 10, section 1.

faces as the returns governing the corn averages (the official calculations of ruling prices) were compiled and the verdict announced. So
could be ended the speculation that was so prominent a feature of the
corn market, and the manipulation that was so frequently hinted at,
or even charged, in the matter of the averages. If prices could be made
more stable and the flow inward of corn made more orderly, this would
promote monetary stability, by reducing the need to send gold abroad
to pay for sudden grain imports.

There were also the arguments in favour of pushing British agriculture to higher efficiency by exposing it to foreign competition. The
Corn Laws seemed to many of those engaged in the debate to be
merely sponsoring stagnation and inefficiency on the land. Moreover,
they protected rents, and on Ricardian lines this meant that the rent
share of the national income must rise as the population increased.[1] But
though hostile to rents, the critics of the Corn Laws suggested to the
tenant-farmers that they might well gain if incomes generally were to
rise in consequence of free trade, for then they would be able to move
away from cereals to a more diversified farming with a range of
products of greater value. It also followed that if the share of the
national income going to rents could be reduced, that accruing to
industry through profits could be sustained or even increased. This, by
improving the capital position, would strengthen the will and ability
to promote new enterprises and so maintain general economic growth.

Closely related to this argument, though not always reconciled with
it, was the claim that if the position of capitalists was improved, *vis-à-vis*
the landlords, then on the lines suggested by Adam Smith, employers
would compete vigorously for labour and so cause the wages share of
the national income to be increased. In contrast, in so far as rents were
protected and landlords' incomes were cherished, no such competition
for labour would result. So the free trade case came back to the simplest
appeal of all, that of higher real wages for industrial workers. In this
way the anti-Corn Law arguments constituted a system promoting
the needs of growth, stability, and justice.

How then could so compelling a case be resisted for so long by so
perceiving a man as Sir Robert Peel, who had already shown such
fiscal enlightenment? To much indeed of the anti-Corn Law case he
gave intellectual consent. But so did he and his supporters to many of
the opposing arguments. The altruism of the members of the League in
working so strenuously for higher real wages through cheaper food

[1] See below, Chapter 10, section 1, p. 386.

was bound to provoke at least some cynical reflection. Many of the Chartists were fiercely outspoken on this matter. As early as 1815 it had been argued that where there was a question of the distribution of gain from cheaper food between workers and employers, the latter would be in a position to determine the outcome, and would not be too nice in the matter; in short, as food got cheaper employers would push money wages down.

On the question of lowering the costs of manufacturing in order to promote exports, Peel might well ask whether the difference in the price of corn due to the operation of the law was really so important in securing the competitive position of Britain. Did it not rather rest upon her enormous development of powered machinery? So far as price stability was concerned, would this really be promoted by a permanent opening of the ports? Indeed, with the British market for food exposed to the world, might not free entry produce the reverse effect? British agriculture, confronted by a bad season, would be unable to compensate itself for smaller output through an increase in the protected home price. Nor was there compelling force in the argument stressing the sudden gold losses that occurred under the protective system when it was necessary to bring corn in. Corn came in in quantity over a fairly short period in the autumn in any case, and the money market would be obliged to adjust itself to a loss of gold even with free trade. Indeed, if such were the case, the losses might well be greater, as dependence upon foreign supply increased. The argument of improved efficiency through competition also had its limitations. Surely the will to improve agriculture depended as much upon stable incomes and a sense of economic and political security as it did upon the need to compete in price with foreign corn? There were indeed signs that British agriculture was improving itself well before the Corn Law debate reached its height. Nor would the implication of the free trade case, namely that some farmers would have to retire because they would no longer be competitive, escape the attention of the farming interest.

On the great question of the expansion of the economy, Peel and his party saw further difficulties. Was unbridled industrial growth really desirable? Was there not still some reality in the idea of control, especially at this most important point, where the state might be able to hold some kind of balance between the new and the old forms of society? Peel's party, containing as it did so many of the landed interest, certainly had much less distaste than did manufacturers for the idea of unproductive consumers, in the form of landlords generously spending to maintain incomes. They also, like the Chartists, doubted that capitalists could be obliged by their very repletion of capital to compete for labour. Rather would the capitalists find the means, by agreement, tacit or otherwise, to avoid this.

There were yet further arguments which were not merely refutations of the repealers. There was the question of the security of the nation's food supply. Once heavily dependent on foreign corn, Britain would be in grave danger of coercion or conquest. Here the repealers had their answer ready: the kind of interdependence between nations that would flow from free markets would mean that other countries would become heavily dependent on British manufactures, and so would have no more interest than Britain in taking to aggression. But the resisters of repeal took the more historical view held by Malthus that passion might be more powerful than a calculation of interest in determining war or peace, and that it was wrong to put the safety of the nation to such a hazard.[1]

Yet ideas, circumstances, and propaganda all converged in favour of repeal. The trend had long been in this direction, for each of the actions taken to simplify trade had brought expansion in the sector concerned. It appeared as though the system of natural liberty, so impressive as an intellectual construction, was being endorsed in the world of real transactions. Moreover, by the mid-forties the industrial sector had developed two characteristics. It was now so large that the external trade balance was overwhelmingly a matter of manufactured exports; similarly both national income and national political stability depended upon the power of the world economy to continue to grow. The idea of a carefully maintained balance between agriculture and industry was already unrealistic, for it was now out of the question to attempt to deny to industry the conditions necessary for expansion, much less to force it into contraction and retreat. Britain had passed the point of no return in the process of industrialization; indeed, she had entered the phase in which her own stability depended upon being able to stimulate growth in other economies. It was now necessary for American cotton, German and Australian wool, Brazilian sugar and coffee, Chinese tea, Baltic and Canadian timber, and Russian and American corn, to come forward in ever-increasing quantities at ever-diminishing real cost to English consumers.

The cause of Corn Law Repeal, the last great legislative barrier to a general extension of trade, found an extraordinary set of advocates in

[1] W. D. Grampp, *The Manchester School of Economics*, Oxford, 1960, p. 24.

the Anti-Corn Law League. Many of their enormous panoply of debating points, taken severally, were vulnerable. The mass of their arguments, taken together, refused to fit into a consistent pattern. But the movement had an intensity, a sharpness of focus, and a sustained vitality so great as not to be resisted. The Leaguers, at least at the height of their campaign, lost their sense of constitutionality, advocating courses of action that were plainly coercive of Parliament and destructive of the representative system: the refusal of taxes, the promotion of strikes on a crippling scale, and the plan to create a liquidity crisis by a concerted demand for gold in return for bankers' notes.

The League agitation was the grand demonstration of the means available to members of the middle class who had found that a Parliament still dominated by landlords was wholly incapable, through its own inspection of the nation's condition, of appreciating the importance of their reasoning. The contrast between the precision of objective and concentration of means characteristic of the League, and the inchoate agitations of the workers, is most striking. But it is worthy of note that the Repealers, in order to carry their case, had to create sufficient disturbance among the masses to frighten the landlords, and to strengthen this trepidation by the threat that even the middle class, if sufficiently resisted, might not be too nice about constitutionality.[1] Though the League encountered organized opposition in the Anti-League, the latter could not match the more powerful propaganda engine.[2]

Finally, to seal the matter, came the potato famine, bringing food shortage of a kind unknown since 1811, especially in Ireland. Peel could support the landed wing of his party no longer: repeal was carried in 1846. Only the insignificant registration duty of one shilling per quarter remained. Calamitous though the intervention of nature had been, it made it possible for an embattled landed interest to retreat with honour; the sharp confrontation of men of business and men of land was ended, and the slow, almost imperceptible ascent to political power of the former was resumed. Even the landed counterattack through the Factory Acts was not to revive the blank opposition produced by this final struggle over corn.

By now the archaic nature of the Navigation Laws was apparent to most legislators: as affecting foreign trade they were repealed in 1849,[3]

---

[1] Norman McCord, *The Anti-Corn Law League, 1838–1846*, London, 1958.
[2] Mary Tancred-Lawson, 'The Anti-League and the Corn Law Crisis of 1846', *H.J.*, 1960; G. L. Mosse, 'The Anti-League', *Econ. H.R.*, 1947.
[3] J. H. Clapham, 'The Last Years of the Navigation Acts', *E.H.R.*, 1910.

the country retaining control only over its coasting trade. The brick excise went in 1850; glass had been freed in 1845.

One last great operation on the tariff remained – that of Gladstone, carried through chiefly in his budgets of 1853 and 1860.[1] Foreign manufactures, including silk, so long defended, came in free. Duties on colonial and foreign sugar were equalized in 1854, and gradually withdrawn thereafter. This was done by successive stages (1864, 1870, and 1874) in order to reduce the shock to the West Indies. Taxes on cleanliness and knowledge were removed by freeing soap and paper in 1853 and 1861. Many other levies were ended, each minor in itself, but crucial in its own area of impact, and together constituting a formidable whole. Gladstone in 1854 ended the protection of the coasting trade by repealing this last relic of the Navigation Laws.

At last free trade was a reality. The Cobden-Chevalier commercial treaty with France in 1860 extended these benefits, bringing easier access between French and British markets, to last until 1872. The Royal Navy and the Revenue ships no longer fought an expensive battle against the smugglers.[2] The Usury Laws as affecting all but two types of transaction were in abeyance; the first exception, made in deference to the landowners, maintaining the old 5 per cent interest rate on loans on real property, went in 1854, the second special case, that of pawnbroking loans, remained under the old law until the Pawnbroking Act of 1872, when such loans were made the subject of special legislation. By 1854 it was possible to refer to 'the universal competition, from which no art can now save individuals or communities'.[3]

This programme of withdrawal by the state was no simple operation of casual repeal. At every step the Chancellor was confronted by powerful interests and arguments. His general task was clear – to carry to its conclusion the long trend toward reducing to a minimum the government's interference in the processes of trade. But complete withdrawal was possible only if a considerable part of the necessary revenues of the state came from direct taxes upon income.

Gladstone, like Peel his mentor, profoundly disliked the idea of requiring free citizens to declare in detail the elements and total of their incomes. As his liberalism grew, so too did his dislike of putting so powerful a tax engine into the hands of politicians. In spite of all the dangers and inconveniences of meddling in particular markets through

---

[1] Buxton, op. cit., vol. I, chapters VI, X.
[2] Neville Williams, *Contraband Cargoes, Seven Centuries of Smuggling*, London, 1959.
[3] 'Wool – The Extension of Trade', *Econ.*, 24 June 1854, p. 671.

indirect taxation, this dislike of direct taxes was shared by all. There was certainly no sympathy for the 'communistic doctrine' that the tax system should be used as a means of a continuous redistribution of wealth.[1] Moreover with indirect taxes the individual, by altering his pattern of consumption, was at liberty to alter his tax liability, an idea attractive to liberals. Accordingly Gladstone provided a classic set of canons to govern indirect taxation.

Pressure was brought to bear upon him by the Financial Reform Association, founded in 1848, of which his brother Robertson was a leading member, in order to convert him to the view that only a tax on incomes was fair and efficient.[2] On the contrary, he persisted in his efforts to abolish the income tax. But success was denied him. From the late forties onward, the costs of state services rose, especially as questions of armaments and war became more pressing. The Crimean War killed Gladstone's programme of 1853, under which the income tax was to end in seven years. As late as 1874 he fought a general election on the proposal to abolish income tax, but was defeated. Thus was the fiscal dilemma solved by events: both direct and indirect taxation were necessary and the Chancellor, as he put it in terms of two charming sisters, had to divide his attentions between them. Yet in the mid-eighties the income tax yielded no more than the tax on the stamps necessary for transactions in property and a few other matters (some £11 million) whereas the very highly taxed consumer goods, tobacco, snuff, and spirits (yielding £13 million) could pay comfortably for the Royal Navy.[3]

Gladstone never ceased to fight against expenditure. A close watch on such matters had been kept by Joseph Hume, whose attacks on even the slightest carelessness had been unrelenting, if sometimes ludicrous. The economy for which Hume stood was elevated by Gladstone into a national virtue; there was to be no waste in the national housekeeping, and there was to be no casual acceptance of new burdens, with their inevitable consequence of higher taxation or increased debt. Cheese-paring was the fiscal principle for the implementation of the minimal state. It was also the way to keep the bribery and corruption of the eighteenth century from returning; the Government Chief Whip was known as the 'Patronage Secretary' well after 1832. But however

[1] 'The Government Scheme for Raising More Revenue from Probate Duties on Personal Estate', *Econ.*, 20 March 1880, p. 325.

[2] W. N. Calkins, 'A Victorian Free Trade Lobby', *Econ. H.R.*, 1960.

[3] William Page, op. cit., vol. II, pp. 41, 67.

the domestic responsibilities of the state were limited, defence costs mounted, especially as science, impeded by antique organization, converted the Admiralty into an immense manufacturing and experimental concern.[1]

With such a restrictive view of the function of the civil service it is hardly surprising that little positive thought went into the provision of administrative judgement and initiative. It was not until after some twelve or fifteen years of routine tasks that a civil servant reached a post demanding any power of decision taking, so that the formerly bright young man emerged from 'the benumbing icehouse of his preliminary service' with his powers much impaired.[2] Yet something of an administrative revolution was taking place as new agencies to deal with new state functions came into being.[3]

Struggle though they might toward non-intervention, mid-century governments found that invention and investment had created new entities that had to be encouraged or controlled. Mail subsidies provided a basis for shipbuilding innovation, especially in the use of steam over long distances, beginning with the P. and O. contract of 1837. By 1859 the mail subsidy to British ships, on the North Atlantic, was running at nearly a third of a million per year, to the outrage of liberal economists and American shipowners.[4] The railway boom culminating in 1847 was a frightening demonstration of what the new kind of construction could mean. A Bill of 1849 for the auditing of railway accounts was killed by the powerful interests concerned. Even earlier, in 1844, Gladstone's Railway Act gave the government power to control the rates of carriage charged on old lines, and even, subject to conditions, to purchase new ones outright, but this latter power was never used.[5] The Act further stipulated, in the interests of the workers, a daily return train on all lines at a penny a mile for third-class passengers — the celebrated parliamentary train. In 1873 the Railway and Canal Commission was established to deal with rates and amalgamations.

In 1869 the government took further action running directly contrary to the general liberal view: it became the owner, through purchase

---

1 'The Solid Work that is before the New Ministry', *Econ.*, 19 December 1868, p. 1446
2 'Tests for the Public Service', *National Review*, 1861, p. 135.
3 David Roberts, *Victorian Origins of the Welfare State*, Oxford, 1960.
4 'The American and British Atlantic Steam Ships', *Econ.*, 21 May 1859, p. 559. See above, Chapter 4, section 5, p. 144.
5 H. W. Parris, *The Regulation of the Railways by the Government in Great Britain: the Work of the Board of Trade and the Railway Commissioners*, Leicester Ph.D. thesis, I.T.A., 1960–61, no. 3188.

at a price highly remunerative to the sellers, of the nation's electric telegraphs. It had become abundantly clear that in the interests of an efficient integrated national system this had to be done; for the first time private enterprise using the new technology had produced a service that only the state could fully implement. By the purchase of the Suez Canal shares in 1875 the government became the 46 per cent owner of the world's greatest international public utility.

So too, with the conditions of work. The elaboration of a system of official control in the mines and factories went steadily on, leading the state ever further in the direction of intervention.[1] The struggle, through the Passenger Acts, and the Colonial Land and Emigrants Commission (set up in 1840) to improve the horrifying and often fatal conditions at sea for emigrants, provides an instructive lesson in the problems of controlling economic behaviour.[2]

But the *laissez-faire* case did not succumb at all points. The opinion, for example, that the state had no responsibility for the pursuit of dishonest or irresponsible promoters impeded the improvement of the Bankruptcy Law, for, in spite of a Committee in 1865, the view persisted that prosecutions should be left to aggrieved creditors, even though it was well known that the victims very often declined to send good money after bad in launching prosecutions, and thus left the defaulter unpunished.

Local and city government produced their own form of the dilemma of intervention.[3] By the mid-century the revolt against central control was at its height: it was hoped that the Municipal Corporations Act of 1835, by releasing new initiatives, would permit the effective abdication of the central government from local affairs. Heavy blows were dealt to the principle of centralization. The powerful Poor Law Commission begun by the 1834 Act was diminished to the relatively feeble Poor Law Board. The Board of Health, set up in 1854, lost much of its power with the removal of its ruler, Edwin Chadwick, and was wound up in 1858. As a substitute for centralization the government sought to discharge its responsibilities for sound local government by passing general Acts setting out local authority powers and duties, in the belief that these statutory powers would be effectively applied without the need for coercion. Though some authorities responded, it was clear

[1] See above, Chapter 7, section 7.

[2] O. Macdonagh, *A Pattern of Government Growth: 1800–1860; the Passenger Acts and their Enforcement*, London, 1961.

[3] Robert M. Gutchen, 'Local Improvements and Centralization in Nineteenth-Century England', *H.J.*, 1961.

that there were grave failures of local initiative. The Sanitary Act of 1866 represented a reversion to the principle of enforcement by the central government on local ones, acting upon complaints. So the centralizing principle returned. But the system continued to rest upon the hope that the necessary local initiatives would be forthcoming.

By the late sixties the towns and cities, in their hectic growth, found the problem of local taxation assuming a new seriousness.[1] To find an adequate basis for local revenues, commensurate with the new claims upon them (health, police, education, poor law), was one of the most baffling of fiscal conundrums. For in spite of the rapid and obvious accumulation of wealth in Britain, the problem of getting a reasonable portion of it into the hands of local government involved a serious clash of principles. Municipal or county bodies could only tax 'visible' assets — lands and houses — under their jurisdiction. All else was beyond their knowledge and their power. The new wealth in business profits — all investment other than houses and land — was untaxed for local purposes.

8. THE UNIONS, COOPERATION AND THE FRANCHISE: 1848–84

With Chartism as an organized movement waning to a remnant in the early fifties the English workers entered upon their first lengthy period of quiescence for some twenty years. For something like thirty years no challenge was to be offered, in ideology or action, to the general principles upon which economic and social life rested. Though in the terrible winter of 1854–55 the number of destitute among the weavers of Coventry approached that of the soldiers suffering in the Crimea, the old obsolescent groups, the hand-loom weavers and other craftsmen whose functions had been superseded by new methods, had been largely purged from society or were incapable of effective protest; the continuous flow of new casualties declined into the residuum, where, increasingly concentrated in urban slums, they passed from the sight not only of the middle class, but of the more successful part of the working class.[2] The latter felt the quickening pulse of economic expansion as it improved their prospects, and in so doing made for better conditions for the army of the unskilled who attended upon them.

[1] E. Cannan, *The History of Local Rates in England*, 2nd ed., London, 1912.
[2] 'The Coventry Weavers', *Econ.*, 22 December 1860, p. 1422; see above, Chapter 7, section 10.

Working-class movements were thus stripped of the mass support that rallied or exploded when times were bad. It was a time of testing for the working men who sought to keep alive the insistence that it was wrong that the mass of men should have no effective voice in the affairs of the nation. Sponsored emigration made some contribution to domestic problems by diminishing the labour force.[1] But most trade union strength was applied in three main directions: cooperation, trade unionism, and pressure for the franchise.

The idea of cooperative production and distribution, for all its buffetings, had not lost its attraction. It was a means, at best, of allowing self-help and mutual aid to work together to reduce conflict within and between humans, and at worst, of reducing the vulnerability to exploitation of at least some of the workers. The success of the Rochdale Pioneers, founding their shop in 1844, represented a new beginning in retail cooperation, where sobriety of aspiration and the close adherence to sound rules tempered but did not destroy idealism.[2] Curiously enough it was a group of middle-class men who, in their anxiety to help the workers, showed greater excitability over what might be possible. The French Revolution of 1848 had brought brief power to the followers of Louis Blanc, whose ideas were reminiscent of those of Owen, seeking to place industry on a new syndicalist basis. The Christian Socialists, men of the middle class, F. D. Maurice, J. M. Ludlow, E. V. Neale, Tom Hughes, Charles Kingsley, and others formed the Society for Promoting Working-Men's Associations in 1850.[3] It was to sponsor cooperative enterprises by urging and advising the workers, and, most important, by providing some at least of the always scarce capital. A second middle-class group, the Positivists, led by Richard Congreve, J. H. Bridges, Frederic Harrison, and Professor E. S. Beesly, following Auguste Comte, believed that social regeneration could be achieved by the application of science to society and politics; they too earnestly sought to come to terms with the workers.[4]

[1] C. Erickson, 'The Encouragement of Emigration by British Trade Unions, 1850–1900', *P.S.*, 1949; R. V. Clements, 'Trade Unions and Emigration, 1840–1880', *P.S.*, 1955.

[2] G. J. Holyoake, *Self-Help by the People, History of the Rochdale Pioneers*, London, 10th ed., 1893; Rev. W. N. Molesworth, 'On the Progress of Cooperation at Rochdale', *R.B.A.*, Section F, p. 225; A. E. Musson, 'The Ideology of Early Co-operation in Lancashire and Cheshire', *Trans. Lancashire and Cheshire Antiquarian Society*, 1960.

[3] Torben Christenson, *Origins and History of Christian Socialism, 1848–54*, Aarhus, 1963.

[4] R. J. Harrison, *The Activity and Influence of the English Positivists on Labour Movements, 1859–1885*, Oxford D.Phil. thesis, I.T.A., 1955–56, no. 671. 'Professor Beesley and the Working Class Movement', in Briggs and Saville, eds., op. cit., Chapter 6.

The long struggle to create effective unions in the various branches of industry continued.[1] By 1872 the railwaymen were able to form their first all grade union.[2] By 1850 the Miners' Association of Great Britain and Ireland was far gone in decline; not until 1858 was it replaced by Alexander MacDonald's National Miners' Association; even then strength was only gathered slowly, accelerating after 1863.[3] The Potters, on the other hand, emerged still powerful from the troubles of the forties. But it was the Engineers who demonstrated the new line of development. The various small unions were brought together in 1851 to form the Amalgamated Society of Engineers.[4] The A.S.E. could take high contributions and pay large benefits, it was closely organized at the centre, and after the famous lock-out of 1852 in which its funds were consumed, it used the strike weapon with careful selectivity. The 'new model' principles were followed by other crafts.

Less striking though continuous and important progress was made in gaining the employers' recognition of the unions in the cotton and hosiery trades, and in building. Though some employers tried to avoid the creation of unions by establishing closer relations with the men, often through spokesmen nominated by the employers, and others fought the unions outright, an increasing number began to come to terms with organized labour, whatever might be the state of the law at the time.[5] Conciliation Boards appeared in a number of trades.[6] At the same time many employers and observers were fearful that too great a capitulation to the workers and their spokesmen would be 'to embrace a future of slavery, discomfort and failures'.[7]

Some degree of national coordination of the Unions was achieved through the London Trades Council after 1860, then by the Trades Union Congress founded in 1868. A strong central control developed under the heads of the leading unions, labelled by the Webbs the

---

[1] H. A. Turner, *Trade Union Growth, Structure and Policy: a Comparative Study of the Cotton Unions*, London, 1962.

[2] Philip S. Bagwell, *The Railwaymen*, London, 1963.

[3] *Transactions and Results of the National Association of Coal, Lime and Iron-stone Miners of Great Britain*, London and Leeds, 1864.

[4] J. B. Jefferys, *The Story of the Engineers, 1800–1945*, London, 1946.

[5] 'To Prevent Strikes', *Econ.*, 26 November 1853, p. 1331; R. A. Buchanan, *Trade Unionism and Public Opinion, 1850–1875*, Cambridge Ph.D. thesis, 1956–57, I.T.A., no. 359.

[6] Sir Rupert Kettle, *Strikes and Arbitrations*, London, 1866; T. J. Dunning, *Trades Unions and Strikes*, London, 1860; Lord Amulree, *Industrial Arbitration in Great Britain*, Oxford, 1929.

[7] 'Labour Attempting to Control Capital', *Econ.*, 6 May 1854, p. 475.

Junta. These men were not without their critics, as the problems of centralization and bureaucracy began to overtake organized labour.[1] In general the unions were composed of craftsmen, with the unskilled without effective organization. The adoption of a more conservative policy by many of the unions did not of course put an end to industrial strife. The Junta could not control all the unions; those outside its influence usually continued in militancy and the rejection of middle-class values.[2]

On the whole the skilled artisans were not greatly interested in levelling ideas. To them went the better type of working-class housing, and among them thrived the idea of respectability.[3] It was they who attempted constructive social thought of an ameliorative rather than a revolutionary kind. The leadership of the new model unions was provided by them. Even when they joined in movements of a radical nature, including Chartism, the skilled artisans were usually to be found on the side of persuasion and moral rather than physical force. The nature of so many public demonstrations among the workers often astonished foreigners. The insistence upon strict order, discipline, and constitutionality, though this was sometimes related to fears of military force, especially in the early years after Waterloo, was in the main due to workers' leaders who even when seeking parliamentary reform, felt themselves to be part of the constitutional process.

During the serious stoppage of 1852 the A.S.E. experimented with the cooperative employment of its own members. But these efforts did not last for long in spite of Christian Socialist financial support; cooperative production had to wait until the later sixties and early seventies for effective revival. These latter years produced cooperative collieries, especially on the north-east coast, cooperative textile mills, the occasional engineering works, and even an Industrial Bank. These enterprises, however, though viable in boom conditions, were largely destroyed by the collapse of the seventies.

Cooperative selling did better. Success with retail shops suggested the further economies of large-scale buying. After regional beginnings

---

[1] S. Coltham, *George Potter and the Bee-Hive Newspaper*, Oxford D.Phil. thesis, I.T.A. 1956–57, no. 698.

[2] G. D. H. Cole, 'British Trade Unionism in the Third Quarter of the 19th Century', *I.R.S.H.*, 1937.

[3] Asa Briggs, *Victorian People*, chapter VII, for Applegarth; for another leader see D. R. Moberg, *George Odger and the English Working-Class Movement, 1860–1877*, London Ph.D. thesis, I.T.A., 1953–54, no. 366.

in Lancashire and Yorkshire, the North of England Cooperative Wholesale Society was formed in 1863.[1] Its successful spread over the whole country caused its title to be simplified in 1872 by the dropping of the territorial prefix. The Industrial and Provident Societies Act of 1852 laid the necessary legal basis for the extension of the movement. The Cooperative alternative, so attractive to workers, to Christian Socialists, and to many philosophical radicals, was gradually seen to have a real role in industrial society. But it was hardly capable of extension to a general principle.

With the craft unions pursuing limited objectives, and the cooperative movement subject to severe limitation, attention turned once more to the franchise. By the late fifties the old aspirations were reviving, and with them the old problem of collaboration with the middle classes. The north-east coast, Yorkshire, and Lancashire all saw the reform banner raised again, these initiatives coalescing into the National Reform Union, with Manchester its centre and middle-class men prominent in leadership. In the Midlands and the South the trade union element was more independent, working through the National Reform League centred in London. After Palmerston's death in 1865 the whole movement accelerated.

Though the Union and the League could work together against common enemies, there were sharp differences about how far the franchise should go. The radical element was prepared to see responsible workers on the electoral roll, and so favoured household suffrage; the working-class leaders wanted manhood suffrage, with the secret ballot. It was clear that both Liberal and Conservative parties were prepared to enlarge the franchise, for both, by the mid-sixties, accepted the notion that the skilled worker at least could no longer be debarred.[2] Short of universal suffrage only one criterion was at all practical, and that was the old one of the value of the house or lodgings occupied. How far down the scale was this to be pushed? Each reduction would bring in an additional group of voters; their geographical distribution and their political leanings had to be taken into account in assessing the possible effect. Here was a great challenge to the political theorists and political arithmeticians. Some, like Walter Bagehot, thought that to enfranchise the workers fully would bring mobocracy, and a disastrous competition between political parties for popular support, the latter manipulated by

---

[1] Percy Redfern, *The Story of the C.W.S. 1863–1913*, Manchester, 1913.
[2] Francis H. Herrick, 'The Second Reform Movement in Britain 1850–65', *J.H.I.*, 1948; John B. Halsted, 'Walter Bagehot on Toleration', *J.H.I.*, 1958.

irresponsible agitators with no concept of the welfare of society as a whole.[1]

As the debate on the franchise was moving to a climax so too was that on the trade unions; they were in fact inextricably mixed. There was a growing uneasiness about the unions, and about the unstable truce between capital and labour. 'What we have to be afraid of here', said an industrialist in the sixties, 'is a strike, not a socialist coup.'[2] The point was made that the whole course of legislation had been to diminish monopolies of one kind or another, yet the workers were seeking the right to create conditions under which they could corporately control the supply of their labour. Moreover this great power might well be given to leaders chosen on the most dubious grounds, without real knowledge of the nation's affairs, and forced by their situation to make outrageous demands; thus would industry fall under the influence of a tiny group of untutored exhibitionists. Such men might find themselves the victims of their own rhetoric when the incendiarists among the workers seized the initiative.

The debate on the franchise coincided with an outbreak of violence that caught the public imagination and revived latent fears of workers' irresponsibility: the 'Sheffield Outrages' of 1866 took the form of violent attacks by members of some of the small unions in the cutlery trades upon fellow-workers who declined to throw in their lot with the union.[3] It was an ugly business but had a disproportionate publicity, coinciding with the Fenian disturbances in Ireland. More subtle was the fear that the workers, if given any real authority over the conduct of industry, would insist upon the control of output and a diminution of competition among enterprises so that each industry would then be able to combine against the public and raise prices. Workers' leaders had in fact pointed out how that the revenue of various branches of industry was much less than it might be if the component firms would abstain from such fierce cutting of prices.

Those radicals who wished the workers to move from the level of mere hirelings were beginning to feel that employers' attitudes left much to be desired. Charles Babbage, as early as 1832, described the engineering masters as concealing the state of their order books lest the

---

[1] See Alastair Buchan, *The Spare Chancellor*, London, 1959; Norman St John-Stevas, *Walter Bagehot: A Study of his Life and Thought*, London, 1959.

[2] Taine, op. cit., p. 232.

[3] Sidney Pollard, 'The Ethics of the Sheffield Outrages', *Hunter Society Transactions*, 1953–54.

workers might use such knowledge as a guide to strike action, with the result that, in a sense, union claims were bound to be irresponsible because there was no data on which to base a judgement.[1] The Social Science Association in 1860 tried to investigate the sources of strikes; the workers were helpful in the presentation of their point of view, but the masters declined to discuss the matter.[2] The theory of the wages fund was wearing thin.[3] On the other hand it was increasingly realized that the workers were now, in many branches of industry, confronted by combinations of capitalists that, though invisible to the law, were powerful.

The year 1867 saw both the extension of the franchise and important changes in trade union law. A great gathering in 1866, having been excluded from Hyde Park, pushed over hundreds of yards of the iron railing of the Park in the presence of the police, a reminder of the dangers of failure to meet the aspirations of the workers. Disraeli's Reform Act extended the vote to cover the £10 householders, lodgers paying £10 rent in the towns, and to £12 householders in the counties. Not all were satisfied with what had been gained, but it was a notable advance over 1832. The artisan in the towns was now virtually enfranchised. In 1872 came the secret ballot, so strongly urged by the Chartists and so soon to produce the Irish members of Parliament who were to consume so much legislative time.

In 1859 the Combinations of Workmen Act (the Molestation Act), the first trade union legislation since 1825, had made explicit the right of peaceful picketing. After 1864 the unions pressed for an amendment of the old law of master and servant, so hostile to effective collective bargaining, and so invidious in its discrimination against workers as parties to contracts of employment. It treated a worker's breach of contract as criminal, though an employer's default was merely a civil offence, with proportionately different penalties. The miners felt the weight of the law with special severity. The Sheffield affair seemed likely not only to stop advance, but even to threaten a deterioration of the legal position. A Royal Commission on the Trades Unions was appointed in 1867, where the workers' leaders could publicly state their views.[4] In this they were aided by their middle-class allies, especially Frederic

---

1 Charles Babbage, *On the Economy of Machinery and Manufacture*, London, 1832, p. 249.
2 Strikes, and the Policy of Employers', *Econ.*, 14 April 1860, p. 392.
3 See below, Chapter 10, section 7.
4 *Reports of the Royal Commission appointed to inquire into the Organization and Rules of Trades Unions and other Associations*, 1867, 1867–68, 1868–69; George Potter, 'The Trade Societies of England from the Workmen's Point of View', *C.R.*, 1870.

Harrison and Tom Hughes. So reassuring an impression was created that the Master and Servant Act of 1867 made certain favourable, though minor, changes in the heavily biased law: a workman could now give evidence on his own behalf and was to be summonsed to answer a charge instead of being arrested. Hardly had this weakness in the workers' position been partly remedied when a new one appeared. The Courts held in the case of *Hornby* v. *Close* that a union could not proceed against a defaulting official to recover stolen funds, thus depriving unions of recourse against their servants. An Act of 1869 rectified this.

The Liberal government in 1871 passed two notable Acts, the one, the Trade Union Act, improving the unions' position, the other, the Criminal Law Amendment Act, detracting from it. The former, at last, gave legal status to the unions, freeing them of the old doctrine of restraint of trade, and allowing them to register as Friendly Societies. But the second Act forbade the unions to picket, thus debarring them from the use of one of their principal weapons. The liberalism of the Gladstone government found the idea of coercion so distasteful that it would not sanction the kind of public demonstration against both employers and non-union workers that picketing might involve.

The newly enfranchised workers not unnaturally turned from their Liberal loyalty in the general election of 1874, bringing in Disraeli's government. From it was obtained the Conspiracy and Protection of Property Act of 1875, legalizing peaceful picketing, though within a fairly precise definition, and prescribing that any act legal for an individual was legal for a trade union. The Employers and Workmen Act of the same year put both parties on the same footing as subscribers to a civil contract, a most important change. In 1876 the unions were brought within the scope of the Friendly Societies Acts. Though the trade unions were not yet invulnerable to those who would restrain or suppress them, starting in 1825 with a status not to be achieved in France until 1884 and in Germany in 1892, they had by 1875 attained a strong position, an advance noted with concern by many English wage-payers.[1]

But the unions too had their problems of continuous adjustment; in 1875 the machine workers in the boot and shoe industry, finding their interests diverging from those of the handsewn craftsmen, broke away to form their own union, a portent of the new claims to be made

[1] Ensor, op. cit., p. 131.

by the less skilled workers.[1] Lower down the scale the gravitation of the unskilled to the great urban centres caused the demand for political and union action to be raised among the casual labourers. On the other hand there was reality in Disraeli's concept of the Tory working man, for many of those to whom he had given the vote in 1867 rallied to his party.[2]

It was time at last to enfranchise the men of the countryside and the miners; this was done by the Reform Act of 1884. Such an extension made possible a change in political outlook in the shires, signalled by the election of Joseph Arch to Parliament in 1885.[3]

British and French labour debates were very different.[4] In Britain discussion turned entirely upon matters of practical legislation, with ideology almost wholly absent; in Britain there were many popular employers, and men of business like Brassey, Mundella, Forster, Chamberlain, and others who constituted a bridge between masters and men.[5] But the idea was dawning upon the more percipient of the middle class that the worker must not merely be heard through philanthropic spokesmen like Hughes or Professor Fawcett. Such men could never hope really to enter into the attitudes of the workers; indeed they were oddly described as having 'the same sort of admiration for the working man that a lover has for his mistress'.[6] It was noted with regret that the new reformed Parliament of 1868 produced not a single artisan to give greater realism to the nation's debates.

The Trades Union Congress had met first in 1868, but had confined itself to industrial matters.[7] The Labour Representation League, founded in 1869, succeeded in winning two of the twelve parliamentary seats they contested in 1874. Two years later the T.U.C. resolved upon the need to get workers into Parliament. The Liberals, in 1885, came to terms with this ambition by allowing trade union candidates a number of clear runs, with the result that eleven, mainly miners, were elected. These men voted with the Liberals and were known as Lib-Labs.

---

[1] Alan Fox, *A History of the National Union of Boot and Shoe Operatives, 1874–1957*, Oxford, 1958.

[2] E. J. Feuchwanger, 'The Conservative Party and Reform after 1867', *V.S.*, 1959.

[3] See above, Chapter 8, section 4.

[4] Frederic Harrison, 'The French Workmen's Congress', *F.R.*, 1878, p. 664.

[5] W. H. G. Armytage, *A. J. Mundella, 1825–1897: the Liberal Background to the Labour Movement*, London, 1951.

[6] 'The Elections', *Econ.*, 21 November 1868, p. 1324.

[7] C. Roberts, *The Trades Union Congress 1868–1921*, London, 1958. But labour leaders had negotiations with Gladstone's Whips — see Royden Harrison, 'The British Working Class and the General Election of 1868', *I.R.S.H.*, 1961.

Though the time had not yet come for Labour to make its own direct challenge for power, the working man had made his parliamentary debut.[1]

9. THE NEW PATERNALISM

Mr Gladstone, after the fall of his first government in 1874, enraged the eager young radicals of his party by appearing to think that the scope for reform was exhausted.[2] There was some warrant for this if attention is concentrated upon what had been achieved rather than upon what remained to be done. The flood of legislation had begun to assume modern proportions. Over the previous two decades, in addition to the extension of the franchise, the further definition of the status of the trade unions, and the remaking of the fiscal system, there had been a formidable series of measures for the regulation of business and society, the catalogue of which astonished men, who like Gladstone, remembered England before the first Reform Bill. Legislative control of working conditions was extended with the Coal Mines Regulation Act of 1860 and the Factory Acts Extension Acts of 1864 and 1867; the employment of chimney-sweeps was further regulated at the same time.[3] In 1867 came the Workshops Regulation Act, bringing protection to workers in small premises in which fewer than fifty persons were employed; in 1872 came a further Coal Mines Act. Grave abuses among casual workers in agriculture, especially women and children, were tackled with the Gangs Act of 1867.[4] The Education Act of 1870, setting up Board Schools on a national basis, though it embodied an inhibiting compromise between radical nonconformist ideas and the conservative Anglican view, was yet a great landmark.[5] The Local Government Act of 1871 began the process of tidying up the central administration for dealing with the counties and boroughs. Between 1868 and 1874 the army was remade, in the teeth of aristocratic opposition.[6] The holding of posts in the universities was freed from religious

---

[1] D. W. Crowley, *The Origins of the Revolt of the British Labour Movement from Liberalism, 1875–1906*, London Ph.D. thesis, I.T.A. 1952–53, no. 662; A. E. P. Duffy, *The Growth of Trade Unionism in England from 1867–1906 in its Political Aspects*, London Ph.D. thesis, I.T.A. 1956–57, no. 712.

[2] Ensor, op. cit., p. 33.

[3] See above, Chapter 7, section 7.

[4] See *Sixth Report on Children's Employment (Agricultural Gangs)*, 1867.

[5] See above, Chapter 7, section 8, p. 258.

[6] See above, Chapter 8, section 2, pp. 285, 286.

requirements by the University Tests Act of 1871. The Licensing Act of 1872 was an attempt to regulate the supplying of alcohol to the public. Such measures involved lengthy debate and much arranging of tactics outside debate. The House of Commons was for the first time being tested as an instrument for regulating an advanced industrial society. No small part of the test fell upon the men concerned, especially those who sponsored the great changes. Moreover Parliament had to face these new constructive obligations at a time when issues of foreign and Irish policy occupied vast tracts of time, in the new ages of mass armies, colonial rivalry, and competitive armaments.

Though brash young radicals might resent Gladstone's complacency in 1874, it was shared by most politicians of the time: the 'languid indifferentists' had now succeeded to 'the sanguine innovators' of a decade or two earlier; 'the new wealth', it was said, 'is, on the whole, satisfied, and wants the world to be as it is'.[1] But new challenges were to confront the nation as it slowly became aware that the basis upon which a quarter of a century of prosperity had rested was changing radically.[2]

It was of course no new thing that vested interests of one kind or another should have its spokesmen in Parliament. In the old days the Tory element, with its landed predominance, had enjoyed this advantage. The Whig landlords began, after 1832, to form a lesser part of the liberal element in the Commons, as business representation grew. By the seventies the Liberal party had the support of most of the new wealth; the Tories had failed to find an effective bridgehead in the industrial sector, though they had done rather better among merchants and shipowners.[3] The brewers and distillers had been kept loyal to the Liberal party by the appeal of economic liberty and social and political emancipation. But the Licensing Act of 1872 sent them into the enemy's camp. Thereafter Tory party funds could benefit greatly, with consequent effects upon the political process; the publican, a man with considerable political influence, passed with his masters from the Liberal to the Conservative party, and the Liberals placed among their fixed targets of abuse the liquor trade and the beer barons. W. H. Smith, entering Disraeli's Cabinet in 1874 was one of the first commercial magnates to hold office as a Tory.

But against this new complication of party structures came something of a simplification of party lines brought by Gladstone's Home

---

[1] 'The Results of Recent Election', *Econ.*, 10 January 1874, p. 30.
[2] See above, Chapter 2, section 8.
[3] Briggs, *A. of I.*, p. 505.

Rule Bill of 1886. Previously the Liberal party had contained among its Whig aristocrats a landed element almost as splendid as that of the Tories: the Bill, with its threat to landed property, caused many of them to change their allegiance. Yet not all the issues of state intervention produced the kind of polarity that accompanied the Licensing laws and Home Rule. There were important matters, like sanitation, that were not party questions but lay between 'interests' that cut across party lines.[1]

In spite of the quite vigorous extension of state intervention under Gladstone there remained plenty for Disraeli's administration to do. An extraordinary flow of legislation passed in 1875: the Conspiracy Act, a Sale of Food and Drugs Act, a Public Health Act systematizing sanitary law, an Artisans' Dwelling Act (the Cross Act, giving local authorities powers to clear slums and build dwellings), an Agricultural Holdings Act, a Land Transfer Act, all concerned with matters that by now were in a thoroughly unsatisfactory state. Though in some directions a good deal of this programme of legislation stopped short of what was really required, in others it represented a new beginning, and in all it marked an increase in public concern. The following year produced Plimsoll's Merchant Shipping Act, intended to stop overloading and the operation of the notorious coffin ships, and Lord Sandon's Education Act, bringing compulsory schooling to the age of twelve.

With the return of the Liberals in 1880 the radicals took the initiative again, but were subject to the maddening obstructionist tactics of the Irish members and the Tory 'Fourth Party' under Randolph Churchill. The Employers' Liability Act of 1880 was a first step toward transferring from the shoulders of the worker to his employer the risks of accident or illness due to the job. A Seamen's Wages Act and a Grain Cargoes Act of the same year were the work of Chamberlain at the Board of Trade. The Married Women's Property Act of 1882 was a great stride toward female emancipation.[2] The Settled Land Act of the same year made land transfer much easier, striking a blow at the long-established defences of hereditary estates. In 1883, at long last, a Bankruptcy Act and a Patents Act sought to remove abuses and protect rights in two important fields. A Housing Act of 1885 was scarcely important as an

---

[1] Ensor, op. cit., p. 125; For the new techniques of party management now arising see H. J. Hanham, *Elections and Party Managements, Politics in the Time of Disraeli and Gladstone*, London, 1959.

[2] See above, Chapter 8, section 9, pp. 321, 322.

effective remedy, though it extended the scope of the Cross Act; not until 1890 was the housing problem to provoke real public action.

This mounting congestion of the Statute Book brought with it a great increase in the number of public servants; the whole trend was contrary to the old liberal aspiration for the minimal state. But just as there had appeared to be no alternative in the earlier part of the century to the dismantlement, piece by piece, of the old system of control, so now the legislature found in one area after another that the need for action, reaching far into the economic and social life of the nation, was proved by practical demonstration. This further involved the rise of the expert administrator, chosen on ability as measured by examinations, a further blow to aristocracy. By 1870, in spite of the argument that to admit the lower classes would cause the higher to withdraw, all departments except the Foreign Office were open to merit.[1]

10. PROGRAMMES OF THE EIGHTIES

In the mid-eighties a new phase in the politics of England was about to begin. The people, at least as adult males occupying reasonable housing, had been enfranchised. The Parliament of 1880 had been elected from a roll of some three million electors, one-third of them of the working class; the Parliament of 1885 issued from a roll of five millions, three-fifths of whom were workers. The economy was now industrial rather than agricultural, the system of restraints upon trade and industry that Adam Smith had condemned was gone, but a new one, for the regulation of industrial society, was far advanced. The liberal ideal of the individualist competitive society had enjoyed a period of ascendency, but as the need for a new kind of regulation grew, the radical mind found its dilemma becoming ever sharper; how were individualism and collectivism to be reconciled?

The men of conservative mind had been obliged to accept a grave diminution of landed predominance, and yet were able to see a considerable future for much of the old way of thinking and acting. Respect for aristocracy was by no means dead, even among the working classes. The conservative mind (like the socialist) had always resisted the atomistic view of society, and so had no difficulty with the idea of using the state in order to operate upon the condition of the people.

[1] See John Morley, *The Life of William Ewart Gladstone*, London, 1903, vol. II, p. 511.

As the *Pax Britannica* weakened, especially after the Franco-Prussian War, and as the great phase of colonial expansion was about to begin with Salisbury's accession to power in 1886, the idea of aristocrats at the head of affairs renewed its old appeal. There was still scope, even though Toryism was beginning to acquire its industrialists, for selective baiting of the factory masters and for posing as protectors of workmen's rights.

But the Conservatives also had their difficulties. The mature industrial society raised problems for which their minds and interests were not greatly suited. Their part in bringing in the Factory Acts and industrial legislation generally had been a matter of fastening a new code upon others. They, like the Liberals, accepted the working of the market mechanism as the mainspring of the economy. To do otherwise would be either to revert to an elaborate system of trade regulations, or to go all the way toward the socialist idea of public ownership and control. Both alternatives were impossible; the latter for self-evident reasons. As to the former, so dependent had Britain become upon international trade that the sons of those who had fought in the forties for the continued protection of corn made no effective move to defend agriculture, from the late seventies onward, against imported foods; no doubt the dependence of aristocratic incomes on industrial expansion had something to do with this passivity.[1] Much less could they be expected to interfere at other points in the economy in order to construct a new system of regulation. Even the monetary mechanism, the point at which general manipulation might be least criticized, was not to be tampered with, despite the vigorous campaign of the bi-metallists and others. There was nothing for it but to accept the market as the basic arbiter of economic affairs.

Thus Liberal and Conservative, so recognizably distinct in the early decades of the century, were drawing together as each had to accept and even elaborate realities that could not be resisted. Both political parties had their old guard who were mainly interested in clinging to the position from which their faction had started, minimizing and even ignoring the incoherent appendages that had forced themselves upon younger minds with a higher sense of contemporary relevance. Each party had its young men, whose minds were concerned with the problem of wooing the new electorate with realistic and attractive

---

[1] B. H. Brown, *The Tariff Reform Movement in Great Britain, 1881–1895*, Oxford, 1944; Sydney H. Zebel, 'Fair Trade: An English Reaction to the Breakdown of the Cobden Treaty System', *J.M.H.*, 1940.

policies. As exemplars, the Liberal side produced Joseph Chamberlain the Tories Lord Randolph Churchill.[1]

But this confrontation of old antagonists in a refurbished and partially false antithesis did not exhaust the political scene of the eighties. New movements were afoot as groups of men and women sought a new formula for society that would be a true alternative to the two great parties in the state. With this the workers had little to do. Not until the later eighties did a new militancy begin to appear among them, expressing itself in the urge toward the formation of unskilled, industrial unions. The renewal of the search for an alternative system for society came mainly from elements of the middle class.

The new interest in socialism was beyond the comprehension and outside the sympathy of many Liberals. Men of the old style like Robert Lowe and Leonard Courtney were still much involved in fears dating from Poor Law reform days. Courtney in 1885 made a long and deeply felt speech against the Medical Relief Bill, a measure to save the worker who sought parish medical relief from losing the franchise; he did so on the ground that such a concession would gravely weaken the character of those who received it.[2] Old style Liberalism was still without sympathy for the trade unions, haunted as it was, in spite of the destruction of the wages-fund theory, by the dread that any attempt to press for higher wages would do damage to profits and hence to entrepreneurial incentives and so to the economy as a whole.

Such old ideas were confirmed by new fears. The new socialism was seen, not as an indigenous product with a long tradition going back to the England of the middle ages, but as a sinister imitation of continental models. The atrocious disorder in which the Paris Commune of 1871 ended gave England a glimpse of the perils of revolution; the organized power of Prussia, strengthened by its collectivist welfare provisions, was equally frightening and equally likely to cause the older Liberals to cling to their hopes for a society in which the minimal state could operate. Alfred Harvey, a leading City banker, and a fine example of the business Liberal, assured his countrymen that 'Socialism, believe me, is the child of militarism and protection, and admirably reproduces the vices of its ancestry'.[3]

---

[1] J. L. Garvin and Julian Amery, *The Life of Joseph Chamberlain*, London, 1923–51, 4 vols; R. R. James, *Lord Randolph Churchill*, London, 1959; W. S. Churchill, *Lord Randolph Churchill*, London, 1906.

[2] G. P. Gooch, *Life of Lord Courtney*, London, 1920, p. 227.

[3] Alfred Spalding Harvey, *Writings*, London, 1907, p. 84.

But a new kind of Liberal was emerging. With the extension of the franchise and the consolidation of the position of organized labour, it was essential, not only on grounds of humanity and ethics, to improve the condition and therefore the character of the workers, but also to meet the new aspirations they were bound to express. John Morley, Mill's disciple, was the prototype of the new Liberal intellectual who, though like his mentor Mill, well aware of the arguments both for freedom and state action, redefined freedom in such terms that it was only possible, in an industrial society, where the state nurtured it with vigour and not with diffidence.[1] Joseph Chamberlain was the type of the new man of business who, having been successful in industry and commerce, turned to politics with a view to practical intervention for the improvement of society. Like Owen he had discovered his own powers while in industry; unlike Owen, as he turned to government he did not lose his sense of what was possible, but became, on the contrary, a master of the politician's art. Perhaps some of the difference arose because Chamberlain was formed in dealing with the Birmingham craftsmen, with their long tradition of self-expressive radicalism, whereas Owen, in manipulating his passive recruits to the New Lanark Mills, formed a different view of society and of himself. With the radicalism of Morley and Chamberlain came the aspiration, not merely to relieve pressures and sufferings at particular points where they had become so bad as to demand public attention, but to use the powers of the state positively, in order to bring a better life to the generality of Englishmen. To these ideas other rallied: Sir Charles Dilke (1843–1911), Sir George Otto Trevelyan (1838–1928), Jesse Collings (1831–1920).

These men were excited by the success of municipal socialism and inspired by the vision of a society capable of lifting all its members to a new level of living through education, judicious state intervention where private initiative was inadequate, and the more effective putting down of monopoly power. The high Whig doctrine, limiting the functions of the state to the preservation of life and property and the enforcement of contracts, was no longer tenable; the radical alternative of the eighties was to regard the state as the agent for moral, material and social reforms, with a special duty to think and act with respect to questions affecting the food, health, housing, amusement, and culture of the working class. They discovered also that sustained effort over a long period meant that political organization and party discipline were necessities. Chamberlain's Liberal Association set up in Birming-

[1] Francis W. Knickerbocker, *Free Minds: John Morley and his Friends*, Oxford, 1943.

ham in 1865 was the herald of new techniques of organization and persuasion that were to spread from civic to national government. The neo-Liberals in defining their new aspirations found themselves confronted by old enemies. Though the House of Commons was now changing in its composition, the Lords was still largely a landed preserve. In the countryside generally there operated what the radicals called a political quadrilateral, formed by the landed squire, the parish clergyman, the farmers, and the publicans. These together could be very powerful when working upon the traditional conservatism of the farm worker. The notorious concentration of land in the hands of nobility and gentry was passively accepted by those who were its victims – a thought that frustrated the eager radical and the revolutionary. The resistance to the extension of education also lay largely in the same quarter, for in addition to general lack of enthusiasm it was the issue of Church Schools that so bedevilled organization, finance and teaching.[1] For all the diminution the landed class had suffered in the nation's economic and political life they were still conceived by the radicals of the eighties as the great enemies of progress.

But the men holding such views as those of Morley and Chamberlain were not numerous; moreover excited though they might be over such concerns, there were others that were even more absorbing, especially the Irish Question, rapidly moving into a phase of crisis that was to make the radical Chamberlain into a conservative. Yet the banner of reform had been raised. Matthew Arnold had been concerned in the sixties and seventies that culture and fullness of life should reach the middle classes; in the eighties a significant group of middle-class men made the claim that the workers too should be embraced in the new aspirations.

This was the political challenge that the aspiring young Tory aristocrat, Lord Randolph Churchill, sought to meet. In the tradition of Disraeli and Young England, Churchill, with his tiny 'Fourth Party', confronting the Liberal majority of 1880, sought for a new appeal that might restore the fortunes of his party. To the horror of his seniors, Churchill clearly saw that with enfranchisement of the commonalty going forward there was no point in hesitating; in spite of the long-standing distrust of the urban masses by the upper classes he proclaimed the doctrine 'Trust the people'.[2] But he did not mean trust a newly educated, close reasoning, and highly responsible people to consider

[1] See above, Chapter 7, section 8, pp. 257, 258.
[2] W. S. Churchill, op. cit., p. 239.

the issues confronting the nation; he meant rather that the old confidence between aristocracy and commonalty should be revived, in the belief that the masses would then continue their affairs in the hands of their traditional leaders. Even Gladstone, whom Churchill so relentlessly harassed, paid tribute to the political skill of this appeal. Gladstone himself had an extraordinary veneration for aristocracy; so too, as he well knew, had the people.[1] Working-class supporters warmed to the approach, in the towns as well as in the country districts. Lancashire produced much Tory support, and in Chamberlain's own fortress of Central Birmingham Churchill ran John Bright very close in the election of 1885. To confirm such an alliance Churchill saw the necessity for Tories to follow the Radicals in the formation of powerful party machinery on a nation-wide basis.

Radical liberalism still had much of its old flavour, calling upon the workers to improve themselves through the newly created opportunities for education, but accepting the middle-class leadership of society. Tory democracy had an air of cajolery and good fellowship that, at election times, burst out over beer and enthusiasm for men who were different from the populace; behind this amiable relationship lay a much more permissive attitude toward the workers, accepting their frailties and calling for no great exertions in the direction of self-improvement.

To other groups within the middle class both attitudes were contemptible, appearing to rest upon an almost frivolous failure to understand the forces at work in society. The Positivists in the sixties and later had hoped for the perfection of a science of society, available for the guidance of an enfranchised and educated people. Philosophers like Green, Bosanquet and Caird, aspired to a more organic union of society.[2] But in contrast to those whose hope lay in social union were the converts of the *Communist Manifesto* and of *Das Kapital*. Here the idea of a society moving toward crisis, with a mounting tension between classes, was starkly expressed. The idea that basic conflict was present between the components of society was revived. In 1881 H. M. Hyndman founded the Democratic Federation, to become in 1884 the Social Democratic Federation.[3] Here was a curious militant movement, of middle-class eccentrics, calling upon the workers to unite for the inevitable and imminent class struggle, incurring the hostility of the

[1] G. W. E. Russell, *Portraits of the Seventies*, London, 1916, p. 197.
[2] See below, Chapter 10, section 10, p. 427.
[3] Chūshichi Tsuzuki, *H. M. Hyndman and British Socialism*, Oxford, 1961.

respectable craft unions with their hectic, exotic, and presumptuous appeal. In 1884 the Socialist League hived off, with William Morris its principal figure. He, like Ruskin, urgently wished to bring more beauty and brotherhood into the lives of the masses, using a form of socialism as the means.

In 1884 the Fabian Society was formed. It was to include an extraordinary variety of people of various shades of brilliance: George Bernard Shaw, H. G. Wells, Graham Wallas, Sidney and Beatrice Webb. They were a group of men and women who, without inhibitions about the sanctity of property and the limitations of state action, were seeking far-reaching changes in society. But they did not propose a general revolutionary challenge.[1] Instead they planned to make close and scientific studies of particular problems, and to prescribe remedies resting on objective demonstration of their correctness. That this would increase the area over which state action was taken was inevitable; this was not to frighten people merely on that account. The Fabians neither held a brief for private enterprise nor sought its indictment: where it worked it would remain, where not it would be modified or dispensed with. But in the mid-eighties the Fabians and the others who described themselves as socialists were a portent rather than an important part of the political scene.

The political evolution of England was a matter of assigning authority in society and providing for the efficient implementation of its decisions through administrative arrangements. But it is clear that political discussion could not proceed very far without the aid of some kind of economic theory, explicit, as with the intellectuals, implicit, as with the practical men.

[1] See below, Chapter 10, section 8, p. 419

CHAPTER TEN

The Effort to Understand

I. THE SYSTEM OF NATURAL LIBERTY

MEN may attempt at many levels to understand their condition: they may aspire to no more than an explanation of how the events of today are related to those of yesterday, or they may seek to assimilate themselves to some great system of thought that embraces the whole of the observable world together with the intangible concepts it has provoked in their minds. Or, more probable than either, they may live by some unassimilated mixture of both.

Our starting point for the understanding of the economic facet of the nineteenth-century mind in Britain must be the system of Adam Smith.[1] Its hold in the nineteenth century was real, but subtle. It was the outcome of an attempt by a moral philosopher to see economic behaviour as a part of behaviour generally and to assimilate it to a unifying system.[2] It was an extraordinary achievement. It seemed to synthesize philosophy and science, so rudely sundered by the speculations of the seventeenth century, when man was first dazzled by the fruits of the new empirical science. Secondly, it provided a philosophic apology for individual initiative. Yet, thirdly it implied that society, for all the disruptive activities of individuals within it, was nevertheless stable and free from any tendency to embark on cumulative accelerating change.

With the dissolution of the medieval schoolmen's view of the universe, as Renaissance and Reformation gathered speed, speculative thought about society had bifurcated. There was concern with man and metaphysics on the one hand, and with nature and science on the other. These became distinct studies, expressed respectively in the psychology of Locke and the science of Newton. But the yearning to restore unity

---

[1] For Smith, Bentham, J. S. Mill, and Marshall, see Jacob Viner, *The Long View and the Short: Studies in Economic Theory and Policy*, Glencoe, 1958; Keitars Amano, *Bibliography of the Classical Economics*, part I, Tokyo, 1961.

[2] See A. L. Macfie, 'The Scottish Tradition in Economic Thought', *S.J.P.E.*, 1955, for the context in which Smith's mind was formed.

of thought was very strong. In a way, the *Wealth of Nations* performed this function.[1] Smith accepted the Lockean view that man was passive – that he was the product of the sensations that impacted upon him from without. This was Locke's escape from the older, churchmen's assumption that man came into the world with a stock of innate ideas. Smith also borrowed from Newton.[2] The natural universe as described by Newton was characterized by perpetual unchanging circularity, with gravity as the controlling and regulating principle. Self-interest could be regarded as the social equivalent of gravity. Thus society was conceived in terms of atomistic individuals, moving in orbits around one another within a closed system, kept from collision by the pursuit of self-interest in a competitive market. This was to think of man as fundamentally passive, without the urge or ability to operate upon his universe of action but living and working within more or less fixed parameters.

Thus Smith's system was in the Lockean tradition of resistance to the claims of authority, made by either Church or State, ostensibly a great apology for individual initiative against the prohibitions of authority, yet it took an extremely limited view of the capacity of man to control or alter his society or his environment. Men of initiative who came after Smith could thus believe that they were free to respond to the stimuli they felt as economic change gathered speed; there was an implied reassurance that the system as a whole would not run away with them. It was of course beyond their powers to realize that their pursuit of self-interest in market situations would carry them and their children into a new kind of life that would alter not only their physical situation but also their own personalities. Smith, a major philosophical thinker, realized that the system of natural liberty did not guarantee the future either of individuals or of societies. He rejected the idea that man was the creator and master of his environment, but he did not go all the way with the opposite view, namely that man was merely its creature: he believed that men could make their own economic history (and their own moral systems), but emphasized that they could never make them quite as they wanted to. But this level of sophistication of thought was

---

[1] Adam Smith, *Wealth of Nations*, Edwin Cannan ed., London, 1904. For a general view see E. Ginzberg, *The House of Adam Smith*, New York, 1930. For recent discussions of Smith see: R. Koebner, 'Adam Smith and the Industrial Revolution', *Econ. H.R.*, 1959; J. Cropsey, *Polity and Economy: An Interpretation of the Principles of Adam Smith*, Nijhoff, 1957.

[2] David B. Hamilton, *Newtonian Classicism and Darwinian Institutionalism: a Study of Change in Economic Theory*, Albuquerque, 1953.

bound to be lost as his system was simplified by pragmatic men of affairs.

Smith was the greatest exponent of the view that, provided there is for each commodity or service a market in which there are a great number of buyers and a great number of sellers, there will emerge from the contending bids an objective price. Such a price is the outcome of the actions of all, but yet is independent of the actions of any. The system is thus free of coercion, yet it operates through the attempt to promote self-interest. The objective price which thus emerges serves to assign the gains or losses of economic activity among the participants. More than this, it serves to allocate the total effort of the community between various lines of production, for profitability depends upon price, and where good profits appear there will men concentrate more investment. In so doing they will be continually altering the pattern of effort away from products that are less wanted toward those for which demand is strong. This set of ideas was bound to have great appeal to men who were seeking to break down the old inhibitive system of regulation and protection that had lasted from the seventeenth to the nineteenth centuries.

There was, however, one great limitation. The conditions of competition must in fact obtain. This meant that, somehow or other, monopoly positions must be stopped from coming into being, or must be destroyed if they do arise. But it seemed plausible that with the growth of new industries composed of many units, and with the rise of a new generation of vigorous traders, the competitive assumption, if not wholly warranted, was sufficiently valid to serve as the basis for theory and policy.

Indeed Smith, in his theory of distribution, provided a kind of safeguard against the power of business men. He had no fear of excess capital accumulation, but rather welcomed the growth of capital in the hands of traders and industrialists, because, in their continuous efforts to use it in investment, they would be driven to reduce prices through competitive selling. The growth of capital thus did not give the business men a mounting control over others, but drove them to share the benefits of their enterprises. These gains could pass to others through reduced prices, but also through increased wages. Business men in receipt of large profits, and looking for a means of reinvesting, were bound to compete with one another for labour, and thus to raise wages. Thus would the labourers thrive with extending commerce and industry, but the traders and industrialists would be subject to a law of mutual control through their competitive attempts to re-apply their former gains. It is of course quite wrong to regard Smith as a simple apologist for business men; on the contrary his plea for the free market system rested on the belief that the only effective control that could be exercised over such men had to be enforced by themselves, though inadvertently. Even his desire to minimize the economic role of the state had the same origin – that by so doing business cliques might be deprived of a source of influence they had so long wielded.

So the case for the dismantlement of the traditional set of economic controls was made. In its place was to come a system requiring no governmental surveillance, but, on the contrary, the abstinence by governments from interference. If the official role was to be thus minimized, it was also essential that governmental spending, and the debts arising from past spending, should be kept as small as possible, in order that the taxation system might also be kept on a minimal basis.

This was the view that was to persist throughout the nineteenth century, in spite of great changes both in circumstances and outlook.[1] It was the economic case for liberal society; it also made possible a solution of the problem of power, simply by making market bids the universal arbiter. This system of ideas was so powerful that it survived long after the assumptions upon which it had rested had been seriously impaired. It blinded many in Britain to the conditions prevailing in other countries and so made them unsympathetic and sometimes obtuse in their dealings with others. It is true that Smith, in his great philosophical work *The Theory of Moral Sentiments* of 1759, wrote in terms of the duties of each member of society to his fellows in the interests of the whole, but this element of his thought, the idea of 'sympathy', was overshadowed in nineteenth-century minds by the more potent concept of self-interest.

David Ricardo, the greatest of nineteenth-century economists, accepted the market approach made by Smith to the economic problem, though he was wholly without Smith's broad grasp and intense interest in the philosophical roots of the system of natural liberty.[2] Ricardo, so far as having a mentor other than in economics was concerned, found

[1] For the attempt by a group of brilliant young men to assimilate and apply Smithian ideas in the early decades of the nineteenth century see John Clive, *Scotch Reviewers: The Edinburgh Review, 1802–1815*, London, 1957.
[2] See Piero Sraffa and M. H. Dobb, eds., op. cit.; Oswald St Clair, *A Key to Ricardo*, London, 1957. For a comprehensive discussion and excellent bibliography see Mark Blaug, *Ricardian Economics; a Historical Study*, New Haven, 1958.

one in Jeremy Bentham, whose ideas were powerfully conveyed to him by James Mill, perhaps Bentham's greatest disciple.[1] The greatest happiness of the greatest number was accepted by Ricardo as the test of sound policy. He was no professional scholar; he was in fact a City man – an enormously successful wartime dealer in government loans, become a country gentleman. His logical faculties were among the most powerful ever applied to economic questions. But to a very considerable degree his mind was parochial, working in terms of the conditions prevalent in his own time and place. His very success in reducing to logical coherence the economic experience of Britain in the early decades of the nineteenth century served to confirm this element of his outlook.

He began with practical problems. In 1810 he was thinking deeply about the monetary confusion that was arising in consequence of the inflation that had followed the suspension of cash payments.[2] By 1813 the discussion of the circulating medium and the exchanges had carried his mind, via the opening or extension of markets, to the problem of profits. This invoked a great debate on the laws regulating the distribution of the product between the contributing social classes, in the form of rent, wages and profits. In turn, out of this theory of the distribution of income between the leading social categories of the day – landlords, labourers, and men of business – came a theory of long-run development. For the ability of the economy to grow depended directly upon how the product was shared.

Mindful of rising population and increasing output, Ricardo reasoned thus: as numbers increased, and with them the demand for food, land of a less and less fertile quality would require to be cultivated, so that the yield of the soil per unit of effort would decline. But the landowner would do well in this situation, for rents would rise as the margin of cultivation was pushed outward; the owners of all intra-marginal acres could charge increased rents, and the tenant farmers would be able to pay them, for if it was worth while to cultivate at the margin as it extended, then all other tilled acres must be yielding higher returns. The individual labourer, however, would find his position unchanged. His wages were more or less determined by the needs of subsistence. If he received less his numbers would fall; if he received more they would increase, and so numbers were in continuous adjustment to the level of subsistence. But though the average labourer neither gained nor

[1] Elie Halévy, *The Growth of Philosophic Radicalism*, London, 1928, esp. part II, chapter 3.
[2] See above, Chapter 6, p. 195.

lost in the long run, the share of the total product of the economy going to the labourers as a class was bound to rise, for the effort required to produce the food supply was becoming an increasing element in the total effort of society. Finally, the profit going to entrepreneurs was treated as a residual. If both the rent share and the wage share of the total product were increasing, the portion left for the reward of business men must diminish. Smith too had argued that the reward per unit of capital must fall as expansion proceeded, but Ricardo gave the idea a new precision.

From this reasoning Ricardo's view of the long-run development of the economy followed directly.[1] Like Smith, he had no sense of a system raised to ever higher levels of wealth on the basis of a continuous renovation of technology. Rather, following Malthus, he believed without question that diminishing returns to effort on the land was inherent, and must serve to end expansion. The implication followed that capitalists, finding the reward to their lending falling as expansion went on, would be less and less inclined to save; entrepreneurs would offer less and less in terms of interest payments as an incentive to lenders, so that eventually there would be no net saving and no net investment. The whole economic process would thus be free of any important renovation, and would revolve in a kind of static circularity.

Of course Ricardo was not unaware of other possibilities. He did not deny the reality of technical progress, but felt that it could not keep ahead, indefinitely, of population growth. Thus, in particular times and places, diminishing returns to effort in agriculture might be suspended or even reversed, but never permanently. Secondly, in the course of debate he realized that a minimal provision for subsistence, conceived purely in physical terms, was unreal, not only as a biological concept but also because it took no account of psychological factors that caused the acceptable minimum to vary. Yet for his system to be tenable he was obliged to couch it in terms of diminishing returns and a physical standard.

By 1815 Ricardo, under the watchful eye of James Mill, was trying to put together a general treatise on political economy. In so doing he found himself 'stopped by the question of price'. He discovered that in order that his thinking should be intellectually satisfying it was necessary to attempt an explanation of the price of particular goods; this in turn demanded a theory of value. He had already shown how the

---

[1] For a useful formal statement of the Ricardian system see Irma Adelman, *Theories of Economic Growth and Development*, Stanford, 1961, chapter 4.

producers of goods were rewarded, in their characters as groups of labourers or entrepreneurs. But now, in order to complete his system, he sought a means of reasoning from the other end, from the goods themselves. For if the Smithian case for the market system as the most effective way of regulating price, and thus incomes, was to stand, then the mystery of price and value had to be penetrated.

Previous to Smith's time the idea of subjective value had occasionally been raised: that goods did not carry value as a quality intrinsic in themselves, but were valuable according to the attitudes that men took toward them, including how much of a given good a man might already possess. But Ricardo found that he could not incorporate this idea into his system. His distribution theory ran in terms of homogeneous groups of men (landlords, capitalists, and labourers) interacting upon each other, with no consideration of subjective differences between them.

Much more appropriate, apparently, to the theory of price and value, was another concept with a long history – the labour theory of value. One good, said Ricardo, subject to certain important qualifications, would exchange for another according to the ratio between the amount of labour embodied in each. Thus he seemed to have found a universal equivalent, embodied labour, which was an objective entity. The fact that he used labour as the key to value was to be taken up by socialist thinkers, who deduced from it the consequence that if the value of a good was determined by the labour-ratio it bore to other goods, then the labourer was entitled to receive the full benefit of his labour, without deduction by landlord or capitalist. Death overtook Ricardo in 1823 while he was still struggling to integrate his theory of labour value with his theory of distribution between classes.

The Ricardian system thus presented a rather puzzling front to contemporaries. It was much harder to comprehend than that of Smith. Moreover, whereas Smith had disposed of the problem of power by committing it to the market mechanism, Ricardo showed how one element in society, the landlords, for long the wealthiest in the kingdom, was to wax even fatter as the need to feed a growing population increased. Thus monopoly was endemic. Moreover, the business men, controlled in Smith's system by their need to reduce prices in market competition and the necessity to bid among themselves for labour, were shown to be incapable of breaking through the ancient barrier to greater wealth, the limitations of agriculture.

But though there were important differences between Smith and Ricardo in the matter of analysis, their policy prescriptions came to the same thing. Smith desired the system of natural liberty in order that the market might replace arbitrary power. Ricardo placed a heavy emphasis upon the need to liberate commerce in order that the restrictive effect of the limited land supply in Britain might be thrown into the distant future by making it possible for Britain to draw upon the food production of the entire world. The total outcome of Ricardo's theory of distribution and of long-run growth, in policy terms, was to place a new, though different, emphasis upon Smithian pleas for a free market system.

This concern for the system of natural liberty extended to the conditions of work. Ricardo and those who followed him, often known as the 'classical economists', took the view that children, not being fully rational beings, were entitled to protection as employees in factories. But adults were not: the terms upon which they laboured lay between them and their employers, and were no proper concern of the state. Indeed this second principle was of greater importance than the first, and if one must suffer in the interests of the other, child protection must go.[1]

The Rev. Thomas Robert Malthus, the third great name of the early industrial period, had serious misgivings about an uncontrolled movement of England from agrarian to industrial society; he was in fact the leading voice of those who were fearful.[2] He was so on two grounds. Malthus had much more sympathy than Ricardo for the old land-based structure of society, and was in consequence anxious to preserve agricultural incomes from drastic decline – hence his support for the Corn Laws.[3] But he had fears for the industrialist too. He envisaged the possibility of capital glut, or in more modern terminology, general unemployment.[4] Was it not inevitable, he argued, that if for one reason or another society attempted to save beyond what could be profitably reinvested, there would be a general breakdown? This surplus of saving was much more likely to occur as industrialization extended; hence Malthus's feeling that the traditional landed class, with its high propensity to spend, was a safeguard for incomes generally.

[1] Mark Blaug, 'The Classical Economists and the Factory Acts – a Re-examination', *Q.J.E.*, 1958, p. 212; Kenneth O. Walker, 'The Classical Economists and the Factory Acts', *J. Econ. H.*, 1941.

[2] For a demonstration of the degree to which value judgements entered into attitudes see R. M. Hartwell, 'Interpretations of the Industrial Revolution in England: A Methodological Inquiry', *J. Econ. H.*, 1959.

[3] T. R. Malthus, *The Grounds of an Opinion on the Policy of Restricting the Importation of Foreign Corn*, London, 1815.

[4] T. R. Malthus, *Principles of Political Economy*, London, 1820. Summarized in Sraffa's edition of Ricardo, vol. II.

Government spending was another possible means of sustaining incomes, but this, he thought, would involve a dangerous extension of the power of the state.[1]

On these matters Ricardo was able to out-argue Malthus. He and his followers urged that farmers' incomes would not in fact fall if free trade in foodstuffs was permitted, but rather, though the margin of cultivation at home would not be pushed out so far as under protection and therefore there would presumably be fewer farmers, each would be better off, for the yield per acre would be higher. Though this would mean less in rents to landlords, their incomes would still be high enough for them to maintain their traditional position. On the matter of gluts, Ricardo adopted what became known as Say's Law: the statement that general glut could not occur, for supply created its own demand.[2] As an optimist in these two important respects, Ricardo downed the last real attempt among political economists to state an anti-Smithian position.[3]

2. THE NUMBER AND QUALITY OF MEN

Discussion of population problems had remained on a pretty rudimentary level until the nineteenth century.[4] There was a general assumption that if numbers were increasing, this was a sign of a healthy society, for how else, ran the implied reasoning, could it be so?[5] Having made this correlation between increase and general well-being, older authors tended to treat those circumstances likely to be adverse to increase as bad. The new beginning came with the publication by Malthus of his *Essay on Population*, first appearing in 1798.[6] For him population increase was no matter for national congratulation, but rather for profound misgiving. The new view of population, substituting an attempt at systematic theory for the many scattered sayings of

[1] B. A. Corry, 'The Theory of the Economic Effects of Government Expenditure in English Classical Political Economy', *Econa.*, 1958; J. J. O'Leary, 'Malthus' General Theory of Employment and the Post-Napoleonic Depressions', *J. Econ. H.*, 1943; R. L. Meek, 'Physiocracy and the Early Theories of Under-Consumption', *Econa.*, 1951, p. 250.

[2] See T. W. Hutchison, 'Some Questions about Ricardo', *Econa.*, 1952, p. 418.

[3] For the policy views of the classical economists see Lionel Robbins, *The Theory of Economic Policy in English Classical Political Economy*, London, 1952.

[4] For population behaviour, see Chapter 2, sections 6 and 9, and Chapter 7, section 3.

[5] James Bonar, *Theories of Population from Raleigh to Arthur Young*, London, 1931.

[6] For Malthus's summary statement of his theory and for a discussion of the historical context see D. V. Glass, ed., *Introduction to Malthus*, London, 1953. Also G. F. McCleary, *The Malthusian Population Theory*, London, 1953.

former times, coincided, perhaps paradoxically, with the new age of industrial expansion.

Malthus developed his ideas over virtually a generation until his death in 1834; he was the dominant figure of the nineteenth-century discussion. The debate ranged far, for it was impossible to say anything very important about population behaviour without becoming involved either directly or by implication in all major aspects of the national life. It was also impossible for the debaters to shake themselves free of all manner of preconceptions and value judgements as they made this first attempt to think systematically about so fundamental a matter.

The immense impact of Malthus's theory upon the thinking of the nineteenth century has often puzzled later observers. It reflected the misgivings of the time. One of the great unknowns in contemporary industrializing society was the future behaviour of the masses. It had been all very well in previous generations to acclaim their increase in times when it was fairly modest, and hardly threatened the structure of society. In pre-industrial conditions, when calamity came in the form of famine or war, the shock could often be assimilated and the numbers briskly replaced by new births. Such increases were associated in men's minds with recovery and expansion, with new opportunities for those who had survived catastrophe. Moreover, such troubles seemed to have been more or less external to the society concerned — they were visitations. But with the new and sustained growth in numbers new attitudes appeared. Society in the new conditions could no longer suffer amputation in this way and yet remain stable. A new sense of precariousness began to develop.

The social and economic changes that were gathering speed had far-reaching political implications. Would the new masses accept and conform to the older view of their place? Were they to be treated as drudges, devoid of intelligence, a subservient element in society kept, as Arthur Young urged, at their tasks by keeping wages down? Or must they be assimilated, as rational responsible beings, to a society placed, in consequence, very much in their hands? The Malthusian debate of the first half of the nineteenth century acted as a focus for all such questions. The new machines and the rest of the new physical apparatus could be exhilarating, but the increase in humans that came with them inspired other thoughts, for numbers on the new scale meant a new society.

Malthus sought to cut through the welter of ideas by a dramatic simplification. He made the whole of population behaviour depend upon the relationship between two rates of change: the growth of human numbers, and the trend in the yield to human effort in agriculture. Men had the power and inclination to increase their numbers in a geometric ratio, doubling them every twenty-five years; the best man could do with agriculture, however hard he might try, was to make its yield increase arithmetically, in the ratio over successive quarter-centuries of 1, 2, 3, 4, 5, etc. The strict mathematical ratios need not be insisted upon; it was enough to argue that a dramatic disparity existed between the potentials for procreation and for production.

Mankind was thus always inviting a visitation from one or all of the positive checks: vice, misery, and war. In the second edition of his *Essay*, published in 1803, and thereafter, Malthus introduced a further possibility – the preventive check. This involved the delaying of marriage until personal income was sufficient to support a family, with continence before marriage. Thus by conscious and calculated choice man might obviate otherwise inevitable calamities.

But though Malthus embodied the possibility of moral restraint in his theory, and argued and propagandized sturdily for it, he does not seem to have had any real confidence that it would act effectively as a preventive check. A strong note of pessimism, that the increase in numbers must end in calamity or corruption, survives through all his writings. For though, over the long debate, he made mention of circumstances that later thinkers were to emphasize about the dampening effect that higher productivity was likely to exert on the birth rate, his principal emphasis, and the basis of his policy prescriptions, remained the fear of excess of people.

Not every one, of course, was prepared to accept the full measure of Malthusian pessimism. Throughout Malthus's lifetime and for much of the nineteenth century, refutations poured in upon his *Essay*.[1] There were many suggestions in the older literature that a society growing in wealth would automatically produce checks to the increase of its own numbers. Many earlier authors, using among other sources the classical texts of Greece and Rome, pointed out that the growth of luxury, or a higher standard of living, was a deterrent to increasing numbers. Urbanization also, it had been frequently argued, was adverse to increase, for the common view, probably well based in fact, had been that on balance towns were consumers rather than creators of population. Many men had unbounded confidence in the fertility of English

---

[1] Kenneth Smith, *The Malthusian Controversy*, London, 1951; J. A. Field in *Essays on Population*, Cambridge, 1931.

soil and were offended by what appeared to be a slight to the national agriculture; others, more intellectual, had been affected by physiocratic ideas about the land being the only true source of wealth. There was also a deep-rooted feeling (encouraged by many churchmen against their brother of the cloth, Malthus) that it was proper in this matter to trust in God, and that to suspend action on the basis of a personal calculation that could be both petty and in error, was wrong. But not only were some Christian co-religionists antagonized by Malthus, so too were those holding the eighteenth-century confidence in the Age of Reason, stressing man's power to control his condition by conscious thought. Romantics also, lauding the principle of free expression of the personality, found ground for damning Malthus. Robert Southey pronounced his condemnation in the roundest terms: Malthus's theory, he thought, compounded 'brute mechanism, blind necessity, and blank atheism'. It is one of the curiosities of the debate that religionists, rationalists, and romantics all criticized Malthus, and one of the lesser ironies of the first third of the nineteenth century that this mild and kindly man, with his colour blindness and cleft palate, harassed by his inability to keep his pupils at Haileybury College in order, should become so great a focus for vituperation.

But the elements that converged in Malthus's mind and were synthesized in his system were powerful enough to carry many others with him, and to make 'the Malthusian devil' a reality.[1] The idea that man must struggle for redemption was deep in the minds of other men; they too were offended by the optimism of William Godwin and those who took a similar view, suggesting, as it did, that mankind was capable of indefinite improvement and was uncursed by original sin. There was something of the Old Testament prophet in Malthus, warning the people against debauchery. The work ethic, too, was powerful; he feared the prospect of mankind gaining material benefits too easily, describing his fellows as inert, sluggish, and averse from labour unless compelled by necessity. His concept of pleasure, at least as it affected the masses, placed more emphasis upon the escape from pain and calamity than upon the positive enjoyment of benefits. Sexual asceticism was a strong component in this general attitude, for the restraint imposed by man's conscious control of his numbers was regarded by Malthus and his followers as intrinsically good. Far from being concerned that neurosis might develop in the struggle between inclination and calculation, he thought that the discipline involved

---

[1] Richard B. Simons, 'T. R. Malthus on British Society', *J.H.I.*, 1955.

would add greatly to the stature of men. His political outlook was that associated with traditional agrarian paternalism; he feared anything that threatened this. In a sense he was in the older tradition of speculation, trying to construct universal theories about man and nature, and in the attempt wrongly assuming that both were homogeneous and stable. Finally, there was the Hellenic preoccupation with harmony and proportion, seen in a static rather than a dynamic light. All these elements were of a traditional, backward-looking kind. In this way the Malthusian case could attract to itself many whose minds were governed by misgivings rather than by confidence.

Alongside this weight of archaism there was a surprising modernism. The Malthusian theory had strong elements of the scientific, capable of persuading men of radical stamp of mind who were attracted by what was undoubtedly a new and fruitful way of thinking. For though Malthus followed others in wedding the Christian concept of man prone to error with the Greek concern for proportion, he pioneered in presenting this union in the form of a quantitative theorem. His two categories of man and nature could be made exhaustive. Though some of the sources he used now seem to have but a slender basis in fact, he carried sociology a great step forward both by the material he presented, and the method he demonstrated. In practical terms, also, Malthus's system commended itself to men who were increasingly concerned with the problem of the poor living in subsidized poverty. He was the major prophet of the return to rigour embodied in the Poor Law Act of 1834.

Yet the theory had serious scientific weaknesses. The mode of operation of the positive checks of war, vice, and misery was left largely unexplored until Darwin, excited by this gap in Malthus, produced his evolutionary thesis of the survival of the fittest.[1] Marx was to attack Malthus with great bitterness because his manner of formulation made over-population inherent in man and nature, thus obviating any possibility of solution through the improvement of technology and institutions.[2]

The theory of population made its impression on thinking men at two levels. There was the theory standing by itself — the simple confrontation of man and his food supply. But there was also the incorpora-

[1] See J. Stassart, *Malthus et la Population,* Liège, 1957; H. J. Habakkuk, 'T. R. Malthus', in *Royal Society of London: Notes and Records,* vol. XIV, no. 1, London, 1959; *Life and Letters of Charles Darwin,* London, new ed., 1902, vol. I, p. 68; *More Letters of Darwin,* London, 1903, vol. I, p. 118.

[2] R. L. Meek, *Marx and Engels on Malthus,* London, 1953.

tion of the theory in the general view of the economy as a whole, as had been done by Ricardo. The law of diminishing returns to effort on the land was a natural outcome of the Malthusian mode of thinking. As numbers increased, and with technological improvement incapable of making much contribution to improved yield, the effort of each new worker must add less to the product. Whatever might happen in industry and trade the land ruled the long-run future.

Secondly, men like Nassau Senior found it possible to construct on Malthusian-Ricardian lines the famous Wages Fund theory.[1] It explicitly stated that the level of wages was determined by the means available for wage payments, divided by the number of employed workers. This, of course, was in a sense a self-evident statement. But by assuming or implying that the fund itself was fed by profits, it was possible to suggest that any attack on profits would diminish the source from which wages were paid. Beyond a certain point this was undoubtedly true – if all profits were dispersed as wages, there would be little hope of increasing wages in the long run through better machines and equipment.

In this way the Malthusian system placed the responsibility for preserving society from the great traditional curses solely upon the workers: vice, misery, famine, and war, were the result of excessive breeding uncurtailed by foresight and continence. No amount of application or ingenuity could relieve men of the limits imposed upon their numbers by nature through agriculture.

But neither Malthus nor Ricardo found it easy to define the subsistence level of wages, a concept crucial for the theory. It proved very difficult to find examples of men whose existence and outlook was determined merely by the conditions of physiological survival. A new concept gradually began to gain ground: that of a standard of living that changed as men changed. The older philosophers had pointed out how the luxuries of one generation became the necessities of a later one. In this way it was seen that the size of the family was a matter determined by a flexible, but conventional, view of family life.

By the late twenties Nassau Senior was making the distinction between necessaries, decencies, and luxuries. For Senior it was the fear among the workers of the loss of decencies that acted as the preventive check on population. Indeed if necessaries were reduced, the upshot was likely to be not a fall in population but an increase, due to the fecklessness that comes with loss of hope. But the prospect of decencies or even luxuries had quite the opposite effect. This was the escape from the

[1] Marion Bowley, *Nassau Senior and Classical Economics*, London, 1937.

population problem. If only men and women could learn to value a higher standard of living more than an increased family, and to act accordingly, the general pressure could be eased and society would not press against the limitations inherent in agricultural productivity. Richard Jones pursued the same line of thought.[1] He went further than Senior in showing that the whole debate about population had failed to take account of historical uniqueness: that societies were always changing, and the views and attitudes of the men who composed them were also in constant flux. Other authors took up the same theme.[2]

In this way social philosophers were moving away from the rather bleak estimate of the commonalty held by Malthus. Not that they took a more optimistic view of the workers' ability to take rational control of his situation, as Malthus demanded but little expected. Rather, men adjusted themselves more by instinct and intuition to their new opportunities. In place of Malthus's soul-building activities of thought and self-discipline, an impersonal social element had appeared to redeem the situation; men simply fell into the way of having smaller families when they began to sense the possibilities of a higher living standard. The problem thus became: how was society to cause a sufficient increase in the money incomes of labourers, and to make available real goods for them to purchase, in such a pattern as to place a premium upon the restraint of family?

It was as well that this new possibility could be entertained, for it would appear that men and women in general were immune to the didactic approach on this matter. One of the difficulties Malthus never solved was the nature of the calculation expected to be made by the ignorant labourer. Once married, he was to have no truck with contraception, nor was he to subject his marriage to the strain of continence in wedlock. This meant that his estimate of his future progeny was very difficult, for there was no scope for adjustment within marriage. The very high death rates of the time added to the problem of foresight, for how was he to tell how many of his children once born, would survive? Useless to tell him to calculate an average of what he saw around him, for he was by assumption proposing more care than that practised at present by his neighbours. It was all very well to appeal to the rational element in man, but to do so in the absence of data is likely to induce frustration.

---

[1] W. Whewell, ed., *Literary Remains of the Late Rev. Richard Jones*, London, 1859.

[2] See W. F. Lloyd, *Two Lectures on the Checks to Population*, Oxford, 1833, for the argument that education can induce reasoning and therefore prudence.

Whatever their origin, the new conclusions about population by men like Senior were optimistic. They were partly the outcome of diminishing fear of the masses, and partly they served in turn to reduce such fears further. Ricardo and the elder Mill had been political radicals, but had not been able to reconcile their pleas for the extension of the franchise with their pessimism about the workers' ability to behave responsibly in planning the affairs of their families. By the mid-century the bogey of numbers through the birth rate was very much reduced and by the eighties it had virtually gone.[1]

The new acceptance of the masses, coinciding with the great quarter-century of economic expansion, was accompanied by new views of the whole social process. Herbert Spencer, much influenced by Darwin, sought a view of society which synthesized with the evolutionary view of the natural world.[2] Spencer argued that as numbers increased, a selection process would operate, for only those types suited to the new congestion and competition would survive. They, in turn, would be of a higher order of complexity and refinement ('individuation'). This condition was inconsistent with high fertility; with the coming of such a society the size of families would contract. Thus the evolution of society into more sophisticated forms would produce a new kind of equilibrium, for the very conditions conducive to high densities (industrialization) also contained the opposite tendency.[3]

In Malthus's system all workers had been treated as similar — they were a homogeneous category. With Spencer came a new idea, or rather a new emphasis upon an old one. It was that men were different from one another and that in a competitive society there would be a constant tendency for the fittest among them to rise to the top. This was a further stage in the attempt to come to terms with the workers and their place in society. In Malthusian minds society was a matter of stratification of an almost caste-like rigidity. Theories of social behaviour tended to reflect this — one for the labourers, one for the middle, one for the upper class. With Spencer the attempt was made to place the whole of society within the same general system of thought.

The evolutionary view was more appropriate to British society in the age of opportunity and expansion. But there remained a common element between Malthus and Spencer. Both regarded those at the

[1] E. Cannan, 'The Changed Outlook in regard to Population, 1831–1931', *E.J.*, 1931.

[2] See Walter M. Simon, 'Herbert Spencer and the "Social Organism" ', *J.H.I.*, 1960.

[3] For a discussion of this and other 'cultural theories' see Sidney H. Coontz, *Population Theories and the Economic Interpretation*, London, 1957. For theories of fertility see D. E. C. Eversley, *Social Theories of Fertility and the Malthusian Debate*, Oxford, 1959.

bottom of society as being there because such a position reflected their true character and ability. Indeed with Spencer this element of deserts was stronger than with Malthus, for the very idea that equality of opportunity was now greater had the effect of emphasizing in a new way the implication that those at the bottom of society were there because they were, in fact, basically inferior. It was beginning to appear that as equality of opportunity increases, inequality of natural endowment is made more apparent.[1] Moreover, there was the fear of the fecundity of the slum dweller, the pauper, and the criminal, with which Spencer's system failed to come to terms. With the increasing concentration of the slums of the great cities, evolution seemed to operate in the direction of deterioration, as was remarked by observers who were struck by the apparent superior ability of those 'with low cerebral development' to multiply.[2] Fears were expressed of the effects of 'interference with nature's selective mechanism' through the lowering of the death rate.[3]

But though thought and debate about the population problem went on with some vigour in Spencer's time and thereafter, the whole matter occupied a much smaller place in men's minds than it had done in Malthus's day. Now population was seen in a larger context — rather than attempt to make all aspects of society stem from a demographic starting point, speculative men were engrossed in the exploration of the new thesis thought by many to embrace the whole of human experience: the principle of evolution. The new way of thinking meant that population debate was submerged in a much more generalized discussion.

Though Malthusian ideas exercised a weaker grip on the minds of thinkers and politicians, they were not purged from the corpus of political economy. The workers found them still part of the canon during the long dominance of J. S. Mill's *Principles*, provoking the comment that the political economists who followed Malthus were 'lowering the whole social life of England by representing marriage as an indiscretion in the youth of the working and middle classes'.[4]

[1] See H. A. Boner, *Hungry Generations: the Nineteenth Century Case against Malthusianism*, Oxford, 1955, p. 4.

[2] Arnold White, *The Problems of a Great City*, London, 1886, chapter III, 'Sterilization of the Unfit'.

[3] Alfred Marshall, *Principles of Economics* (9th variorem ed., C. W. Guillebaud, ed.), London, 1961, vol. I, p. 201.

[4] G. J. Holyoake, *John Stuart Mill as some of the Working Class Knew Him*, London, 1873, p. 17.

Miss Harriet Martineau's tales serving as *Illustrations of Political Economy and Taxation* which appeared in the thirties, insisting upon the responsibility of the workers to adjust the proportion of population to capital, set a tone that continued long after.

Yet many groups of workers were prepared to adopt, through emigration, a solution to the Malthusian problem of numbers. Labour unions sponsored schemes whereby the old principle of mutual aid could be applied to the transfer of workers overseas.[1] Malthus himself had accepted emigration as a remedy of some efficacy in the short run, while empty lands were still available, though of course such a redistribution of men was no permanent solution.[2]

During and since Malthus's time there had been those who had argued that the natural result of intercourse might be controlled by artificial means. Abortion and even infanticide had been canvassed in the earlier phase of the debate, but without any positive advocacy. But in the twenties, contraception had found powerful protagonists in Francis Place and Richard Carlile, who underwent fierce social ostracism; James Mill suffered in the same cause, so too did Robert Dale Owen.[3] In the fifties George Drysdale urged this solution and continued to do so for many years. Charles Bradlaugh was the leading advocate in the sixties, and the question became the centre of national notoriety in the famous Bradlaugh-Besant Trial of 1877, arising from the reissue by the defendants of Charles Knowlton's *Fruits of Philosophy*, first published in the early thirties.[4] This was the first general public debate; it signalled the long decline of the English birth rate not least by causing the circulation of the offending work to leap from a few hundred to over one hundred thousand a year. Malthus, of course, had been revolted by such a solution, yet the movement for the propagation of birth control, founded in 1879, bore the name of the Malthusian League.

A violent collision occurred between the neo-Malthusians and the Church, both in its Protestant and Catholic branches. It was, of course, no accident that those who first advocated so radical a remedy for the

---

[1] W. S. Shepperson, *British Emigration to North America: Projects and Opinions in the Early Victorian Period*, Oxford, 1957.

[2] James Bonar, *Malthus and his Work*, London, 1924.

[3] Norman E. Himes, 'Jeremy Bentham and the Genesis of Neo-Malthusianism', *E.H.*, 1936; 'J. S. Mill's Attitude Towards Neo-Malthusianism', *E.H.*, 1929; 'The Place of J. S. Mill and Robert Owen in the History of English Neo-Malthusianism', *Q.J.E.*, 1928.

[4] J. A. Banks and Olive Banks, 'The Bradlaugh-Besant Trial and the English Newspapers', *P.S.*, 1954.

population problem were those who had abandoned much of the system of ideas upon which Malthus's case had rested; Bradlaugh was the most notorious atheist of his day. The Established Church produced a fierce condemnation of what was proposed. Among other fears there was the belief that family limitation might strike at the stability of the family itself.

Closely related was the question of the status of women.[1] Material plenty, generated by the industrial society, was bringing a greater measure of freedom here also. As women developed an increasing sense of their bargaining position, and began to revolt against continuous pregnancy, the case for birth control could be construed as an element in the attack upon male supremacy. One of J. S. Mill's reasons for supporting female emancipation was that childbearing could pass, at least to some degree, under women's control. Another, that brought him close to Wordsworth, was the need of the human to be able to escape from the presence of his fellows into open country.[2]

3. THE CONSERVATIVE PROTEST

Those of conservative mind were bound to react strongly to the ideas of Smith and Ricardo. Indeed, the position of Malthus in this respect is particularly interesting. He had contributed much to Ricardo's re-fashioning of the Smithian system, yet he could not accept its full implications. He had much sympathy for the outlook of the landed men, who were traditionalist and were unwilling to face the implications of the new industry, or when forced to do so favoured measures that would arrest its rapid spread.

The landed interest still dominated Parliament as it had always done.[3] When the ideas of the political economists were aired at Westminster they could be sure of hearing landed growls and encountering landed votes. *Blackwood's Magazine* described the doctrines of political economy as 'in their nature democratic and republican, hostile to aristocracy and monarchy; and . . . generally taught by people who virtually confess themselves to be republicans'.[4] The nobility and gentry found it very difficult to argue in the new *a priori* way; indeed most of them regarded the whole mode of thought as pernicious. It seemed both mistaken and

---

[1] See above, Chapter 8, section 9, p. 321.
[2] Basil Willey, *Nineteenth-Century Studies: Coleridge to Matthew Arnold*, London, 1949, p. 160.
[3] See above, Chapter 9, section 1.
[4] 'Brougham on the Education of the People', *Blackwood's Magazine*, 1825, p. 542.

dangerous to try to reduce the situation to such simple terms as were embodied in the Ricardian mode of thought; much better to trust in a general acceptance of reality, with such modifications as were unavoidable.

Nor were the agricultural labourers without their protagonists. The greatest of these was William Cobbett. Although he had plenty to say in criticism of landlords and farmers who did not discharge their responsibilities, he looked back to the old agrarian society that was now threatened, and felt that the mass of men would be safer and happier if not subjected to the uncontrolled impact of the mounting forces of industry. Cobbett's vituperation against Ricardo was notorious, including the designation of 'muck-worm'. But Cobbett was also loud against Malthus, detesting him for the population doctrine but quite failing to see the common ground shared with him.[1]

A most important section of the *literati* of the day held much of Cobbett's attitude, though without his violence of expression.[2] The Lake Poets — Wordsworth, Southey, Coleridge — had all as young men welcomed the French Revolution as a new birth of freedom, but had lost heart at the excesses that the violent rupture of society produced.[3] After Waterloo they were active in their attacks on those who sought to base future policy on system, especially a system incorporating the hated Malthusian ideas. Moreover, the romantic movement was gaining great strength at this time, especially in Germany; through Coleridge and others it found its expression in Britain.

The conservatism of Thomas Attwood and his followers was of a different kind. For him all the ills of society were attributable to one great defect — the faulty monetary system.[4] Given correct principles at this single point, then all else was capable of self-adjustment. Yet he felt compelled to take a prominent part in both the great movements of his day — Reform and Chartism — for only so could he hope to

---

[1] Charles H. Kegel, 'William Cobbett and Malthusianism', *J.H.I.*, 1958, p. 348. See also G. D. H. Cole and M. Cole, eds., *The Opinions of William Cobbett*, London, 1945; G. D. H. Cole, *The Life of William Cobbett*, 3rd ed., London, 1947.

[2] See above, Chapter 8, section 6, pp. 305, 306; C. E. Pulos, 'Shelley and Malthus', *Pubs. of Mod. Languages Assoc. of America*, 1952; W. P. Albrecht, 'Hazlitt and Malthus', *Modern Language Notes*, 1945.

[3] Crane Brinton, *The Political Ideas of the English Romanticists*, Oxford, 1927; F. W. Bateson, *Wordsworth: A Re-interpretation*, London, 1954; Geoffrey Carnall, *Robert Southey and his Age*, Oxford, 1960; J. A. Colmer, *Coleridge: Critic of Society*, Oxford, 1959; W. F. Kennedy, *Humanist Versus Economist: The Economic Thought of Samuel Taylor Coleridge*, Berkeley, 1958.

[4] S. G. Checkland, 'The Birmingham Economists, 1815–1850', *Econ. H.R.*, 1948.

find support. But from neither agitation did it come in adequate measure.

Early in the century, in France, the St Simonians had been anxious to replace the aristocracy of birth, especially as it formerly showed itself in the military and ecclesiastical establishments, with one chosen on the basis of success in business and in science. In England this concept of the pyramid, refurbished with a new aristocracy, found its most dramatic and uncompromising advocate in Carlyle. Industrialism had created perversion at all levels of society, he argued, from the labourer, through the millocracy, to the traditional aristocracy. A new élite of 'Aristocrats of Nature' and 'Captains of Industry' should preside over a system of disciplined social castes. Even the analogy with the old cosmogony of the medieval schoolmen was there, for, said Carlyle, the universe itself was a 'Monarchy and a Hierarchy'.[1] It should be said that Carlyle, like Plato, insisted that his new aristocracy were to be chosen on merit whatever their level of origin in the pyramid.[2]

But Carlyle's view of society, like that of the schoolmen, was incapable of producing a reasoned economics. It sought escape from the perplexities of economic and social analysis, through delegation to a new aristocracy. The enormous vogue Carlyle enjoyed showed how deeply rooted, in spite of Locke, Adam Smith, and the *laissez-faire* case, was the ancient view of society, yearning for leadership, not reason. But Carlyle, and those who felt as he did, could not solve the problem of the basis of selection for the new aristocracy. Some hoped, as Galtonian genetics gained ground, that society could solve its economic and social problems through it; Galton wrote in 1873 of 'a golden book of natural nobility'.[3] Earlier, the phrenologists had held the same hope in a different form, seeking the clue to the social pyramid in cranial bumps.[4]

Men of conservative mind often felt themselves in a dilemma from which they saw no escape. They could complain of J. R. McCulloch, one of the great advocates of the system of natural liberty, that 'he was ready to turn the whole country into one vast manufacturing district, filled with smoke and steam-engines and radical weavers'.[5] The later

---

[1] See J. S. Schapiro, 'Thomas Carlyle, Prophet of Fascism', *J.M.H.*, 1945, p. 100; F. A. von Hayek, ed., John Stuart Mill, *The Spirit of the Age*, Chicago, 1942.

[2] See above, Chapter 8, section 6, p. 307; Emery Neff, *Carlyle*, London, 1932.

[3] Karl Pearson, *The Life, Letters and Labours of Francis Galton*, Cambridge, 1930, vol. IIIA, p. 264.

[4] George Combe, *A System of Phrenology*, 3rd ed., Edinburgh, 1830.

[5] *Political Economy Club Minutes of Proceedings*, etc., vol. VI, London, 1921, p. 234.

thirties and the forties produced much evidence of nostalgia, especially at the upper end of society, for an idealized medievalism, some of it ludicrous, as with the abortive tournament rained out at Eglinton Castle in 1839, some of it an attempt to restore older relations through politics, via the 'Young England' movement.[1] But the new trend could not be stopped, for the rapidly expanding nation had come to depend on a continued increase in output. Even the aristocracy, to whom societies in other times and places looked to control the impact of blind change, could do nothing; the Frankenstein monster of steam was too powerful – the system had moved beyond comprehension and control.

To add to such misgivings, certain overtones of the organic and evolutionary approach, gaining ground from the fifties onward, were distinctly frightening. Though evolution had its individualist aspect, it could also be made to run in terms of the history and attitudes of the masses. Events could now be regarded as the outcome of elemental trends and stresses engendered at a level of society of which the middle classes were ignorant and frightened. *The Economist* remarked in 1853 how the belief in great men as the ostensible controllers of society had yielded place to the realization that 'now the gregarious multitude carries the day, and continually gives a new form to society'.[2] The new view suggested the transitoriness of existing arrangements, not least in the holding of property. This was the more alarming when accompanied by the fear that much of society was without mentors or discipline, and a prey to the unscrupulous agitator. This prompted many to lay a new emphasis upon organic gradualism, stressing the 'harmonious development of all the constituent parts' of society, and condemning any proposal likely to damage such a process.[3]

4. THE SOCIALIST PROTEST

For those who felt a profound disgust at the impact of events on the mass of Englishmen it was impossible to begin the analytical task at once.[4] It was necessary first to restate the assumptions that lay behind consideration of the economic system. There were three great themes

---

[1] Briggs, *A. of I.*, p. 300; C. Whibley, *Lord John Manners and his Friends*, Edinburgh and London, 1925.

[2] 'Further Census Details', *Econ.*, 18 June 1853, p. 675.

[3] Sir Edward Bulwer Lytton, *Address to the Associated Societies of Edinburgh*, Edinburgh and London, 1854, p. 18.

[4] For general treatment see M. Beer, *A History of British Socialism*, London, 1929.

that worked on the minds of such men: rationalism, the natural or inherent rights of men, and a mixture of compassion and indignation. The confidence in rationalism embodied the belief that men, by exercising their thinking faculties upon their social condition, could both understand it and improve it. The belief in natural rights meant that the share which each man enjoyed of the national income ought not to be the outcome of a bargaining process, with the counters often weighted with gross unfairness, but should be determined by ethical considerations: a man had a natural right to the product of his labours. Compassion, of course, came with the observation of the plight of many (perhaps most) workers in town and country, and indignation was its accompaniment.

Rationalism was the dominant element in the work of William Godwin, principally embodied in his *Political Justice*, published in 1793. Rousseau was his master, for Godwin believed in the almost infinite power of man for self-improvement, once he was able to base his society upon correct principles. Godwin had complete confidence in man's ability, through his reasoning faculties, to arrive at such principles. Yet he was an arch-individualist, his ideas approaching anarchy. So strong was his individualism that he objected to orchestral playing as damaging to self-reliance.

The combination of rationalism and individualism led to a dilemma. It was very difficult to see how correct principles were to be brought to operation if man was to eschew the state. Godwin, in fact, believed in anarchy combined with communism — no state, and no insistence upon individual property rights. The coercion of men was wrong; their conduct could only be legitimately altered by appeal to their reasoning faculties. Perhaps, indeed, Godwin deserves a place, not among the disturbers of society, but among the great quietists. But this view of him was to be falsified in the longer run, for elements of Godwin's approach were to form an important part of the creed of militant socialists. His confidence in the power of intellect was in strong contrast to the views of the Malthusian school, for it asserted both the power to increase productivity, and the ability to control the course of society.

It was from the men sometimes called agrarian socialists that the plea for natural right came most plainly. These men, headed by Thomas Paine, their minds formed in a pre-industrial society, began their thinking from the ancient problem of the right of the individual to a share in the possession of the land. Such men knew full well that the entire position a man held in society could depend upon this — his power, his wealth, his ability to perpetuate his family, his ability to survive catastrophe of one kind or another. The idea that the land had been filched from the people was very ancient; taken together with the new abstract idea of natural right, it made a powerful impact.[1] The existing pattern of claims to property in land was no longer a thing unquestioned; it required to be established in equity. For the liberal school, derived from Locke, the sanctity of private property was the basis of freedom; from Paine's point of view it was the basis of the denial of freedom to the commonalty. From this attitude men of active mind easily took the step of questioning the whole of the process of distribution, especially as the product grew at the new accelerating rate.

But it was also necessary for someone to cast a questioning eye over industrializing society, and to note its immediate effects upon the lives of the workers. This Charles Hall did.[2] His medical duties in the west of England gave him opportunity for close social observation. With it came compassion and indignation. From these in turn came, not frothy condemnation, but an attempt at analysis. The first socialist capable of scientific procedures had appeared. Hall made explicit the class structure of his day, and insisted that attitudes and power were related to wealth. Even elements that wore the public face of neutrality – the judiciary, the executive, and the legislature, could be subject to strong class bias. The operation of the market was not neutral, as those following Smith claimed or implied, but some bargainers were much more powerful than others. Moreover, the market tended to increase rather than to diminish inequalities.

Between Hall and the next group of early socialists, William Thompson, John Gray, Thomas Hodgskin, and John Bray, came the writings of Ricardo.[3] He had seemed to say that conflict between the three classes, landlords, labourers, and capitalists, was inherent.[4] In addition,

---

[1] Christopher Hill, 'The Norman Yoke', in John Saville, ed., *Democracy and the Labour Movement*, London, 1954, chapter I.

[2] Charles Hall, *The Effects of Civilization in European States*, London, 1805, new ed., 1850.

[3] R. K. P. Pankhurst, *William Thompson, 1775–1833: Britain's Pioneer Socialist, Feminist, and Cooperator*, London, 1954; John Gray, *A Lecture on Human Happiness*, London, 1825, reprinted 1931; Janet Kimball, *The Economic Doctrines of John Gray, 1799–1883*, Washington, 1948; Thomas Hodgskin, *Labour Defended Against the Claims of Capital*, London, 1825, reprinted 1922; Elie Halévy, *Thomas Hodgskin, 1786–1869*, London, 1903, new ed., A. J. Taylor, ed., London, 1956; John Francis Bray, *Labour's Wrongs and Labour's Remedy*, Leeds, 1839, reprinted, London, 1931; *A Voyage from Utopia*, M. F. Lloyd-Pritchard, ed., 1957; H. J. Carr, 'John Francis Bray', *Econa.*, 1940.

[4] H. Foxwell, Introduction to Anton Menger, *The Right to the Whole Produce of Labour*, London, 1899, pp. xl–xli; E. Lowenthal, *The Ricardian Socialists*, New York, 1911.

his labour theory of value appeared to imply that labour was the sole source of value. From this approach stemmed the Ricardian socialists' claim that to labour should belong the whole produce of its efforts, or, at any rate, a much greater proportion of it. Natural Right thus took on a new content. The ideas of Godwin, Paine, and Hall could now be given a new precision, borrowed from the great apologist for the competitive system. In 1814 there had also come to hand Patrick Colquhoun's *Treatise on the Wealth, Power, and Resources of the British Empire*, with its dramatic demonstration, in quantitative terms, of inequality.

But though the socialist reasoners could extract support from Ricardo's system, they did not make any great general theoretical progress in consequence. They could not accept the Ricardian stationary state as the eventual outcome, for though, as John Stuart Mill implied, it could bring many of the elements valued by socialists, including a great easing of competitive pressures, they could not accept the notion of a cosy and mellowed capitalism.

Certain difficulties, long present in socialist thinking, now became more explicit. How was equality to be reconciled with freedom? For full freedom to those of ability and opportunity might well cause them to acquire a very large share of output. How was a fair distribution, approximating to equality, to be reconciled with the provision of capital for further improvement? On the first problem there appeared no alternative but to argue that rational man, properly educated, would be too soundly based to be infected with the lust for wealth and power. On the second, the socialist writers in general had no real success in working out a theory of capital that made it possible to judge how much was needed for new investment, and how it was to be provided.

The four leading Ricardian socialists struggled with these problems, each in his fashion. Thompson tried hard to adapt the Benthamite standard. The greatest happiness of the greatest number was the object; the problem was one of calculating how a policy was to be measured in this sense. Thompson was certain that competition would not serve the desired end. Gray paid great attention to the problem of industrial breakdown, through a discrepancy between demand and output, and stressed the elements of exploitation operating in landholding and in banking. Thomas Hodgskin, especially, showed theoretical insight. He argued that capital was supplied not from some stored-up fund, but by varying the distribution of the current flow of output between investment and consumption, through the banking system and wage bargaining. For him growth did not depend upon capital, as with the Ricardians, but upon labour. Hodgskin's rather impracticable remedy was combination among workers against capital, in order to maximize the workers' total share, but competition within the labour force to determine the allocation between workers.

In spite of their good intentions, these men could not supply the baffled workers with an analysis and a programme. Hodgskin, for all his theoretical acuteness, was too much infected by the anarchistic individualism of Godwin. Bray proposed a National Joint-Stock Scheme, to organize the workers into companies. Financed by the creation of new money by state action, each such company, confined to a single trade, would place the productive and distributive process in the hands of the workers. But force was to have no place in the change — it must be carried through by a convinced and willing community. Thompson and Gray did not wish to wait upon this universal acceptance of the need for change, but neither did they wish to see coercion employed. Their solution took the form of voluntary communities on the Owenite model, capable of demonstrating within themselves a new way of communistic life and a new mode of cooperative production. So clear would be the advantages that such communities would increase and become general, remaking society without offering an overt challenge to authority. None of these men were prime movers in the field of action.

But Robert Owen was. With his immense drive toward human amelioration, he sponsored large-scale experimentation and led the movement of political protest for at least a decade down to 1834.[1] Alone among the early socialists Owen had preceded theorizing by an astonishingly successful career as a man of business. The young Calvinist from Wales had been pious and earnest to an extraordinary degree. A deadly seriousness approaching obsession accompanied all he did; this was reinforced by the authority which naturally came with business success. 'Thou needst be very right', said David Dale, his lay preacher father-in-law, 'for thou art very positive'. Effective action in so immature a cause as the challenging of capitalism depended upon a leader possessed of extraordinary qualities of self-assurance and persistence. Yet the very conditions that had made him so powerful, also served to weaken him.

Owen was one of the first of industrialists to learn how profitable it

[1] See above, Chapter 9, section 6, pp. 345-9.

could be to nurture his workers as he cherished his machines. His experiments in management and industrial psychology at New Lanark near Glasgow helped to make him a fortune; they also taught him how human behaviour could be altered by varying the conditions under which working and living were done. In 1813 Owen published his *New View of Society,* or *Essays on the Principle of the formation of the Human Character.*[1] In it his view of man was expressed in all its simplicity: his character is made for, and not by him. With this as a cardinal doctrine, Owen could not fail to be horrified by the kind of society that was emerging in much of urban and industrial Britain. If nothing was done, the workers would be debased from childhood, and the employers corrupted, for it was the means by which each acquired his income that determined his character. Owen's breach with the churches was not merely wilful extravagance – it too derived directly from his environmentalist view; the Church blamed evil upon individual sin, but for Owen evil arose from a misconceived society. Owen was one of the great prophets of the idea that man was a prisoner – not of nature, for the new industry showed his mastery there – but of the kind of society that the new industrial methods had produced.

There was of course a strong strain of élitism, if not messianism, in Owen's thinking. If the mass of men were creatures of environment, then they could only be redeemed by the thought and action of the very few who could break free from their conditioning. Godwin's rationalism had been a concept that extended to all men – all had the potential to think out their condition; not so with Owen.

His approach was thus radically different from that of his socialist precursors, though he borrowed freely from them all. His ideas stemmed directly from his experience. His failure stemmed from the inadequacy of his theoretical equipment and his inability to choose a political course that would bring his ideas into effective operation.

It is hardly surprising that the universal manager, Owen, and at least two of the leading socialist theorists, Thompson and Gray, should converge in the founding of communistic communities.[2] Such entities provided Owen with a means of extending his principles; they also seemed to offer to the Ricardian socialists an escape from the necessity for revolutionary action, with its inevitable violence. The communities were not intended as retreats from the world, but rather

[1] See also, Robert Owen, *Life of Robert Owen,* M. Beer, ed., New York, 1920; G. D. H. Cole, *Robert Owen,* London, 1925; M. I. Cole, *Robert Owen of New Lanark,* Oxford, 1953.
[2] See above, Chapter 9, section 6, p. 345.

were to be dramatic demonstrations both of superior production methods and of a richer social life. For the failure of the movement Owen bore a heavy responsibility which all his generous financial support could not discharge. He seems not to have realized that those who entered the cooperative communities would be the unregenerate creatures of the system he condemned, so that the greatest skill and devotion on the part of the leaders was called for. But the industrial genius that had created New Lanark could not be brought to bear in the cooperative colonies. Partly this was because Owen now conceived himself in the grander role of prophet, and partly it derived from the inherent contradiction between the community ideal and the rigours of effective management.

The Chartist socialists of the activist wing went back to Charles Hall and his view of the class structure of society (strengthened as it had been by the Ricardian analysis). For Bronterre O'Brien the clashes between class interests were so great and so inherent that it was useless to try, as Owen had done, to seek the aid of the middle and upper class to remake society. In taking this view O'Brien was using a form of Owen's own basic principle: environment had operated so effectively upon the wealthier classes that no appeal to them to liquidate their own position could hope to succeed. With the class view came a restatement of the behaviour of capitalists that was a step toward Marx. Output was being held back by the competitive principle, so much less productive than the cooperative alternative. Those in command of trade and industry were concerned for the success of their own enterprises, and not with social ends. Moreover, under the competitive impulse machinery was used to replace men, weakening their bargaining position and lowering their wages. It was of little avail to organize in trade unions, so long as the position of the employer was left intact — hence political action took precedence over union organization.

Between the conservative protest and the socialist one the distinction was by no means absolute. The Christian Socialists, with Frederick Denison Maurice and Charles Kingsley as their chief exponents, were men of the middle class who were deeply disturbed at the social injustice and damage being done to the workers.[1] But they were not revolutionaries: they sought the means, through assistance to the cooperative movement and the trade unions, whereby the workers could evolve their own institutions that would safeguard their place in

---

[1] See above, Chapter 9, section 8, p. 364; J. M. Ludlow and Lloyd-Jones, *Progress of the Working Class, 1832–1867*, London, 1867.

society. Among young Tory politicians in the forties there was a similar urge to improve the lot of the masses accompanied by a somewhat romantic wish to bring together the upper and lower classes in a relationship of mutual confidence and respect.[1] But though both Christian Socialists and Young England could help forward the trend toward amelioration, their thoughts did not run to any radical alteration of society.

Thus it was that Britain passed the mid-century with the socialist effort to understand society still little more than a collection of fragments of various origin, and encumbered by a sense of failure. The great expansion of the economy then beginning considerably improved the conditions of the mass of men so that protest from the lower orders fell away.

5. BENTHAMITE COLLECTIVISM

Among the radical-intellectuals of the nineteenth century the concept of natural liberty did not stand alone. There was also the apparently inconsistent idea that the state must play a positive role through carefully considered legislation and administration, to bring about, where necessary, an artificial identity of interests.[2] The Benthamites did not believe that the market mechanism was basically defective, that it suffered from inherent tendencies to break down. But they did believe it capable of omissions, and these of a serious kind. Indeed only an anarchist could imagine that the principle of freedom could solve all the problems of society.

The possibility of omissions had been provided for by Adam Smith in the third great duty he assigned to government.[3] The state was to provide security against foreign enemies, and the internal administration of justice. But in addition it was to make itself responsible for those tasks which, though of the greatest importance in the life of the community, would not be discharged on the basis of the unaided profit motive.

The great work of Bentham and his associates was precisely to the end of elaborating the duties and procedures of the ruling power.[4]

---

[1] See Benjamin Disraeli, *Coningsby*, London, 1844.

[2] For the debates to which this difference has given rise see J. B. Brebner, 'Laissez-faire and State Intervention in Nineteenth Century Britain', *J.E.H.* Supplement, 1948; Willson H. Coates, 'Benthamism, Laissez-faire and Collectivism', *J.H.I.*, 1950.

[3] *Wealth of Nations*, vol. II, pp. 184–5.

[4] For his economics see W. Stark, ed., op. cit.; Elie Halévy, *The Growth of Philosophic Radicalism*, London, 1928.

The Benthamites sought the extension of the franchise partly in order that a Parliament might come into being which correctly perceived not only the merits of the market, but also the need to keep government alive to its further range of duties. Chadwick, the most notable implementer of Benthamite policies, engrossed himself in the Poor Law, the Factory Acts, sanitation, and public health.

Benthamism meant identifying the urgent tasks of society and prescribing the means for their discharge; it meant specific legislation, with inspectors in the field and administrators in centralized offices.[1] It meant Members of Parliament who thought, as Bentham did, in terms of 'agenda'. The fight for the Factory Acts meant an *ad hoc* alliance with the landed interest – emotion and hostility were allied to implementation to put down self-interest as it showed both in factory masters and in the parents of wage-earning children.

It was natural that men of Chadwick's kind, adept at organization, should place emphasis upon the positive tasks to be done, and perhaps minimize the concept of the automatic economic system. But the minds of many economists, including Nassau Senior and John Stuart Mill, were dominated by the need to allow the springs of private enterprise to work freely. It was essential that the sources of profit, and therefore of employment and economic expansion, should not be impaired. So arose a dilemma. The new services were costly, involving increased taxation the burden of which had to lie somewhere; the enforcement of minimal conditions under which working and living went on often meant raising the business man's costs; any attempt to alter the grossly unequal distribution of incomes between classes could not fail to have effects upon entrepreneurial incentives, or so it seemed.

6. THE POST-RICARDIANS

From the thirties onward it was not only the conservative minded, the socialists, and perhaps some collectivists, who made inroads into the Ricardian system. So too did the professional economists. The 'vulgar economists', as Marx called them, had succeeded Ricardo. They abandoned, either implicitly or explicitly, the Ricardian theories of distribution and value. The theory of distribution was very difficult to handle. Partly this was because the attempt to alter the idea of a minimum level of subsistence into a psychological rather than a physio-

---

[1] David Roberts, 'Jeremy Bentham and the Victorian Administrative State', *V.S.*, 1958–59.

logical concept had deprived it of precision. Partly it was because a system of thought that made the whole course of events depend upon what happened in agriculture seemed increasingly unreal. Finally, though the idea of diminishing returns on the land was still part of the formal theoretical apparatus, the old contrary view, held by Horner and others, namely that there was no need to fear limitations of 'the prolific virtue of the soil, as well as the effective power of industry', was recovering its support.[1] Ricardo's value theory suffered, either because the next generation found the labour theory intellectually barren, or because they found it politically dangerous.[2] Even such theoretical coherence as Ricardo had managed to achieve, after so great an effort, was gravely impaired.

The attitude to Britain's international position was also changing. Ricardo had felt no fear of excess capital at home. But as the mid-century approached, with capital accumulation in Britain proceeding apace, the adequacy of free trade alone as a means of saving profits from decline was questioned.[3] J. S. Mill, Robert Torrens, E. G. Wakefield, and others began to develop a plea for a policy of capital exports and emigration.[4] This would make it possible for both capital and men to find better opportunities in the Empire and in America. The difficulties of the forties could not fail to give rise to such a line of thought; it was becoming increasingly evident that Britain's ability to continue to industrialize and expand depended on aggressive action to exploit the idle resources of other parts of the world.

John Stuart Mill published his *Principles* in 1848; it was to be the most highly regarded treatise on the subject down to the eighties. It ensured that political economy retained its public importance: the Prince Consort and the Queen submitted the royal princes and princesses to a course of lectures on the subject in 1858 and rewarded the giver with a very elegant silver inkstand.[5] It is unlikely that the royal children appreciated how great were the conflicts within the *Principles*.

Mill was confronted throughout his adult lifetime with the task of

[1] F. W. Fetter, ed., *The Economic Writings of Francis Horner in the Edinburgh Review, 1802–6*, London, 1957.

[2] See T. W. Hutchison, 'Some Questions about Ricardo', *Econa.*, 1952, p. 424.

[3] G. S. L. Tucker, *Progress and Profits in British Economic Thought: 1650–1850*, Cambridge, 1960, chapter VIII.

[4] Lionel Robbins, *Robert Torrens and the Evolution of Classical Economics*, London, 1958; Frank Whitson Fetter, 'Robert Torrens: Colonel of Marines and Political Economist', *Econa.*, 1962; Edward Gibbon Wakefield, *A Letter from Sydney and Other Writings*, with Introduction by R. C. Mills, London, 1929.

[5] W. B. Hodgson, *Life and Letters*, J. M. D. Meiklejohn, ed., Edinburgh, 1883, p. 376.

reconciling the Ricardian orthodoxies in which he had been raised with the need for a view more appropriate to rapidly changing conditions.[1] He saw that it was wrong to view other societies simply in English terms. It was hopeless to try to promote in Ireland a capitalist agriculture on the English basis, with substantial tenants freely renting from landlords; account had to be taken of the cottier system and the tenant who was little more than a labourer.[2] New countries, desirous of developing their industries, were bound to feel that there was a case for protection against the products of British factories, at least while their own development was in the 'infant' stage. Mill made concessions to this argument in the case of Australia which got him into a great deal of trouble. Wakefield's scheme for planned emigration, with a controlled market for land, also gained his support. He made a considerable movement in the direction of socialism, amending his *Principles* in collaboration with Mrs Taylor.[3]

The programme favoured by the radicals who regarded Mill as their guide necessarily consisted of a rather various set of measures. Freedom of international trade would ensure peace between nations by substituting for the confrontation of sovereign states a system of mutual interdependence.[4] General education was essential so that all adults could attain full stature. Peasant proprietorships or small holdings were needed so that the ownership of land might no longer be the monopoly of the few, so that the romantic urge of many radicals for a restoration of healthy contact with the soil might be met, and so that individual initiative could operate to best effect in agriculture.[5] Profit-sharing schemes in industry, by allowing the worker a share in the gains, would make him feel himself a part of the enterprise and not something merely hired by it. Cooperative projects in producing and selling would demonstrate a healthy alternative to private enterprise. Savings banks would provide a means whereby the workers might attain some measure of financial independence based upon their own foresight.

Each of these proposals was seen as being consistent with both liberal

---

[1] John Stuart Mill, *Autobiography*, London, 1873, pp. 244–8; Michael St John Packe, *The Life of John Stuart Mill*, London, 1954.

[2] R. D. Collison Black, *Economic Thought and the Irish Question, 1817–1870*, Cambridge, 1960.

[3] F. A. Hayek, *John Stuart Mill and Harriet Taylor: Their Correspondence and Subsequent Marriage*, London, 1951; H. O. Pappe, 'The Mills and Harriet Taylor', *Political Science*, 1956.

[4] F. R. Flournoy, 'British Liberal Theories of International Relations, 1848–1898', *J.H.I.*, 1946.

[5] See W. T. Thornton, *A Plea for Peasant Proprietors*, London, 1848.

and socialist ideas: the radical programme in the sixties and after was in fact obliged to meet these two mutually unassimilated criteria. The upshot was a new emphasis upon pragmatism: Jevons, toward the end of his life, in 1882, expressed the opinion that no tests of a general nature could be established that would assign the form and limits of proper state action: 'we can lay down no hard and fast rules, but must treat every case in detail on its merits'.[1]

7. THE WAGES FUND AND THE SIXTIES

Perhaps the most important subject of theoretical debate in the sixties was wages. The increasing strength of the unions and the mounting pressure for the extension of the parliamentary franchise focused attention very strongly upon the effect of labour's increasing power. In Australia, it was said, capital was dependent upon labour – the worker, in short supply, enjoyed the upper hand; in Britain the reverse was true: it was labour that depended upon capital. But better union organization in Britain was having its effect. This meant that for the well-being of all, including labour itself, capital must be protected against excessive claims by labour, for if capital were damaged all would suffer. Hitherto capital had managed to protect itself to a very real degree through its influence in the legislature, especially in its jealous view of trade unions and of taxes upon capital. Now there seemed to be real danger that if the control of Parliament fell into the hands of the workers, capital would be seriously impaired. The signs were multiplying that the workers were placing an increasing emphasis upon their own solidarity and security.

There were two ways of meeting this situation. An attempt could be made to continue to 'contain' the worker – both by resisting his admission to the franchise, and by providing theories of society that would convince him of his duty to accept the limitations implied in the wages fund theory.[2] Alternatively, the implications of democracy had to be faced – the political participation of the workers welcomed, and a more satisfactory economic theory found, which could reconcile the interests composing a society undergoing fundamental change.

On the whole the trade unionists rejected outright the body or

[1] Lionel Robbins, 'The Place of Jevons in the History of Economic Thought', *Proc. Manchester Statistical Society*, 1936, p. 19.
[2] See above, Chapter 10, section 2; W. R. Hopper, 'An Iron-Masters' View of Strikes', *F.R.*, 1865, p. 753.

thought labelled political economy, derived from Smith, Malthus, and Ricardo, as being quite without validity.[1] To its purported natural laws they opposed the claim of natural right, a concept independent of any systematic view of the economy, and instinctive in origin.

It was at this point (1869) that John Mill 'recanted', conceding that the Wages Fund was an artificial concept. 'The doctrine hitherto taught by all or most economists (including myself), which denied it to be possible that trade combinations can raise wages . . . is deprived of its scientific foundation, and must be thrown aside.'[2] There were valid theoretical reasons for its abandonment. With the idea of excess of capital gaining ground it seemed perverse to insist that labour should make sacrifices to cherish the Wages Fund. Mill had always held that though the laws of production were universal and immutable, those of distribution were within human control: the Wages Fund had clashed with this. But there was also the practical necessity of getting to terms with the workers on this basic problem. It had been clear for a long time that the theory had no meaning for them. It was time to move away from such a crude formulation, but without forgetting that growth would be arrested if too much was dispersed from current output in wage payments. Moreover, the old doctrine, like the Ricardian class distribution scheme, was an invitation to think in terms of masses — to set the worker over and against his employer. It ran in terms of aggregates, and so involved a dangerous mental habit, at least in the eyes of those who still believed in atomistic individualism. The fear of the mass agitator, and of the full-time labour spokesman, was also very real — the way to deal with them was to keep wages bargaining on the basis of particular masters and their men, and not to raise questions of the general wages share in total output.

Masters and men knew from long experience of the decennial trade cycle that the workers were obliged to press for increased money wages as prices and productivity rose, and to resist downward pressure as they fell. Bargaining, in fact, had become accepted as necessary and right over a good deal of industry. Neither masters nor men conducted such negotiations in the light of a generalized Wages Fund, but as a specific sharing of the gains or losses of particular transactions. When Brassey, one of the greatest of entrepreneurs, talked of wages, it was in

---

[1] R. V. Clements, 'British Trade Unions and Popular Political Economy, 1850–1875', *Econ. H.R.*, 1961.

[2] John Stuart Mill, 'Thornton on Labour and its Claims', *F.R.*, 1869, p. 517; Sir L. Stephen, *The English Utilitarians*, London, 1900, vol. III, pp. 203 ff.

terms of the relationship between payments and productivity as they worked out in the execution of a job, and not in terms of a universal theorem. At long last orthodox political economy was obliged to abandon a theory that had played a large part in employers' insistence that combinations of workers to raise wages were not only incapable of achieving their aim, but were positively damaging. Yet there remained truth in the idea that without increases in productivity there were severe limits to the workers' power, through coercion, to raise wages. The real contest had to do with the continuous adjustment of the proportions in which the benefits of increased output or the losses due to contraction were distributed between capital and labour.[1]

But though there were good grounds for releasing wages from the Wages Fund theory, the result was a worrying hiatus. Business men had been able for more than a generation to argue, on the highest authority, that 'in the case of labour, as in the case of wheat and sugar, the fluctuations of price must rest upon natural causes'. They were now deprived of theoretical sanction.[2]

8. NEO-CLASSICS AND NEO-SOCIALISTS

Though the theoretical disintegration of the Ricardian system had not greatly troubled those in charge of politics and business, there were others who, from the seventies onward, became increasingly dissatisfied with the lack of any real synthesis of economic experience. As the professional scientist was emerging, so too, now, was the professional economist, mainly in the universities. These men were bound to feel that a further effort was now called for in order to restore the scientific claims of political economy.

Here two great names are to be noted: W. S. Jevons and Alfred Marshall. Ricardo, in spite of his Benthamism, with its criterion of the greatest happiness of the greatest number, had, after a brief reference, relegated to secondary importance any consideration of the feelings of the individual as they might affect demand. Others had strongly resented this, and a steady flow of subjective value theorists continued through the years. But it required the genius of Jevons to make explicit

---

[1] For the problem of containing the pressures on resources due to the needs of rapid capital formation see S. Pollard, 'Investment, Consumption and the Industrial Revolution', *Econ. H.R.*, 1958.

[2] 'The Tyranny of the Strike', *Econ.*, 17 September 1859, p. 1036; 'The "Law" of Supply and Demand', *Econ.*, 20 October 1866, p. 1221.

the concept of marginal utility, making value dependent upon the psychological state of the buyer rather than upon some intrinsic quality of the good (in Ricardo's scheme, embodied labour).[1] To give precision to his analysis Jevons made his reasoning depend upon the effect on the demander's attitude toward a particular good of the acquisition of one further unit. The differential calculus was used to demonstrate this. Marshall worked along the same lines, generalizing the approach, applying it to the supply side in his *Economics of Industry* in 1881, presenting it in its classic form in his *Principles of Economics* of 1890. The seller's behaviour could be understood in similar terms, his attitude to giving up one additional unit being crucial to the explanation. In the longer run, the marginal principle could be used to explain with precision how capital was distributed between alternative projects, investment being so applied that, assuming the degree of risk to be equal, the return in the form of profits would be equal, at the margin, between all the contending projects.

By this kind of thinking a new kind of precision was given to the case for the system of natural liberty. Each buyer in each market, including that for labour, would so allocate his spending that he would get equal benefit, at the margin, from all his purchases. This would be true of both sides of the market, for each buyer could be regarded also as a seller, yielding up his money for some other good. Capitalists would behave in the same way, so that investment too would distribute itself between the various lines of production to optimal effect. Labour too would follow the same principle in distributing itself among alternative employments.

So, it might seem, a great step was taken to restore the coherence of the classical system. But there were two great difficulties. The neo-classical system was one of partial equilibrium; that is to say it envisaged the general parameters of the system to be fixed. Thus spenders were regarded as having a given income, leaving open the problem of how they would behave if their incomes changed. Equilibrium thinking also meant that the question of economic growth was virtually taken off the agenda. Secondly, neo-classical thinking rested, as had Smith's system, upon competitive assumptions. But by the eighties the elements of monopoly, that might be more or less ignored in the years of expansion, had now to be considered seriously. The question

[1] See Richard S. Howey, *The Rise of the Marginal Utility School, 1870–1889*, London, 1960; W. Stanley Jevons, *Theory of Political Economy*, London and Oxford, 1871, 2nd ed., 1879.

arose again, were they such as to render irrelevant a system, that, however theoretically brilliant, did not fit the realities of the time?

Socialist thinking had gone into something like abeyance since the forties or early fifties. But it too had never lacked protagonists. Mill, as we have seen, had much sympathy, though it has been strongly argued that Mill's effect upon socialist thinking was 'soporific'.[1] He had failed to analyse the early English socialists as they deserved, and had deprived the movement of protest of its vitality by his equivocal acceptance. In Britain, in the seventies, socialism was without any systematic statement.[2] The political economists, even had they wished, would have found it difficult to judge its credentials. But abroad, especially in Germany, where there was much less diffidence about the role of the state, socialist theory had been developing. By the early eighties, the ideas of the early British socialists were returning to their country of origin, incorporated in German systems.

Marx had given the world the outline of his views in the *Communist Manifesto* issued in 1848. These he had built into a system during long hours spent in the Reading Room of the British Museum.[3] The first volume of his great work, *Das Kapital*, appeared in German in 1867. But it was not translated into English until 1887.[4] His ideas, in fact, though known to a considerable circle of intellectuals in the early eighties in Britain, made little real immediate impact on general opinion.[5] But those who were seeking an explanation of the behaviour of an economy that had reached an advanced stage of industrialization found much that was stimulating in his writings. The neo-classical economists, so far as their general system was concerned, thought mainly in equilibristic terms, trying to locate the norm around which a system of limiting assumptions would vacillate, and studying the processes of adjustment to the norm. The kind of theory of which Marx was the great exponent ran in terms of the continuous renovation of societies, seeking its clue in the dynamics of change.

This type of thinking made a strong appeal to the historically minded, who resented the new flowering of the classical system in the marginal

---

[1] Foxwell, op. cit., p. lxxviii.

[2] T. W. Hutchison, *A Review of Economic Doctrines, 1870–1929*, Oxford, 1953, p. 294.

[3] Isaiah Berlin, *Karl Marx, His Life and Environment*, 2nd ed., London, 1949; Leonard Krieger, 'Marx and Engels as Historians', *J.H.I.*, 1953.

[4] Karl Marx, *Capital*, trans. from 3rd German ed. by S. Moore and E. Aveling and ed. by F. Engels, London, 1887.

[5] John Rae, *Contemporary Socialism*, London, 1884. See also W. Cunningham, 'The Progress of Socialism in England', *C.R.*, 1879.

analysis. There had emerged in Britain in the seventies and eighties an historical school of economists, including John Kells Ingram, Cliffe Leslie, and Arnold Toynbee, who shared with German and American scholars the view that economic studies had been stultified by attempts at abstract theorizing and that the truth could only be known by the close study of real experience unrolling itself continuously over time.[1] Nothing was to be gained, said Thorold Rogers, by browsing on 'the thorns and thistles of abstract political economy'.[2]

The Fabians, by drastic simplification, cut loose from the dilemma that had haunted Mill throughout his life: how to reconcile the basic conditions necessary for the working of the free enterprise economy with the state action necessary to repair its omissions.[3] The Fabians simply took no serious account of the system as a whole: whether, if the market mechanism were abandoned over one area after another, it would cease to be effective as its scope diminished, with the final result that there would be no alternative but to choose a new general principle upon which to rest the system. The Fabians, in fact, like the Benthamites, rather unscientifically assumed the general outlines of society to be stable, and saw their role to be that of modifiers who could safely rely upon a basic continuity.[4] Upon the Fabians there converged all kinds of protestors. In part the deliberate repudiation of a systematic theory arose from the same situation that had given the People's Charter its form — the need to find unity in an agreed programme of action.

Those of conservative and traditional mind had never had any confidence in generalized theories of economic and social behaviour. They were, from this point of view, no better and no worse off in the situation of the eighties, and might indeed have found a wry satisfaction in the confusion now appearing among the theorists. Moreover, for all the changes going forward in the organization of England, it was still possible to believe that change, if inevitable at all, would be evolutionary rather than hectic, so that the conservative mind could

---

[1] A. W. Coats, 'The Historist Reaction in English Political Economy, 1870–90', *Econa.*, 1954.

[2] James E. Thorold Rogers, *The Economic Interpretation of History*, London, 1888, p. vi.

[3] See above, Chapter 9, section 10, p. 381; Alain Maine McBriar, *Fabian Socialism and English Politics, 1884–1918*, Cambridge, 1963; G. B. Shaw, et al., *Fabian Essays*, Intro. by Asa Briggs, 6th ed., 1962.

[4] H. G. Wells, *Experiment in Autobiography*, London, 1934, p. 196, et seq. For the relationship between the Benthamites and the Fabians see Mary Peter Mack, 'The Fabians and Utilitarianism', *J.H.I.*, 1955. See also William Irvine, 'George Bernard Shaw and Karl Marx', *J. Econ. H.*, 1946.

continuously glide over the decades without being exposed to the kind of shock that destroys the old landmarks and leaves some kind of system as the only guide.

In the eighties the old-style liberal could still insist upon the great merits of free markets at home and abroad, and though the more radical minded might sense that Britain's position had undergone great but unappreciated changes, even he was unwilling to abandon a basic confidence in the market mechanism. Socialist thought had pushed many radicals away from liberalisam, yet being unable to carry a full conversion, left them empiricists. The new socialist groups placed their emphasis upon the failures of the system rather than its successes, especially the deplorable condition of the least fortunate in society, but they too were incapable of a systematic explanation of the past or a guide to the future, and so fell back upon judging each case for reform on its merits. Thus liberals and socialists drew together, meeting as proponents of pragmatic improvement.

9. THE PROBLEMS OF GROWTH AND STABILITY

Observation of the working of the economy as a whole called attention to two great problems: on what conditions did growth depend, and what was it that caused the continuous cycle of prosperity and depression?

The growth problem, of course, was not new: many thinkers both before and after Adam Smith had considered the conjunction of circumstances that caused an economy and society to expand, and the possible limits of such expansion. Ricardo had developed his theory of the stationary state. But though it had enjoyed some degree of intellectual assent, it was not of a kind to gain general approval, especially in the face of economic expansion. There was a feeling, expressed by Walter Bagehot and Sir Henry Maine, that Western nations in their modern phase were very different from the rest of the world, in strong contrast to the older civilizations of the East that had got stuck — India, Japan, China.[1] There was a tendency to emphasize the historical uniqueness of Western experience since the renaissance. This implied that the advanced nations had a cumulative advantage over the rest of the world where the principle of originality and the idea of progress had no place. Yet the idea that economies, like the living things in nature,

[1] *The Works . . . of Walter Bagehot*, Mrs Russell Barrington, ed., London, 1915, vol. VIII, pp. 27-35.

must undergo a life cycle involving maturity and decline lurked in the back of many minds.

In practical terms, however, the general belief was that given freedom of world commerce an ever widening circle of producers and consumers would be created. This was the essence for the case for free trade. But it rested heavily upon the assumption that new land would be continuously available for the production of food and raw materials, and that the nations of the new world would be content with the role of primary producers. On the whole there was no great preoccupation in England with the theory of economic growth until the seventies and eighties. Then came a good deal of questioning about the rise of industrial rivals like the United States, Germany, and France, and the conditions upon which the economy of Britain might continue to expand.

There was also a vigorous debate on the relationship between economic growth and the world supply of monetary metals, a tempting theme, for whereas the falling prices and general difficulties of the first half of the century were accompanied by a static gold production, the subsequent general expansion from the mid-century onward coincided with gold discoveries in California and Australia, and, finally, contraction and falling prices in the seventies were related in time to the general demonetization of silver and a flagging world production of gold.[1]

It was in the seventies that men began to think of invention and innovation as an important and even primary element in change. The older liberals had for long claimed that the major contribution to growth had been the dismantling of the state apparatus. Gladstone in 1879 sounded the new note: 'The great salient feature of the age is in its first aspect the constant discovery of the secrets of nature, and the progressive subjugation of her forces to the purposes and will of man.'[2] The concept of man as a creature of his environment, studying to propitiate it (Smith's prime instinct of self-preservation) and the Malthus-Ricardo emphasis upon the irrefragable limits imposed on output by nature had given place to the concept of man the master of nature. But, Gladstone went on, the conquest of nature, though it had raised wages, had produced baleful results. Capital had benefited more than labour; luxury more than industry. The governing and business classes, the 'plutocracy', were being increasingly deranged by the appetite and pursuit of wealth, made possible by their disproportionate

[1] W. T. Layton and G. Crowther, *An Introduction to the Study of Prices*, London, 1938.
[2] W. E. Gladstone, *Rectorial Address, University of Glasgow*, London, 1879, p. 10 *et seq.*

share in the fruits of science. To Gladstone, the contrast between the new potency of man in his manipulation of nature and his impotence in making use of his gains, was becoming more stark. This was indeed a long way from the beneficence of the system of natural liberty.

The question of the stability of the system was a matter of much more continuous concern.[1] Malthusian fears of the possibility of excess saving were expressed by others, including Lauderdale, Chalmers, Cazenove, Richard Jones, and William Whewell.[2] These ideas had been largely irrelevant to the situation after 1815, for two reasons: they were discussed at a high level of abstraction, and they had no immediate bearing upon post-war slump. The under-consumption case did find a champion in Thomas Attwood, who argued that a failure of spending had occurred and had been much aggravated by the government's deflationary policy in returning to gold.[3] From this, Attwood's ideas developed to the point where he was arguing for a managed currency, with the supply of money and credit vigorously manipulated by the government to stimulate or control demand as the state of the economy required. Engrossed in the polemics necessary to obtain political attention, he failed to develop an adequate theoretical basis for his proposals, though many keen insights appear in his writings, overborne by his somewhat strident policy proposals.

An enormous amount of discussion had centred upon the banking system down to the Act of 1844.[4] As the idea of physical and tariff controls lost ground there was only one point at which the government could manipulate the economy, but that, in spite of some exaggeration, was a very powerful one. Whatever diminution Ricardo's doctrines had suffered in other directions, his adherence to the Quantity Theory of Money had a powerful effect on contemporary minds.[5] The behaviour of prices, and through it the general level of activity, was seen as depending in large measure upon the quantity of money in circulation, or more exactly upon the changes in its quantity. But the state was not free, as Attwood seemed to imagine, to vary this at will.

[1] B. A. Corry, *Money, Saving and Investment in English Economics, 1800–1850*, London, 1962.
[2] F. A. Fetter, 'Lauderdale's Oversaving Theory', *A.E.R.*, 1945. For the theory of effective demand at this time see D. H. Macgregor, *Economic Thought and Policy*, Oxford, 1949, chapter 4.
[3] For Attwood and others see Robert G. Link, *English Theories of Economic Fluctuations, 1815–1848*, Oxford, 1959.
[4] See above, Chapter 6, pp. 198–201.
[5] R. S. Sayers, 'Ricardo's Views on Monetary Question,' in *Papers in English Monetary History*, T. S. Ashton and R. S. Sayers eds., Oxford, 1953.

For the domestic price level was the key to international trade. If it went too high exports would fall and imports rise, causing great difficulties with the balance of payments.[1] If no control were present there would be no disciplinary element in the system to cause unsound ventures to be purged. There were also grave political dangers in creating such an immense discretionary power. It was necessary that the monetary system be based upon universally recognized principles which embodied a built-in control that would make it possible for the state to withdraw from this aspect of the economy as from most others. The Bank Act of 1844 put the seal on the system whereby the economy was obliged, when it had expanded as far as the Bank of England's reserves would permit, to undergo selective diminution. The fact that the Act might be suspended in the face of panic liquidation was however a practical recognition that the economy had become so complex that the Bank of England, in deciding who was entitled to support in crisis, must ease the liquidation process.

But though most public discussion of stability down to 1844 centred upon the operation of the Bank Acts, the idea of a fatal flaw of a non-monetary kind in the mechanism of expansion was by no means dead. John Stuart Mill in his *Principles* of 1848 reasserted Say's Law as he had learned it from his father and from Ricardo, dismissing the idea of general glut.[2] Yet he was obliged to take account of the successive crises that were by now so patent. General eagerness to buy, Mill observed, and general reluctance to buy, succeed one another at intervals. He began therefore to try to reason from reality. He came to the conclusion that because barter had given way to an elaborate monetary system allowing individuals to hold wealth in the form, not of real goods, but in money or bank balances, buying and selling had become separate acts. There might well occur, at particular times, a very general inclination to sell, accompanied by an equally general inclination to defer purchases.[3] This might produce an accelerating effect until all were seeking to become as liquid as possible, demanding money from creditors and refusing goods. But Mill, as in other matters, remained ambivalent between the ideas he had inherited and those suggested by experience.

The under-consumption case found other advocates: W. F. Lloyd,

---

[1] For the technical debate on monetary and trade phenomena see Jacob Viner, *Studies in the Theory of International Trade*, New York, 1937, chapters III, IV, and X.

[2] John Stuart Mill, *Principles of Political Economy*, London, 1848, chapter V, section 3.

[3] Hutchison, op. cit., p. 351.

William Sargant, John Lalor, E. G. Wakefield, R. S. Moffatt and a considerable list of others refused to allow the idea of a demand failure through excess savings to die.[1] Lalor believed that speculative fevers were the result of an excess of funds seeking investment: their owners, dissatisfied with the low return available, turned to speculation in the commodity markets and to dubious industrial ventures.[2] Because of excess capital the whole investment process ceased to be one of close calculation, but became largely irrational. In this situation a certain kind of talent could thrive. In normal times, wrote one observer with a fine disregard for consistency of metaphor, 'facts are a good horse to ride; but the knights who tilt in the markets find now that opinion is their most trenchant weapon'.[3]

Yet an alternative approach was gaining ground. The immense creation of physical capital in the railway booms had set James Wilson, founder of *The Economist*, and others, thinking of the situation likely to arise when a society sought, at great speed, to create an enormous addition to its fixed capital. Such conjectures gained further point from the railway collapses of 1836 and 1847. It seemed clear that in phases of great optimism the economy might well seek to create equipment for future use on such a scale that breakdown must come because there were simply not enough resources available to complete the task. So strained might the situation become that no possible lessening of current consumption could meet the case. Bonamy Price, the Oxford professor, gave perhaps the most influential statement of this over-spending, over-consumption, point of view. In this vital matter, wrote Price, there is only one way to escape injury: not to make more fixed capital beyond the amount of savings.[4]

The idea of miscalculation in expansion received further elaboration. If an increase in activity was to be generalized, it was necessary that it not merely bring existing capacity into employment; it must cause new investment in new plant. This would create excessive optimism, with the result that 'over-trading' became general, with traders entering into dangerous bargains based upon high prices. But the response of industrialists was also too late, so that the new industrial capacity delivered goods in excess of demand, causing a fall in prices and employment.

Trade unionists in the seventies, like Alexander Macdonald and

---

[1] William Lucas Sargant, *The Science of Social Opulence*, London, 1856.
[2] John Lalor, *Money and Morals: A Book for the Times*, London, 1852.
[3] Arthur Ellis, 'Influence of Opinion on Markets', *E.J.*, 1892, p. 116.
[4] Bonamy Price, *Chapters on Practical Political Economy*, London, 1878, p. 120.

George Potter, argued along similar lines. The competitive system forced employers into a position in which rational behaviour was impossible. In bad times they often had no alternative but to push wages down, for this was the only way they could gain some part of the shrinking market. In good times they competed so vigorously to extend output, especially in coal and iron, that they induced a fever of activity that made relapse inevitable.

In the sixties the idea was becoming more explicit that the forces at work subjected the economy to what was in fact a cycle. What could be the timing mechanism? Perhaps it was the psychological instability of business men who were subject to cumulative and mutually sustaining phases of optimism and pessimism. John Mills, Manchester banker, argued from his own observations of business behaviour that economic fluctuations were 'cognate with the science of mind'.[1] Alternatively the psychological cycle had its roots in physiology: it was the younger men who were most speculative, it was they who sent the market up as they bid for goods and labour, but, chastened by the calamities brought by their excesses they yielded the control of business to their successors, who within ten years or so were ready for another burst of optimism.[2] Yet again, it could be argued, as with Marx, that fixed capital had an average life of about ten years, so that any bunching of investment in one period was bound to have repercussions of a periodic kind, with the system as a whole becoming cumulatively more unstable.

Jevons sought the clue not in man or the machine but in nature.[3] Was it possible that some overpowering exogenous cause was working upon human society from without? Variations in sunshine due to sunspots seemed to show the appropriate periodicity; to Jevons it seemed that this must induce great cycles in the world's agricultural yield, a pulse powerful enough to control the whole. Though Jevons did not find many to support his sunspots explanation, there were many theorists and practical men who realized how important variations in harvests could be, causing international movements of gold and a decline in incomes that would soon show in workers' budgets. James Wilson had theorized before Jevons on the importance of harvest fluctuations and his creation, *The Economist*, continued to stress their

[1] John Mills, 'On Credit Cycles and the Origin of Commercial Panics', *Trans. Manchester Statistical Society*, 1867–68.

[2] William Purdy, *The City Life: Its Trade and Finance*, London, 1876, p. 10.

[3] W. S. Jevons, 'The Periodicity of Commercial Crises and its Physical Explanation', from *Investigations in Currency and Finance*, H. S. Foxwell, ed., London, 1884.

importance.[1] The manufacturers of shoes were well aware that scarce food meant that footwear simply was made to last longer.

The City preferred its own way of dealing with crisis.[2] There was earnest discussion on the direction of the late over-extension, and account of the course of the crisis would appear. The City would then place a label upon the breakdown: excessive foreign loans, railway investment, post-war restocking, speculation in the Eastern trades, gold discoveries. This would go out to the country as the City view. There, so far as the City in general was concerned, the matter would rest. As recovery came there was usually a tendency to look for the symptoms associated with the previous collapse (and possibly one collapse removed), and, if these were weak or absent, to conclude that the expansion was sound.

No unification of theory occurred, much less did a set of policy prescriptions for the elimination of cycles appear. The Bank of England had the duty to keep a general eye on the state of trade and give warning through a rise in Bank Rate that expansion had gone far enough. The Bank, in fact, did well in minimizing commercial crisis after 1866. But though hectic liquidation was avoided, the economy continued to show a roughly decennial pulse, bringing fortunes and failures, and irregular earnings and employment, affecting to greater or lesser degree all its members. On the whole these were accepted as the costs of expansion in a market economy.

10. UNDERSTANDING THE MATURE ECONOMY

The ideas of Smith, Malthus, and Ricardo, were deeply rooted in the general matrix of philosophical and scientific speculation that had been developed in post-Renaissance, pre-industrial, society. By the eighties the content of these basic ideas had changed greatly as the new industrial society produced its new view of reality.

The physics of Newton remained undisturbed in the sphere in which it was first enunciated. But Newtonianism had lost a good deal of its hold over the minds of students of society. The scientific limelight had passed from astrophysics to the biology of Darwin, and to the physico-chemical-electrical discoveries of Faraday, Mayer and Joule, Kelvin, Rayleigh, Ramsay, Clerk-Maxwell, and others.

---

[1] See Link, op. cit., pp. 103–26; 'The Great Fall in the Price of Wheat', *Econ.*, 29 August 1874, p. 1046.
[2] See above, Chapter 6, p. 211.

The evolutionary idiom of Darwin certainly affected economic and social thinking in Britain. Moreover, with the work of Darwin and Galton came a reassertion of the vitalist principle, so long eclipsed by Newton. They spoke of 'sports' and discontinuous variations, and asserted the power of the social organism to contain within itself an intrinsic vitality. Even the thought of Marx was unable to extinguish this element.[1] Economists, especially at Oxford, including Travers Twiss, Thorold Rogers, and Bonamy Price, showed signs of wishing to escape from mere mechanistic equilibrium, though they were not too clear about alternatives.[2]

But Darwinism, in spite of the efforts of Spencer, could not promote a new economics that was generally acceptable. Nor were the new discoveries in other fields of physical inquiry, so vigorously debated in the last decades of the century, any more capable of providing a new idiom for social thinking.

Philosophy and metaphysics had also changed. The older view of man as a stable, homogeneous entity, rationally reacting to external stimuli, was no longer acceptable. Indeed, the vogue for individualist, mechanistic explanations of human conduct was now at a heavy discount. The new idealism of Oxford and Glasgow was making a strong appeal. T. H. Green, Bernard Bosanquet, and F. H. Bradley, in their pursuit of the Absolute, had come to the conclusion that men are not isolated entities on an atomic model, but members one of another in their aspiration, fears, and actions.[3] Moreover, it was beginning to become apparent that a civilization based upon science has an inherent tendency to create monopoly in industry and trade.[4] Even economic man was no longer an individual.

But philosophy was no more capable than science of sponsoring a new reasoning system about economic relationships. Green and his successors reintroduced all the impalpables about man and society from which political economy with its concept of 'economic man' had abstracted. Thinking which is concerned with wholes and the Absolute is destructive of categories, and accordingly cannot produce articulated reasoning. But though idealism could not produce a new system of

---

[1] See Soloman F. Bloom, 'Man of his Century: A Reconsideration of the Historical Significance of Karl Marx', *J.P.E.*, 1943, p. 500.

[2] S. G. Checkland, 'Economic Opinion in England as Jevons Found It', *M.S.*, 1951, pp. 154–8.

[3] For the effect upon the education debate see W. S. Fowler, 'The Influence of Idealism upon State Provision of Education', *V.S.*, 1961.

[4] 'The Monopolies of Civilization', *Econ.*, 8 August 1868, p. 897.

economics it could certainly damage the old one; the new idealism coming, not as an attack upon political economy but as the outcome of renewed philosophical speculation aimed at conceiving man's place in the universe, had a powerful impact upon the acceptance of the older economics: it altered thinking about society at its philosophical root. The idealists said, believe in the common humanity of the economic actors and you have the principal clue to their behaviour; you need go no further with 'system'. In the face of monopoly even Marshall fell back upon 'economic chivalry'.[1] By the eighties Edward Caird, successor to Adam Smith's Chair at Glasgow, was leading a movement to organize women workers in self-defensive and self-educative organizations; a bitter critic accused him of ignoring the laws of supply and demand, and teaching the 'new or sympathetic economy'.[2]

Thus the acceptability of the older economics was eroded, almost imperceptibly, by the passing of the cosmogony and metaphysics upon which it rested, and no effective substitutes were found. From the side of society, the flow of Blue Books, reviving from the seventies onward, brought a flood of data, including mounting evidence of monopoly and irrationality, that could not be assimilated to any available theoretical scheme.

While in this state of increasing vulnerability, economic thinking was overtaken by the perplexities of the depression of the seventies and eighties. The economy of Britain, after a quarter of a century of hectic expansion, now seemed to many observers to be incapable of sustaining the previous rate of growth. Yet the academic economists, though they made valuable contributions to the discussion of contemporary problems by giving evidence to public bodies and by their less formal writings, were mainly concerned with an attempt to revive and perfect the formalism of the older thinking. In so doing they achieved a superlative reasoning structure, the 'neo-classical' economics, restating the Smithian system in marginal terms. But this was one which seemed to most contemporaries to be irrelevant. A splendid efflorescence of ratiocination served only to annoy those who were worried and perplexed.

Why was there, at such a time, so much concern with the allocative and equilibrating mechanism, and so little with the problems of growth? There was a certain fatalism in the attitude of most observers toward the great long-run trends of which they were beginning to become

[1] A. C. Pigou, ed., *Memorials of Alfred Marshall*, London, 1925, part II, chapter XVII.
[2] Sir Henry Jones and John Henry Muirhead, *The Life and Philosophy of Edward Caird*, Glasgow, 1921, p. 118.

aware – they were too elemental to be meddled with. Vitalism, for all the appeal it might make to biologists, had, from the social point of view, little attraction for a generation shocked by the faltering of a system that had formerly behaved so vigorously. Paradoxically, Darwinism had played a part in this fatalistic reaction, for depression could be construed as a period of healthy purging of inept elements. Charles Booth, whose humanity drove him to his great inquiry into the state of the London poor, saw cyclical movement in the economy in just such terms – as a cathartic and an agent of re-invigoration.[1] It was natural enough that economic thinking should return to the problem of scarcity, with its emphasis upon the limits within which choice and decision may take place, and its application of the principle of adjustment at the margin by both consumers and producers. Indeed this reaction was an academic analogue to the situation in British industry: just as men of business in the seventies and eighties were seeking salvation by the lowering of costs through greater efficiency and better use of resources within the firm, so academics elaborated a system of optimum allocation of resources, not of growth.

The fear of social unrest reinforced the urge to think in such terms. Marshall's retreat into partial equilibrium was an attempt to solve the problems of group conflict by a form of reasoning that resolved all behaviour into the same pattern – choice, by individuals, at the margin. It was an effort to go back to Newton, to think once more in individualist terms, and to argue about behaviour within fixed assumptions. The relative isolation of British economists from the rest of the world perhaps helps to explain how this could happen.[2]

Thought about society had sunk to a low ebb by the eighties. Spencerian individualism, as a means of blocking state action, bore some responsibility for this; but so too did the indiscriminate fulminations of Carlyle and others against economic man. The movements of protest – socialism and land nationalization, were largely intellectual, but were so in a special sense. They came from thinking men who, despairing of any generalized light upon society, joined such movements because they were activist, and might, given support, lead to constructive thinking and action. The 'true' intellectuals, frightened by the proliferation of fractional truths, often uncertain in their own thoughts, and fearful of the effects of theories of single causation on

---

[1] Margaret Cole, *Beatrice Webb*, London, 1945, p. 58.
[2] T. W. Hutchison, 'Insularity and Cosmopolitanism in Economic Ideas, 1870–1914', *American Economic Review*, 1955, Papers and Proceedings, p. 1.

minds increasingly feeling the intoxication of public education, discountenanced strong pleas of any kind.

The worker had more or less abandoned political economy at the theoretical level. In the eighties he was astonished and delighted at the sight of Henry George, an untutored man, accepting full battle with the academic and political doctrinaires.[1] In Britain the major and distinctive role of 'the little American rooster' was to help to break the heavy spell of impotence.

The urge to find a new sense of identity and solidarity was reflected in the outburst of imperial enthusiasm heralded by Disraeli as early as 1872, even in the midst of mounting prosperity. Many in Britain were no longer attracted by the Cobdenite cosmopolitan competitive world; though Gladstone might protest against the new imperial mystique, many sought to derive a new outlook from the diverse elements of empire, acquired piecemeal. Such men warmed to Disraeli's flattering reference to 'the sublime instinct of an ancient people'.[2]

Society had entered upon a phase of experience in which no systematic comprehensive theory of the kind sought since the eighteenth century was available. No satisfactory explanation of economic behaviour in all its aspects and in all its interrelations was possible, much less a system which assimilated the theory of society to that of nature. The members of the first of the great industrial communities were obliged to seek their way forward without the benefit of unifying system. But then, the majority of Englishmen had never had any great confidence in the grand synthesis, especially if it was made explicit, and had always been content to settle their affairs with the aid of what Marshall called 'short chains of reasoning'.

---

[1] Elwood P. Lawrence, *Henry George in the British Isles*, East Lancing, 1957; Charles Albro Barker, *Henry George*, New York, 1955, chapters XII, XIII.

[2] For the ideology of empire see A. P. Thornton, *The Imperial Idea and its Enemies: a Study in British Power*, London, 1959; C. A. Bodelsen, *Studies in Mid-Victorian Imperialism*, London, 1924.

Bibliography

GENERAL WORKS

ASHTON, T. S. *The Industrial Revolution, 1760–1830.* Oxford, 1948.
BRIGGS, ASA. *The Age of Improvement.* London, 1959.
CHAMBERS, J. D. *The Workshop of the World, British Economic History from 1820 to 1880.* London, 1961.
CLAPHAM, SIR JOHN. *An Economic History of Modern Britain*: Vol. I, *The Early Railway Age, 1820–1850*; Vol. II, *Free Trade and Steel, 1850–1886.* Cambridge, 1926, 1932.
CLARK, G. KITSON. *The Making of Victorian England.* London, 1962.
COURT, W. H. B. *A Concise Economic History of Britain from 1750 to Recent Times.* Cambridge, 1954.
FAY, C. R. *Great Britain from Adam Smith to the Present Day.* London, 1928.
HAYEK, F. A. *ed. Capitalism and the Historians.* London, 1954.
REDFORD, ARTHUR. *The Economic History of England, 1760–1860.* 2nd edn. London, 1960.
ROSTOW, W. W. *British Economy of the Nineteenth Century.* Oxford, 1948.
WOODWARD, E. L. *The Age of Reform, 1815–70.* 2nd edn. Oxford, 1962.
YOUNG, G. M. *Early Victorian England, 1830–1865.* 2 vols. London, 1934.
—— and HANDCOCK, W. D. *English Historical Documents 1833–74.* Vol. III, Part I. Oxford, 1956.

Annual Register.
Dictionary of National Biography.
London Gazette.
Official Reports of Debates in Parliament (Hansard).
Literary Reviews: *The Edinburgh, The Quarterly, The Westminster, Fortnightly, Contemporary, Nineteenth Century.*
Newspapers: *The Times, The Economist.*
Historical and Economic Journals: *see* list of Abbreviations, pp. xiii, xiv.

BIBLIOGRAPHIES

Economic History Review, since 1934–35.
FORD, P. and FORD, G., eds. *Hansard's Catalogue and Breviate of Parliamentary Papers, 1696–1834.* Oxford, 1953.
—— *Select List of British Parliamentary Papers, 1833–1899.* Oxford, 1953.
Victorian Studies, since 1957.

431

Index to Theses Accepted for Higher Degrees in the Universities of Great Britain and Ireland. London, 1950 – in progress.

Dissertation Abstracts, Ann Arbor.

Index to Economic Journals, Homewood, Vol. I, *1886–1924*, 1961 – in progress.

International Bibliography of Economics, U.N.E.S.C.O., 1955 – in progress.

International Index: A Guide to Periodical Literature in the Social Sciences and Humanities: Vol. I, *1907–1915*, New York, 1916 – in progress.

GENERAL STATISTICAL SOURCES

(a) *Official*

Statistical Abstract for the United Kingdom, in *Sessional Papers* annually from 1854 to 1939–40.

Reports of the Census of England and Wales, 1801, in progress. As a guide see *Census Reports of Great Britain 1801–1931*, No. 2 in the series *Guides to Official Sources*, H.M.S.O. London, 1951.

Tables of Revenue, Population, Commerce, etc. known as Porter's Tables in *S.P.* in 1833 and annually from 1835–54.

Miscellaneous Statistics in *S.P.* annually from 1855–82.

Report of the Royal Commission on Depression of Trade and Industry. (*S.P.* 1866, XXI, *S.P.* 1886, XXII, and *S.P.* 1886, XXIII).

(b) *Others*

GAYER, A. D., ROSTOW, W. W. and SCHWARTZ, A. J. *The Growth and Fluctuations of the British Economy, 1790–1850.* Oxford, 1953.

MCCULLOCH, J. R. *A Descriptive and Statistical Account of the British Empire.* 2 vols. London, 1854 edn.

MITCHELL, B. R., with the collaboration of Phyllis Deane. *Abstract of British Historical Statistics.* Cambridge, 1962. (Provides a splendid bibliography for each chapter.)

PAGE, WILLIAM. *Commerce and Industry: a Historical Review of the Economic Conditions of the British Empire.* 2 vols. London, 1919. Vol. II, Statistical Tables.

PORTER, G. R. *The Progress of the Nation in its various Social and Economic Relations, from the beginning of the Nineteenth Century.* London, 1847 edn.

There is much miscellaneous statistical material in the *Journal of the Royal Statistical Society*, and *The Economist Newspaper* together with its annual *Commercial History and Review*, from 1864.

VISUAL

Anon. *Life in England in Aquatint and Lithography, 1770–1860.* London, 1953.

DORÉ, GUSTAVE and JERROLD, BLANCHARD. *London, a Pilgrimage.* London, 1872.

HITCHCOCK, RUSSELL HENRY. *Early Victorian Architecture in Britain*. 2 vols. London and New Haven, 1954.

KLINGENDER, FRANCIS D. *Art and the Industrial Revolution*. London, 1947.

REYNOLDS, GRAHAM. *Painters of the Victorian Scene*. London, 1953.

SITWELL, SACHEVERELL. *Narrative Pictures, A Survey of English Genre and its Painters*. London, 1937.

SMITH, ALBERT, ed. *Gavarni in London*. London, 1849.

Periodicals: *The Illustrated London News, Punch.*

See also under Chapter 8, section 9.

CHAPTER TWO: *The Growth and Stability of the System*

ASHWORTH, WILLIAM. *An Economic History of England, 1870–1939*. London, 1961.

BEALES, H. L. 'The Great Depression in Industry and Trade', reprinted in E. M. Carus-Wilson, ed., *Essays in Economic History*. London, 1954.

CAIRNCROSS, A. K. *Home and Foreign Investment, 1870–1913*. Cambridge, 1953.

—— and WEBER, B. 'Fluctuations in Building in Great Britain, 1785–1849', *Econ. H.R.*, 1956.

CARRIER, N. H. and JEFFERY, J. R. 'External Migration, 1815–1950, a Study of the Available Statistics', *Studies on Medical and Population Subjects*, no. 6, 1953.

CONNELL, K. H. *The Population of Ireland, 1750–1845*. Oxford, 1950.

COONEY, E. W. 'Long Waves in Building in the British Economy of the Nineteenth Century', *Econ. H.R.*, 1960.

DEANE, PHYLLIS. 'Contemporary Estimates of the National Income in the First Half of the Nineteenth Century', *Econ. H.R.*, 1956.

—— 'The Implications of Early National Income Estimates for the Measurement of Long-term Growth in the United Kingdom', *Economic Development and Cultural Change*, vol. IV, no. 1, 1955.

—— 'Contemporary Estimates of National Income in the Second Half of the Nineteenth Century', *Econ. H.R.*, 1957.

—— and COLE, W. A. *British Economic Growth, 1688–1959*. Cambridge, 1962.

DOUGLAS, PAUL. 'An Estimate of the Growth of Capital in the U.K., 1865–1909', *Journal of Economic and Business History*, 1930.

FEINSTEIN, C. H. 'Income and Investment in the United Kingdom, 1856–1914', *E.J.*, 1961.

GAYER, A. D., ROSTOW, W. W. and SCHWARTZ, A. J. *The Growth and Fluctuations of the British Economy 1790–1850*. Oxford, 1953.

—— and FINKELSTEIN, I. 'British Share Prices, 1811–1850', *Review of Economic Statistics*, 1940.

GIFFEN, SIR ROBERT. *Economic Inquiries, and Studies*. London, 1904.

GOSCHEN, G. J. *Essays and Addresses on Economic Questions, 1865–1893*. London, 1905.

HABAKKUK, H. J. *Cambridge History of the British Empire*. Cambridge, 1940. Vol. II, chap. XXI, 'Free Trade and Commercial Expansion'.

HOFFMANN, W. G. *British Industry 1700–1950*, trans. by W. O. Henderson and W. H. Chaloner. Oxford, 1955.

HUGHES, J. R. T. *Fluctuations in Trade, Industry and Finance: A Study of British Economic Development, 1850–1869*. Oxford, 1960.

IMLAH, ALBERT M. *Economic Elements in the Pax Britannica*. Cambridge, Mass., 1958.

JENKS, LELAND HAMILTON. *The Migration of British Capital to 1875*. New York, 1927, reprinted 1963.

JONES, G. T. *Increasing Return*. Cambridge, 1933.

KONDRATIEFF, NIKOLAI D. 'The Long Waves in Economic Life', *Review of Economic Statistics*, 1935, reprinted in *Readings in Business Cycle Theory*, American Economic Association Series, London, 1950.

KRAUSE, J. T. 'Changes in English Fertility and Mortality, 1781–1850', *Econ. H.R.*, 1958.

LAYTON, SIR WALTER and CROWTHER, GEOFFREY. *An Introduction to the Study of Prices*. London, 1935.

MARSHALL, T. H. 'The Population Problem during the Industrial Revolution', in Carus-Wilson, *ed.*, *op. cit.*

MATTHEWS, R. C. O. *A Study in Trade Cycle History: Economic Fluctuations in Britain 1833–42*. Cambridge, 1954.

—— 'The Trade Cycle in Britain, 1790–1850', *O.E.P.*, 1954.

MUSSON, A. E. 'The Great Depression in Britain, 1873–1896', *J. Econ. H.*, 1959.

PREST, A. R. 'National Income of the United Kingdom, 1870–1946', *E.H.*, 1948.

ROSTOW, W. W. *The Stages of Economic Growth*. Cambridge, 1960.

—— ed., *The Economics of Take-Off into Sustained Growth*. London, 1963.

ROUSSEAUX, P. *Les Mouvements de fond de l'economie anglaise, 1800–1913*. Louvain, 1938.

SAUL, S. B. *Studies in British Overseas Trade, 1870–1914*. Liverpool, 1960.

SAVILLE, JOHN. *Rural Depopulation in England and Wales*. London, 1957.

SCHLÖTE, WERNER. *British Overseas Trade from 1700 to the 1930's*, trans. W. O. Henderson and W. H. Chaloner. Oxford, 1952.

SHANNON, H. A. 'Migration and the Growth of London, 1841–91', *Econ. H.R.*, 1934–5.

SILBERLING, N. J. 'British Prices and Business Cycles, 1779–1850', *Review of Economic Statistics*, 1923.

SMART, WILLIAM. *Economic Annals of the Nineteenth Century, 1801–1830*. 2 vols. London, 1910, 1917.

THOMAS, BRINLEY. *Migration and Economic Growth, A Study of Great Britain and the Atlantic Economy*. Cambridge, 1954.

TOOKE, T. and NEWMARCH, W. *A History of Prices 1793–1856*. 6 vols. London, 1838–57.

VAN VLECK, G. W. *The Panic of 1857: an Analytical Study*. Oxford, 1943.

WARD-PERKINS, C. N. 'The Commercial Crisis of 1847', *O.E.P.*, 1950.

WILLIAMSON, JEFFREY G. 'The Long Swing: Comparisons and Interactions between British and American Balance of Payments, 1820–1913', *J. Econ. H.*, 1962.

CHAPTER THREE: *The Men of Invention*

BABBAGE, CHARLES. *On the Economy of Machinery and Manufactures*. London, 1832.

BERNAL, J. D. *Science and Industry in the Nineteenth Century*. London, 1953.

BESSEMER, SIR HENRY. *Autobiography*. London, 1905.

CARDWELL, D. S. L. *The Organisation of Science in England*. London, 1957.

COTGROVE, S. T. *Technical Education and Social Change*. London, 1958.

CROWTHER, J. G. *British Scientists of the Nineteenth Century*. London, 1935.

DERRY, T. K. and WILLIAMS, TREVOR I. *A Short History of Technology: from Earliest Times to A.D. 1900*. Oxford, 1960.

GERNSHEIM, HELMUT and GERNSHEIM, ALISON. *The History of Photography*. Oxford, 1955.

GIBBS-SMITH, C. H. *Sir George Cayley's Aeronautics, 1796–1855*. London, 1963.

HABAKKUK, H. J. *American and British Technology in the Nineteenth Century*. Cambridge, 1962.

HUDSON, D. and LUCKHURST, K. W. *The Royal Society of Arts, 1754–1954*. London, 1954.

NASMYTH, JAMES. *James Nasmyth, Engineer: an Autobiography*. London, 1883.

POLE, SIR WILLIAM, ed. *The Life of Sir William Fairbairn, partly written by Himself*. London, 1877.

—— *The Life of Sir William Siemens*. London, 1888.

PRITCHARD, J. LAURENCE. *Sir George Cayley: the Inventor of the Aeroplane*. London, 1961.

RENNIE, SIR JOHN. *Autobiography of Sir John Rennie*. London, 1875.

Report on Technical Education, Schools Inquiry Commissioners, 1867.

R.C. Reports on Technical Instruction, 1882–84.

RICHARDS, J. M. *The Functional Tradition in Early Industrial Building*. London, 1959.

SINGER, CHARLES, *et al.*, eds. *The History of Technology*: Vol. IV, *The Industrial Revolution c. 1750 to c. 1850*. Oxford, 1958.

SMILES, SAMUEL. *Lives of the Engineers*. London, 1861–62.

—— *Industrial Biography*. London, 1863

—— *Men of Invention and Industry*. London, 1884.

THOMPSON, LILLIAN G. *Sidney Gilchrist Thomas*. London, 1940.

TRENEER, ANNE. *The Mercurial Chemist: a Life of Sir Humphrey Davy*. London, 1963.

TREVITHICK, FRANCIS. *Life of Richard Trevithick*. London, 1872.

UBBELOHDE, A. R. *Man and Energy*. London, 1954.

URE, ANDREW. *The Philosophy of Manufactures*. London, 1835.

CHAPTER FOUR: *The Men of Business*

1. The Business Community

EVANS, GEORGE HERBERTON, Jr. *British Corporation Finance 1775–1850; a Study of Preference Shares*. Baltimore, 1936.

FORMOY, R. R. *The Historical Foundations of Modern Company Law*. London, 1923.

HUNT, B. CARLETON. *The Development of the Business Corporation in England 1800–1867*. Cambridge, Mass., 1936.

JEFFERYS, J. B. 'The Denomination and Character of Shares', in E. M. Carus-Wilson, ed., *Essays in Economic History*. London, 1954. Vol. I.

LITTLETON, A. C. and YAMEY, B. S., eds. *Studies in the History of Accounting*. London, 1956.

RAYNES, H. E. *A History of British Insurance*. London, 1948.

SHANNON, H. A. 'The Coming of General Limited Liability' and 'The Limited Companies of 1866–1883', in Carus-Wilson, op. cit.

SMITH, J. G. *Organized Produce Markets*. London, 1922.

STACEY, NICHOLAS A. H. *English Accountancy 1800–1954*. London, 1954.

WETENHALL, *Stock Exchange Lists*. London, 1811–1907, twice weekly.

2. The Traders

BUCK, N. S. *The Development of the Organization of the Anglo-American Trade 1800–1850*. New Haven, 1925.

FOXBOURNE, H. R. *English Merchants, Memoirs in Illustration of the Progress of British Commerce*. London, 1866.

HYDE, F. E., with the assistance of J. R. Harris. *Blue Funnel: A History of Alfred Holt and Company, 1865 to 1914*. Liverpool, 1956.

JEFFERYS, J. B. *Retail Trading in Britain, 1850–1950*. Cambridge, 1954. See especially chapter 1.

JOHN, A. H. *A Liverpool Merchant House*. 1959.

MARRINER, SHEILA. *Rathbones of Liverpool 1845–73*. Liverpool, 1961.

PASDERMADJIAN, H. *The Department Store, its Origins, Evolution and Economics*. London, 1954.

REDFORD, ARTHUR. *Manchester Merchants and Foreign Trade 1850–1939*. 2 vols. Manchester, 1934, 1956.

3. The Textile Manufacturers

ARMYTAGE, W. H. G. 'A. J. Mundella and the Hosiery Industry: the Liberal Background to the Labour movement', *Econ. H.R.*, 1948.

—— *A. J. Mundella, 1825–1897*. London, 1951.

BAINES, E., Jr., *History of the Cotton Manufacture in Great Britain*. London, 1835.

BLAUG, M. 'The Productivity of Capital in the Lancashire Cotton Industry during the Nineteenth Century', *Econ. H.R.*, 1961.

CRUMP, W. B. *The Leeds Woollen Industry 1780–1820*. Leeds, 1931.

ELLISON, THOMAS. *The Cotton Trade of Great Britain*. London, 1886.

FELKIN, W. *History of the Machine Wrought Hosiery and Lace Manufactures*. London, 1867.

FITTON, R. S. and WADSWORTH, A. P. *The Strutts and the Arkwrights, 1758–1830; a Study of the Early Factory System*. Manchester, 1958.

OWEN, ROBERT. *The Life of Robert Owen by Himself*. New edn. 1857.

PREST, JOHN. *The Industrial Revolution in Coventry*. Oxford, 1961.

RIMMER, W. G. *Marshalls of Leeds, Flax Spinners, 1788–1886*. Cambridge, 1960.

SIGSWORTH, ERIC. *Black Dyke Mills*. Liverpool, 1958.

SMELSER, NEIL J. *Social Change in the Industrial Revolution: An Application of Theory to the Lancashire Cotton Industry, 1770–1840*. London, 1959.

TAYLOR, A. J. 'Concentration and Specialisation in the Lancashire Cotton Industry, 1825–1850', *Econ. H.R.*, 1949.

THOMAS, JOHN. 'A History of the Leeds Clothing Industry', *Y.B.*, Occasional Papers no. 1, 1955.

UNWIN, G. *Samuel Oldknow and the Arkwrights*. Manchester, 1924.

URE, ANDREW. *The Cotton Manufacture of Great Britain*. London, 1836, 2nd ed. London, 1961.

4. Engineers, Contractors and Promoters

ARMYTAGE, W. H. G. *A Social History of Engineering*. London, 1961.

CHALONER, W. H. *The Social and Economic Development of Crewe, 1780–1923*. Manchester, 1950.

DEVEY, JOSEPH. *Life of Joseph Locke, Civil Engineer*. London, 1862.

ELLIS, C. HAMILTON. *British Railway History; an Outline from the Accession of William IV to the Nationalisation of the Railways: Vol. I, 1830–1876*. London, 1954.

—— *British Railway History: Vol. II, 1877–1947*. London, 1959.

HELPS, ARTHUR. *Life and Labours of Mr. Brassey, 1805–1870*. London, 1872.

LAMBERT, R. S. *The Railway King, 1800–1871: a Study of George Hudson and the Business Morals of his Time*. London, 1934.

MARSHALL, C. F. DENDY. *A Centenary History of the Liverpool and Manchester Railway*. London, 1930.

ROBERTSON, W. A. *Combination Among Railway Companies*. London, 1912.

ROLT, L. T. C. *Isambard Kingdom Brunel*. London, 1957.

—— *George and Robert Stephenson*. London, 1960.
—— *Thomas Telford*. London, 1958.
WARDLE, D. B. 'Sources for the History of Railways at the Public Record Office', *J. Tpt. H.*, 1955–56.

5. *Makers and Owners of Ships*

GRAHAM, G. S. 'The Ascendancy of the Sailing Ship, 1850–85', *Econ. H.R.*, 1956.
HUGHES, J. R. T. and REITER, STANLEY. 'The First 1945 British Steamships', *Jr. American Statistical Assoc.*, 1958.
JARVIS, RUPERT C. 'Sources for the History of Ships and Shipping', *J. Tpt. H.*, 1958.
JONES, R. J. CORNEWALL. *The British Merchant Service*. London, 1898.
LINDSAY, W. S. *History of Merchant Shipping*. London, 1876. Vol. IV.
MAYWALD, K. 'The Construction Costs and Value of the British Merchant Fleet, 1850–1938', *S.J.P.E.*, 1956.
NORTH, DOUGLASS. 'Ocean Freight Rates and Economic Development, 1750–1913', *J. Econ. H.*, 1958.
SAVAGE, C. I. *An Economic History of Transport*. London, 1959.
THORNTON, R. H. *British Shipping*. Cambridge, 1939.

6. *The Lesser Users of Metals*

ALLEN, G. C. *The Industrial Development of Birmingham and the Black Country, 1860–1927*. London, 1929.
COURT, W. H. B. *The Rise of the Midland Industries, 1600–1838*. Oxford, 1938.
FOX, ALAN. 'Industrial Relations in Nineteenth Century Birmingham', *O.E.P.*, 1955.
GILL, CONRAD and BRIGGS, ASA. *History of Birmingham*. 2 vols. Oxford, 1952.
MINCHINTON, W. E. *The British Tinplate Industry*. Oxford, 1957.
PAYNE, P. L. *Rubber and Railways in the Nineteenth Century*. Liverpool, 1960.
POLLARD, SIDNEY. *A History of Labour in Sheffield*. Liverpool, 1959.
SCOTT, J. D. *Vickers, a History*. London, 1962.

7. *The Metal Makers*

ADDIS, J. P. *The Crawshay Dynasty: A Study in Industrial Organization and Development, 1765–1867*. Cardiff, 1957.
ASHTON, T. S. *Iron and Steel in the Industrial Revolution*. Manchester, 1924.
BURN, D. L. *The Economic History of Steelmaking 1867–1939*. Cambridge, 1940.
BURNAM, T. H. and HOSKINS, G. O. *Iron and Steel in Britain 1870–1930*. London, 1943.
CAMPBELL, R. H. *Carron Company*. Edinburgh, 1961.
CARR, J. C. and TAPLIN, W. *History of the British Steel Industry*. Oxford, 1962.
ELSAS, MADELEINE, ed., *Iron in the Making: Dowlais Iron Company Letters, 1782–1860*. Cardiff, 1960.

ERICKSON, CHARLOTTE. *British Industrialists: Steel and Hosiery 1850–1950.* Cambridge, 1959.

HUNT, ROBERT. *British Mining.* 2nd edn. London, 1887.

JEANS, W. T. *The Creators of the Age of Steel.* London, 1884.

JOHN, A. H. *The Industrial Development of South Wales, 1750–1850,* Cardiff, 1950.

Mineral Statistics of the U.K. of Great Britain and Ireland.

8. The Coal Masters

CHANDLER, D. and LACEY, A. D. *The Rise of the Gas Industry in Britain.* London, 1949.

GALLOWAY, R. L. *Annals of Coal Mining.* First series. London, 1898.

JEVONS, W. S. *The Coal Question, an Inquiry concerning the Progress of the Nation, and the Probable Exhaustion of our Coal Mines.* London and Cambridge, 1865.

MORRIS, J. H. and WILLIAMS, L. J. *The South Wales Coal Industry 1841–1875.* Cardiff, 1958.

SPRING, DAVID. 'The English Landed Estate in the Age of Coal and Iron: 1830–1880', *J. Econ. H.,* 1951.

—— 'The Earls of Durham and the Great Northern Coal Field', *Canadian Historical Review,* 1952.

STIRLING, EVERARD. *History of the Gas, Light and Coke Company.* London, 1949.

TAYLOR, A. J. 'Labour Productivity and Technological Innovation in the British Coal Industry, 1850–1914', *Econ. H.R.,* 1961.

—— 'The Sub-Contract System in the British Coal Industry', in L. S. Pressnell, ed., *Studies in the Industrial Revolution presented to T. S. Ashton.* London, 1960.

VANE-TEMPEST-STEWART, EDITH. Marchioness of Londonderry, *Frances Anne: The Life of Frances Anne, Marchioness of Londonderry, and her husband Charles, Third Marquess of Londonderry.* London, 1958.

WILLIAMS, J. W. *The Derbyshire Miners: A Study in Industrial and Social History.* London, 1962.

9. The Builders

BARKER, T. C. *Pilkington Brothers and the Glass Industry.* London, 1960.

CHADWICK, GEORGE F. *The Works of Sir Joseph Paxton, 1803–1865.* London, 1962.

COONEY, E. W. 'The Origins of the Victorian Master Builders', *Econ. H.R.,* 1955.

DAVEY, NORMAN. *A History of Building Materials.* London, 1961.

DODD, A. H. *The Industrial Revolution in North Wales.* 2nd edn. Cardiff, 1951.

DYOS, H. J. *Victorian Suburb: A Study of the Growth of Camberwell.* Leicester, 1961. pp. 122–37.

HUDSON, KENNETH. *Industrial Archaeology.* London, 1963.

MIDDLEMAS, R. K. *The Master Builders.* London, 1963.
SUMMERSON, JOHN. *Georgian London.* London, 1945.

10. The Producers of Chemicals

BARKER, T. C. and HARRIS, J. R. A. *A Merseyside Town in the Industrial Revolution: St. Helens, 1750–1900.* Liverpool, 1954.
CLOW, A. and CLOW, N. *The Chemical Revolution; a Contribution to Social Technology.* London, 1952.
COHEN, J. M. *The Life of Ludwig Mond.* London, 1956.
GOSSAGE, W. *A History of Soda Manufacture.* Liverpool, 1870.
HABER, L. F. *The Chemical Industry in the Nineteenth Century.* Oxford 1958.
HARDIE, D. W. F. *A History of the Chemical Industry in Widnes.* Liverpool, 1950.

11. The Exploiters of Electricity

GARRATT, G. R. M. *One Hundred Years of Submarine Cables.* London, 1950.
MARTIN, THOMAS. *Faraday's Discovery of Electro-Magnetic Induction.* London, 1949.
SABINE, R. *The Electric Telegraph.* London, 1867.
SCOTT, J. D. *Siemens Brothers, 1858–1958.* London, 1958.

12. Various Men of Parts

COLEMAN, D. C. *The British Paper Industry, 1495–1860.* Oxford, 1958.
MATHIAS, P. *The Brewing Industry in England, 1700–1830.* Cambridge, 1959.
WADSWORTH, A. P. 'Newspaper Circulations, 1800–1954', *M.S.*, 1954–55.
WOODRUFF, W. *Rise of the British Rubber Industry.* Liverpool, 1958.

CHAPTER FIVE: *Food Producers and Rent Receivers*

BOVILL, E. W. *English Country Life, 1780–1830.* Oxford, 1962.
BUTTRESS, F. A. *Agricultural Periodicals of the British Isles, 1681–1900, and Their Location.* Cambridge, 1950.
CAIRD, JAMES. *English Agriculture in 1850–51.* London, 1851.
—— *High Farming under Liberal Covenants, the Best Substitute for Protection.* London, 1848.
—— *The Landed Interest and the Supply of Food.* London, 1878.
COPPOCK, T. J. 'The Statistical Assessment of British Agriculture', *Ag. H.R.*, 1956.
DRESCHER, L. 'The Development of Agricultural Production in Great Britain and Ireland from the Early Nineteenth Century', *M.S.*, 1955.
FLETCHER, T. W. 'The Great Depression of English Agriculture, 1873–1896', *Econ. H.R.*, 1961.
HOSKINS, W. G. *The Making of the English Landscape.* London, 1955.
HUNT, J. G. 'Landownership and Enclosure, 1750–1830', *Econ. H.R.*, 1959.

Owners of Land. England and Wales, Return for 1872-3, Cmd. 1097.

PROTHERO, R. E. (Lord Ernle), *English Farming Past and Present.* 6th edn. with Introduction by G. E. Fussell and O. R. McGregor. London, 1961.

R.C. on Agricultural Depression, 1882.

THOMSON, F. M. L. 'English Great Estates in the 19th Century, 1790-1914', *First International Conference of Economic History, Contributions and Communications,* 1960.

—— *English Landed Society in the Nineteenth Century.* London, 1963.

TROW-SMITH, R. *A History of British Livestock Husbandry, 1700-1900.* London, 1959.

—— *English Husbandry.* London, 1951.

CHAPTER SIX: *The Suppliers and Manipulators of Capital*

BAGEHOT, WALTER. *Lombard Street, a Description of the Money Market.* New edn. London, 1919.

BASTER, A. S. J. *The Imperial Banks.* London, 1929.

BROADBRIDGE, S. G. 'The Early Capital Market: The Lancashire and Yorkshire Railway,' *Econ. H.R.,* 1955.

CHECKLAND, S. G. 'The Mind of the City, 1870-1914', *O.E.P.,* 1957.

CLAPHAM, SIR JOHN. *The Bank of England: a History 1694-1914.* Cambridge, 1944.

CRAMP, A. B. *Opinion on Bank Rate, 1822-60.* London, 1962.

CRICK, W. F. and WADSWORTH, J. E. *A Hundred Years of Joint Stock Banking.* London, 1936.

DAVIS, L. E. and HUGHES, J. R. T. 'A Dollar-Sterling Exchange, 1803-1895', *Econ. H.R.,* 1960.

DUN, J. *British Banking Statistics.* London, 1876.

FEAVERYEAR, A. E. *The Pound Sterling.* 2nd edn, rev. by E. Victor Morgan. London, 1963.

FENN, C. *A Compendium of the English and Foreign Funds and the Principal Joint Stock Companies.* London, 1837 successive editions.

GREENBERG, MICHAEL. *British Trade and the Opening of China, 1800-1842.* Cambridge, 1951.

GREGORY, T. E. *An Introduction to Tooke and Newmarch's A History of Prices, . . .* London, 1928.

—— *Select Statutes Documents and Reports Relating to British Banking 1832-1928.* London, 1929.

—— and HENDERSON, ANNETTE. *Westminster Bank through a Century.* Oxford, 1936.

HIDY, R. W. *The House of Baring in American Trade and Finance; English Merchant Bankers at Work, 1763-1861.* Harvard, 1949.

HORNE, H. OLIVER. *A History of Savings Banks.* Oxford, 1947.

JEVONS, W. S. *Investigations in Currency and Finance*. London, 1884.

JOSLIN, DAVID. *A Century of Banking in Latin America*. Oxford, 1963.

KING, W. T. C. *History of the London Discount Market: with an Introduction by T. E. Gregory*. London, 1936.

MORGAN, E. VICTOR. *The Theory and Practice of Central Banking 1797–1913*. Cambridge, 1943.

—— and THOMAS, W. A. *The Stock Exchange: Its History and Functions*. London, 1962.

NIEBYL, K. *Studies in the Classical Theories of Money*. Oxford, 1946.

PALGRAVE, R. H. I. *Bank Rate and the Money Market, 1844–1900*. London, 1903.

PRESSNELL, L. S. *Country Banking in the Industrial Revolution*, Oxford, 1956.

RAE, GEORGE. *The Country Banker: his Clients, Care and Work*. London, 1885.

S.C. on the High Price of Gold Bullion, 1810.

SAYERS, R. S. *Central Banking after Bagehot*. Oxford, 1957.

—— *Lloyds Bank in the History of English Banking*, Oxford, 1957.

SMITH, K. C. and HORNE, G. F. 'An Index Number of Securities, 1867–1914', *London and Cambridge Economic Service Special Memorandum*, no. 37.

SYKES, JOSEPH. *The Amalgamation Movement in English Banking, 1825–1924*. London, 1926.

VINER, JACOB. *Studies in the Theory of International Trade*. New York and London, 1937.

WITHERS, H. *The National Provincial Bank, 1833–1933*. London, 1933.

WOOD, ELMER. *English Theories of Central Banking Control, 1819–58*. Harvard, 1939.

See also references under Chapter 10, section 9.

CHAPTER SEVEN: *The Lower Orders*

1. Men, Women and Jobs

BOOTH, C. 'Occupations of the People of the U.K., 1801–1881', *J.S.S.*, 1886.

DAY, CLIVE. 'The Distribution of Industrial Occupations in England 1841–61', *Trans. of the Connecticut Academy of Arts and Sciences*, New Haven, 1927.

HOOKER, R. H. 'On Forty Years Industrial Changes in England and Wales', *T.M.S.S.*, 1897–98.

REDFORD, ARTHUR. *Labour Migration in England, 1800–50*. Manchester, 1926.

SAVILLE, JOHN. *Rural Depopulation in England and Wales, 1851–1951*. London, 1957.

2. The Worker as Entrepreneur

GOSDEN, P. H. J. H. *The Friendly Societies in England, 1815–75*. Manchester, 1961.

MAYHEW, HENRY. *London Labour and the London Poor*. London, 1851–62.

Bibliography

TAYLOR, A. J. 'The Sub-contract System in the British Coal Industry', in L. S. Pressnell, ed., *Studies in the Industrial Revolution*, London, 1960.

R.C. on Employment of Children in Agriculture, 1868–9.

Report of the R.C. on Friendly Societies, 1871–74.

3. Birth, Marriage, and Death

FARR, WILLIAM. *Vital Statistics*, ed. by N. A. Humphreys. London, 1885.

Reports and Abstracts of the Census of Great Britain, 1801–1891.

Annual Reports of the Registrar General of Births, Deaths and Marriages, continuous from 1839.

See also population references under Chapter 2.

4. The Workers' Rewards

BOWLEY, A. L. 'Changes in Average Wages (Nominal and Real) in the United Kingdom between 1860 and 1891, *J.R.S.S.*, 1895.

—— 'Comparison of the Rates of Increase in Wages in the United States and in Great Britain, 1860–1891', *E.J.*, 1895.

—— 'The Statistics of Wages in the United Kingdom during the last Hundred Years. (Part IV) Agricultural Wages', *J.R.S.S.*, 1899.

—— *Wages and Income in the United Kingdom since 1860*. Cambridge, 1937.

—— *Wages in the United Kingdom in the Nineteenth Century. Notes for the Use of Students*. Cambridge, 1900.

—— and WOOD, G. H. 'The Statistics of Wages in the United Kingdom during the Nineteenth Century', *J.R.S.S.*, 1899 to 1906 inc.

FOX, ARTHUR WILSON. 'Agricultural Wages in England and Wales during the last Fifty Years', *J.R.S.S.*, 1903.

GIFFEN, R. *The Growth of Capital*. London, 1889.

—— 'The Progress of the Working Classes in the Last half of the Century', reprinted in *Essays in Finance*. 2nd edn., 2nd series. London, 1880–86.

HOBSBAWM, E. J. and HARTWELL, R. M. 'The Standard of Living during the Industrial Revolution: A Discussion', *Econ. H.R.*, 1963.

PHELPS-BROWN, E. H. and HOPKINS, SHEILA V. 'Seven Centuries of Consumables, compared with Builders' Wage-rates, *Econa.*, 1956.

POLLARD, SIDNEY. 'Wages and Earnings in the Sheffield Trades, 1851–1914', *Y.B.*, 1954.

TUCKER, RUFUS J. 'Real Wages of Artisans in London, 1729–1935', *Journal of the American Statistical Assoc.*, 1936.

WOOD, G. H. 'The Course of Average Wages between 1790 and 1860', *R.B.A. Section F*, 1899.

—— 'Real Wages and the Standard of Comfort since 1850', *J.R.S.S.*, 1909.

—— 'Some Statistics relating to Working Class Progress since 1860', *J.R.S.S.*, 1899.

Report of the Industrial Remuneration Conference, London, 1885.

5. The Worker and his Wife as Spenders

HEWITT, MARGARET. *Wives and Mothers in Victorian Industry*. London, 1958.

HILTON, G. W. *The Truck System, including a History of the British Truck Acts, 1465–1960*. Cambridge, 1960.

Report on the Sanitary Conditions of the Labouring Population of Great Britain. Supplementary Report . . . on Interment in Towns, 1842.

Report of the R.C. on Truck, 1871.

6. The Housing Problem

CAIRNCROSS, A. K. and WEBBER, B. 'Fluctuations in Building in Great Britain, 1785–1849', *Econ. H.R.*, 1956.

CHADWICK, E. *Report of the Sanitary Condition of the Labouring Population*, 1842.

HILL, OCTAVIA. *Homes of the London Poor*. London, 1875.

MAYWALD, K. 'An Index of Building Costs in the U.K., 1845–1938', *Econ. H.R.*, 1954.

PALGRAVE, R. H. I. 'On the House Accommodation of England and Wales, with Special Reference to the Census of 1871', *J.S.S.*, 1869.

SINGER, H. W. 'An Index of Urban Land Rates and House Rents in England and Wales, 1845–1913', *Econometrica*, 1941.

Report of the R.C. for Inquiring into the State of Large Towns & Populous Districts, 1844, 5.

R.C. on Housing of the Working Classes, 1884–5.

7. Conditions of Working

DICEY, A. V. *Lectures on the Relation between Law and Public Opinion in England during the Nineteenth Century*. London, 1905.

DRIVER, C. H. *Tory Radical: the Life of Richard Oastler*, Oxford, 1946.

HAMMOND, J. L. and HAMMOND, B. *Lord Shaftesbury*, London, 1924.

WARD, J. T. *The Factory Movement, 1830–1855*, London, 1962.

WEBB, R. K. 'A Whig Inspector: H. S. Tremenheere', *J.M.H.*, 1955.

R.C. on Children's Employment, Second Report, 1843.

Reports of Inspectors of Factories, semi-annually from 1835 to 1877, thereafter *Annual Reports of the Chief Inspector of Factories and Workshops*.

8. Conditions of Living

ASHWORTH, W. *The Genesis of Modern British Town Planning*. London, 1954.

BARKER, T. C. and HARRIS, J. R. *A Merseyside Town in the Industrial Revolution: St Helens, 1750–1900*. Liverpool, 1954.

BOOTH, GENERAL WILLIAM. *In Darkest England and the Way Out*. London, 1890.

BRIGGS, ASA. *Victorian Cities*. London, 1963.

CHALONER, W. H. *The Social and Economic Development of Crewe, 1780–1923*. Manchester, 1950.

Bibliography

CHECKLAND, S. G. 'The British Industrial City as History: the Glasgow Case', *Urban Studies*, 1964.

CRUICKSHANK, MARJORIE. *Church and State in English Education: 1870 to the Present Day*. London, 1963.

CURTIS, S. J. *History of Education in Great Britain*. 2nd edn. London, 1950.

DYOS, H. J. *Victorian Suburb: A Study of the Growth of Camberwell*. Leicester, 1961.

ENGELS, F. *The Condition of the Working Class in England*, trans. and ed. by W. O. Henderson and W. H. Chaloner. Oxford, 1958.

FRAZER, W. M. *A History of English Public Health, 1834–1939*. Baltimore, 1951.

HUTCHINS, B. L. *The Public Health Agitation, 1833–1848*. London, 1909.

POLLARD, SIDNEY. *A History of Labour in Sheffield*. Liverpool, 1959.

REDFORD, ARTHUR and RUSSELL, I. S. *The History of Local Government in Manchester*. London, 1939–40.

SIMON, SIR JOHN. *English Sanitary Institution*. London, 1890.

—— *Public Health Reports*, ed. by Edward Seaton, 2 vols. London, 1887.

WOODHAM-SMITH, CECIL. *Florence Nightingale, 1820–1910*. London, 1950.

The British Association Handbooks, marking visits to British cities, are useful.

S.C. Report on the Health of Towns, 1840.

R.C. on the State of Large Towns, 1844, 1845.

Report on the Sanitary Condition of the Labouring Population, Poor Law Commissioners, 1842.

R.C. on the Sanitary Laws, 1868, 1871.

R.C. on the State of Popular Education in England, 1861.

R.C. to Inquire into the Elementary Education Acts (England and Wales), 1886–1888.

9. The Impact on Personality

ALTICK, R. D. *The English Common Reader: A Social History of the Mass Reading Public, 1800–1900*. Cambridge, 1957.

ASHBY, M. K. *Joseph Ashby of Tysoe, 1859–1919*. Cambridge, 1961.

CLARK, G. KITSON. *The Making of Victorian England*. London, 1962.

JAMES, LOUIS. *Fiction for the Working Man, 1830–50*. Oxford, 1963.

READ, DONALD. *Press and People, 1790–1850, Opinion in Three English Cities*. London, 1961.

SMELSER, NEIL J. *Social Change in the Industrial Revolution* (as above).

THOMPSON, E. P. *The Making of the English Working Class*. London, 1963.

WEBB, R. K. *The British Working-Class Reader, 1790–1848*. London, 1955.

10. The Residuum

BRUCE, MAURICE. *The Coming of the Welfare State*. London, 1961.

DORÉ, GUSTAVE and JERROLD, BLANCHARD. *London. A Pilgrimage*. London, 1872, Chapters XVII and XVIII.

HOWARD, D. L. *The English Prisons*. London, 1960.

JONES, KATHLEEN. *Lunacy, Law and Conscience, 1744–1845: The Social History of the Care of the Insane.* London, 1955.

—— *Mental Health and Social Policy, 1845–1959.* London, 1960.

MAYHEW, HENRY. *London Labour and the London Poor* (as above).

—— and BINNEY, JOHN. *The Criminal Prisons of London.* London, 1862.

ROSE, GORDON. *The Struggle for Penal Reform.* London, 1961.

WEBB, SIDNEY and WEBB, BEATRICE. *English Local Government,* 9 vols., 1906–29, Vols. VII, VIII, IX, *English Poor Law History.*

R.C. *for Inquiring into the Administration and Practical Operation of the Poor Laws,* 1834 (reprinted 1905, Cd. 2728).

CHAPTER EIGHT: *The Other Orders*

2. The Landed Elite

BOND, BRIAN. 'Recruiting the Victorian Army', *V.S.,* 1962.

CHADWICK, OWEN. *Victorian Miniature.* London, 1960.

DALZIEL, MARGARET. *Popular Fiction 100 Years Ago.* London, 1957.

FORTESCUE, J. W. *A History of the British Army.* London, 1902. Vol. XI, 1923, Vol. XII, 1927, and Vol. XIII, 1930, for the nineteenth century.

GIBBS, VICARY, et al., eds., *The Complete Peerage.* 12 vols. London, 1910–59.

HANHAM, H. J. 'The Sale of Honours in Late Victorian England', *V.S.,* 1960.

—— 'British Party Finance, 1868–80', *Bulletin of the Institute of Historical Research,* 1954.

HOLLINGSWORTH, T. H. 'A Demographic Study of the British Ducal Families', *P.S.,* 1957.

PUMPHREY, RALPH E. 'The Introduction of Industrialists into the British Peerage; A Study in Adaptation of a Social Institution', *A.H.R.,* 1959.

TAINE, HIPPOLYTE ADOLPHE. *Notes on England,* ed. by Edward Hyams. London, 1958.

TURBERVILLE, A. S. *The House of Lords in the Age of Reform, 1784–1837.* London, 1958.

See also references under Chapter 5.

3, 4, and 5. The Middle Ground, The Middle Orders and The Lower Middle Class

ADBURGHAM, ALISON. *A Punch History of Manners and Modes, 1841–1940.* London, 1961.

ANNAN, N. G. 'The Intellectual Aristocracy', in J. H. Plumb, *ed., Studies in Social History.* London, 1955.

BAMFORD, T. W. *Thomas Arnold.* London, 1960.

BANKS, J. A. *Prosperity and Parenthood.* London, 1954.

ESCOTT, T. H. S. *England, Its People, Polity and Pursuits.* London, 1879.

GARTNER, LLOYD P. *The Jewish Immigrant in England, 1870–1914.* London, 1960.

NEWSOME, DAVID. *Godliness and Good Learning*. London, 1961.

MUSGROVE, F. 'Middle Class Education and Employment in the Nineteenth Century', *Econ. H.R.*, 1959.

OGILVIE, VIVIAN. *The English Public School*. London, 1957.

ROACH, J. P. C. 'Victorian Universities and the National Intelligentsia', *V.S.*, 1959.

6. *The Problem of Emotional Acceptance*

ARNOLD, MATTHEW. *Culture and Anarchy*. London, 1869.

BRIGGS, ASA and SAVILLE, JOHN, eds., *Essays in Labour History*. London, 1960.

FAIN, JOHN TYREE. *Ruskin and the Economists*. Nashville, 1957.

GILL, JOHN CLIFFORD. *The Ten Hours Parson: Christian Social Action in the Eighteen Thirties*. London, 1959.

HARRISON, J. F. C. 'The Victorian Gospel of Success', *V.S.*, 1957.

HOUGHTON, WALTER E. 'Victorian Anti-Intellectualism', *J.H.I.*, 1952.

SMILES, AILEEN. *Samuel Smiles and His Surroundings*. London, 1956.

SMILES, SAMUEL. *Self Help, with Illustrations of Conduct and Perseverance*, with a centenary introduction by Asa Briggs. London, 1958.

TILLOTSON, KATHLEEN. *Novels of the Eighteen-Forties*. Oxford, 1954.

WARBURG, JEREMY, ed., *The Industrial Muse: the Industrial Revolution in Poetry*. Oxford, 1958.

7. *The Problem of Social Action*

AUSUBEL, HERMAN. *In Hard Times: Reformers Among the Late Victorians*. New York, 1960.

BOSANQUET, HELEN. *Social Work in London, 1869–1912: a History of the Charity Organisation Society*. London, 1914.

BRIGGS, ASA. *Victorian Cities*. London, 1963.

CLARK-KENNEDY, A. E. *The London. A Study in the Voluntary Hospital System*. 2 vols. London, 1962, 1964.

DRIVER, C. H. *Tory Radical: the Life of Richard Oastler*. Oxford, 1946.

FINER, S. E. *The Life and Times of Sir Edwin Chadwick*. London, 1952.

KELLY, THOMAS. *George Birkbeck, Pioneer of Adult Education*. Liverpool, 1957.

LEWIS, R. A. *Edwin Chadwick and the Public Health Movement, 1832–1854*. London, 1952.

MOWAT, CHARLES LOCH. *The Charity Organization Society, 1869–1913; Its Ideas and Work*. London, 1961.

MURISON, W. T. *The Public Library*. London, 1955.

RODGERS, B. 'The Social Science Association', *M.S.*, 1952.

SIMEY, T. S. and SIMEY, M. B. *Charles Booth: Social Scientist*. Oxford, 1961.

WOODROOFE, KATHLEEN. *From Charity to Social Work in England and the U.S.A.* London, 1963.

YOUNG, A. F. and ASHTON, E. T. *British Social Work in the Nineteenth Century.* London, 1956.
See also references to Chapter 7, sections 7 and 8.

8. Homes and Habits

DUTTON, RALPH. *The Victorian Home.* London, 1954.
BELL, QUINTIN. *The Schools of Design.* London, 1963.
EASTLAKE, CHARLES L. *Hints on Household Taste in Furniture, Upholstery.* London, 1868.
EDWARDS, R. and RAMSEY, L. G. G. *The Regency Period, 1810–1830.* London, 1958.
—— *The Early Victorian Period, 1830–1860.* London, 1958.
GERNSHEIM, ALISON. *Fashion and Reality.* London, 1963.
GIBBS-SMITH, CHARLES H. *The Fashionable Lady in the 19th Century.* London, 1960.
GLOAG, JOHN. *Victorian Comfort.* London, 1961.
STEEGMAN, JOHN. *Consort of Taste 1830–1870.* London, 1950.

9. The Integrity of the Family

ABEL-SMITH, BRIAN. *A History of the Nursing Profession.* London, 1960.
BUTLER, A. S. G. *Portrait of Josephine Butler.* London, 1954.
DUNBAR, JANET. *Early Victorian Woman: Some Aspects of Her Life, 1837–57.* London, 1953.
JAEGER, MURIEL. *Before Victoria.* London, 1956.
MCGREGOR, O. R. *Divorce in England.* London, 1957.
MILL, JOHN STUART. *The Subjection of Women.* London, 1869.
QUINLAN, M. J. *Victorian Prelude: a History of English Manners.* New York, 1941.
SIMEY, MARGARET. *Charitable Effort in Liverpool in the Nineteenth Century.* Liverpool, 1951.
THOMAS, KEITH. 'The Double Standard', *J.H.I.*, 1959.

CHAPTER NINE: *The Politics of an Industrializing Society*

1. The Distribution of Power in 1815

ASPINALL, A. *The Early English Trade Unions: Documents from the Home Office Papers.* London, 1949.
COLE, G. D. H. and TILSON, A. W. *British Working Class Movements: Select Documents, 1789–1875.* London, 1951.
See also references under Chapter 8.

Bibliography

2. The Pressures Contained

DARVALL, F. O. *Popular Disturbance and Public Order in Regency England.* Oxford, 1934.

HALÉVY, ELIE. *History of the English People in the Nineteenth Century.* Vols. I and II. London, 1924–26.

HAMMOND, J. L. and HAMMOND, B. *The Skilled Labourer, 1760–1832.* London, 1919.

READ, DONALD. *Peterloo: the Massacre and its Background.* Manchester, 1958.

WHITE, R. J. *Waterloo to Peterloo.* London, 1957.

3. The Economic Role of the State

BUXTON, SYDNEY. *Finance and Politics: an Historical Study, 1783–1885.* London, 1888.

HARGREAVES, E. L. *The National Debt.* London, 1930.

PORTER, G. R. *The Progress of the Nation.* New edn. London, 1847.

SHEHAB, F. *Progressive Taxation: a Study in the Development of the Progressive Principle in the British Income Tax.* Oxford, 1953.

SMART, WILLIAM. *Economic Annals of the Nineteenth Century, 1801–1830.* 2 vols., 1910, 1917.

4. The Attack on Oligarchy

BRIGGS, ASA. 'Social Structure and Politics in Birmingham and Lyons, 1825–1848', *B.J.S.*, 1950.

HALÉVY, E. *A History of the English People in the Nineteenth Century:* 2nd edn. London, 1950. Vol. III.

MATHIESON, W. L. *British Slavery and its Abolition, 1823–1838.* London, 1926.

READ, DONALD. *Press and the People, 1790–1850.* London, 1961. Chapter IV.

See also references in Chapter 7, sections 7, 8, and 10.

6. The Workers' Search for a Formula

ARMYTAGE, W. H. G. *Heavens Below: Utopian Experiments in England, 1560–1960.* London, 1961.

BRABROOK, E. W. *Provident Societies and Industrial Welfare.* London, 1898.

BRIGGS, ASA, ed., *Chartist Studies.* London, 1959.

CLARK, G. KITSON. 'Hunger and Politics in 1842', *J.M.H.*, 1953.

COLE, G. D. H. *Attempts at General Union: A Study in British Trade Union History, 1818–1834.* 2nd edn. 1953.

LOVETT, WILLIAM. *Life and Struggles of William Lovett in his pursuit of bread, knowledge and freedom.* London, 1876.

MACASKILL, JOY. 'The Chartist Land Plan', in Asa Briggs, ed., *Chartist Studies.* London, 1959.

MATHER, F. C. *Public Order in the Age of the Chartists.* Manchester, 1959.

MILIBAND, RALPH. 'The Politics of Robert Owen', *J.H.I.*, 1954.

MORTON, A. L. *The Life and Ideas of Robert Owen.* London, 1962.

POLLARD, SIDNEY. 'Nineteenth Century Cooperation: from Community Building to Shopkeeping', in Asa Briggs and John Saville, *eds., Essays in Labour History.* London, 1960.

READ, D. and GLASGOW, E. Feargus O'Connor, *Irishman and Chartist.* London, 1961.

SAVILLE, JOHN, *ed. Ernest Jones: Chartist, Selections from the Writings and Speeches.* London, 1952.

SCHOYEN, A. R. *The Chartist Challenge: A Portrait of George Julian Harney,* London, 1958.

SHEPPERSON, WILBUR S. *British Emigration to North America. Projects and opinions in the Early Victorian Period.* Oxford, 1957.

WEBB, S. and WEBB, B. *The History of Trade Unionism, 1666–1920.* London, 1920.

WILLIAMS, DAVID. *The Rebecca Riots: A Study in Agrarian Discontent.* Cardiff, 1955.

7. The Open Economy: the Forties and After

BROWN, LUCY. *The Board of Trade and the Free Trade Movement, 1830–42.* Oxford, 1958.

CALKINS, W. N. 'A Victorian Free Trade Lobby', *Econ. H.R.*, 1960.

CANNAN, E. *The History of Local Rates in England.* 2nd edn. London, 1912.

CLAPHAM, J. H. 'The Last Years of the Navigation Acts', *E.H.R.*, 1910.

CLARK, G. KITSON. 'The Repeal of the Corn Laws and the Politics of the Forties', *Econ. H.R.*, 1951.

GRAMPP, W. D. *The Manchester School of Economics.* Oxford, 1960.

MCCORD, NORMAN. *The Anti-Corn Law League, 1838–1846.* London, 1958.

MACDONAGH, O. *A Pattern of Government Growth: 1800–1860; the Passenger Acts and their Enforcement.* London, 1961.

ROBERTS, DAVID. *Victorian Origins of the Welfare State.* Oxford, 1960.

SPRING, DAVID. 'Earl Fitzwilliam and the Corn Laws', *A.H.R.*, 1954.

TANCRED-LAWSON, MARY. 'The Anti-League and the Corn Law Crisis of 1846', *H.J.*, 1960.

8. The Unions, Cooperation, and the Franchise: 1848–84

ARMYTAGE, W. H. G. *A. J. Mundella, 1825–1897: the Liberal Background to the Labour Movement.* London, 1951.

CHRISTENSON, TORBEN. *Origins and History of Christian Socialism, 1848–54.* Aarhus, 1963.

CLEMENTS, R. V. 'Trade Unions and Emigration, 1840–1880', *P.S.*, 1955.

ERICKSON, C. 'The Encouragement of Emigration by British Trade Unions, 1850–1900', *P.S.*, 1949.

FEUCHWANGER, E. J. 'The Conservative Party and Reform after 1867', *V.S.*, 1959.

HERRICK, FRANCIS H. 'The Second Reform Movement in Britain 1850–65', *J.H.I.*, 1948.

JEFFERYS, J. B. *The Story of the Engineers, 1800–1945*. London, 1946.

ROBERTS, C. *The Trades Union Congress 1868–1921*. London, 1958.

TURNER, H. A. *Trade Union Growth, Structure and Policy: a Comparative Study of the Cotton Unions*. London, 1962.

Reports of the Royal Commission appointed to inquire into the Organization and Rules of Trades Unions and other Associations, 1867, 1867–68, 1868–69.

10. Programmes of the Eighties

BROWN, B. H. *The Tariff Reform Movement in Great Britain, 1881–1895*. Oxford, 1944.

GARVIN, J. L. and AMERY, JULIAN. *The Life of Joseph Chamberlain*. 4 vols. London, 1923–51.

JAMES, R. R. *Lord Randolph Churchill*. London, 1959.

TSUZUKI, CHŪSHICHI. *H. M. Hyndman and British Socialism*. Oxford, 1961.

ZEBEL, SYDNEY H. 'Fair Trade: An English Reaction to the Breakdown of the Cobden Treaty System', *J.M.H.*, 1940.

CHAPTER TEN: *The Effort to Understand*

1. The System of Natural Liberty

AMANO, KEITARS. *Bibliography of the Classical Economics*, Part I. Tokyo, 1961.

BLAUG, MARK. 'The Classical Economists and the Factory Acts—a Re-examination', *Q.J.E.*, 1958.

—— *Ricardian Economics; a Historical Study*. New Haven, 1958.

CLAIR, OSWALD ST. *A Key to Ricardo*. London, 1957.

CROPSEY, J. *Polity and Economy: An Interpretation of the Principles of Adam Smith*. Nijhoff, 1957.

HALÉVY, ELIE. *The Growth of Philosophic Radicalism*. London, 1928. See esp. Part II, Chapter 3.

HAMILTON, DAVID B. *Newtonian Classicism and Darwinian Institutionalism: a Study of Change in Economic Theory*. Albuquerque, 1953.

KOEBNER, R. 'Adam Smith and the Industrial Revolution', *Econ. H.R.*, 1959.

MALTHUS, T. R. *Principles of Political Economy*. London, 1820.

ROBBINS, LIONEL. *The Theory of Economic Policy in English Classical Political Economy*. London, 1952.

SMITH, ADAM. *Wealth of Nations*, ed. by Edwin Cannan. London, 1904.

SRAFFA, PIERO, and DOBB, M. H., eds. *The Works and Correspondence of David Ricardo*. 9 vols. Cambridge, 1951, in progress.

WALKER, KENNETH O. 'The Classical Economists and the Factory Acts', *J. Econ. H.*, 1941.

2. *The Number and Quality of Men*

BONAR, JAMES. *Malthus and his Work.* London, 1924.
BONER, H. A. *Hungry Generations: the Nineteenth Century Case against Malthusianism.* Oxford, 1955.
BOWLEY, MARION. *Nassau Senior and Classical Economics.* London, 1937.
CANNAN, E. 'The Changed Outlook in regard to Population, 1831–1931', *E.J.*, 1931.
COONTZ, SIDNEY H. *Population Theories and the Economic Interpretation.* London, 1957.
EVERSLEY, D. E. C. *Social Theories of Fertility and the Malthusian Debate.* Oxford, 1959.
GLASS, D. V., ed. *Introduction to Malthus.* London, 1953.
MEEK, R. L. *Marx and Engels on Malthus.* London, 1953.
SIMON, WALTER M. 'Herbert Spencer and the "Social Organism" ', *J.H.I.*, 1960.
SIMONS, RICHARD B. 'T. R. Malthus on British Society', *J.H.I.*, 1955.
SMITH, KENNETH. *The Malthusian Controversy.* London, 1951.
STASSART, J. *Malthus et la Population.* Liège, 1957.

3. *The Conservative Protest*

BRINTON, CRANE. *The Political Ideas of the English Romanticists.* Oxford, 1927.
COLE, G. D. H. and COLE, M., eds. *The Opinions of William Cobbett,* London, 1945.
KEGEL, CHARLES H. 'William Cobbett and Malthusianism', *J.H.I.*, 1958.
KENNEDY, W. F. *Humanist versus Economist: The Economic Thought of Samuel Taylor Coleridge.* Berkeley, 1958.
NEFF, EMERY. *Carlyle.* London, 1932.
SCHAPIRO, J. S. 'Thomas Carlyle, Prophet of Fascism', *J.M.H.*, 1945.

4. *The Socialist Protest*

BEER, M. *A History of British Socialism.* London, 1929.
BRAY, JOHN FRANCIS. *Labour's Wrongs and Labour's Remedy.* Leeds, 1839; reprinted, London, 1931.
COLE, G. D. H. *Robert Owen.* London, 1925.
COLE, M. I. *Robert Owen of New Lanark.* Oxford, 1953.
FOXWELL, H. Introduction to Anton Menger, *The Right to the Whole Produce of Labour.* London, 1899.
GRAY, JOHN. *A Lecture on Human Happiness.* London, 1825; reprinted 1931.

Bibliography

HALÉVY, ELIE. *Thomas Hodgskin, 1786–1869*. London, 1903; ed. by A. J. Taylor, London, 1956.

HILL, CHRISTOPHER. 'The Norman Yoke', in John Saville, *ed.*, *Democracy and the Labour Movement*, London, 1954. Chap. I.

HODGSKIN, THOMAS. *Labour Defended Against the Claims of Capital*. London, 1825; reprinted 1922.

KIMBALL, JANET. *The Economic Doctrines of John Gray, 1799–1883*. Washington, 1948.

LOWENTHAL, E. *The Ricardian Socialists*. New York, 1911.

OWEN, ROBERT. *Life of Robert Owen*, ed. by M. Beer. New York, 1920.

PANKHURST, R. K. P. *William Thompson, 1775–1833: Britain's Pioneer Socialist, Feminist, and Cooperator*. London, 1954.

5. Benthamite Collectivism

BREBNER, J. B. 'Laissez-faire and State Intervention in Nineteenth Century Britain', *J. Econ. H.* Supplement, 1948.

COATES, WILLSON H. 'Benthamism, Laissez-faire and Collectivism', *J.H.I.*, 1950.

HALÉVY, ELIE. *The Growth of Philosophic Radicalism*. London, 1928.

ROBERTS, DAVID. 'Jeremy Bentham and the Victorian Administrative State', *V.S.*, 1958–59.

STARK, W., *ed., Jeremy Bentham's Economic Writings*. 3 vols. London, 1952–54.

6. The Post-Ricardians

BLACK, R. D. COLLISON. *Economic Thought and the Irish Question, 1817–1870*. Cambridge, 1960.

FETTER, FRANK WHITSON. 'Robert Torrens: Colonel of Marines and Political Economist', *Econa.*, 1962.

VON HAYEK, F. A., *ed. John Stuart Mill, The Spirit of the Age*. Chicago, 1942.

MILL, JOHN STUART. *Autobiography*. London, 1873.

—— *Principles of Political Economy, with some of their Applications to Social Philosophy*. London, 1848; ed. by W. J. Ashley, 1909.

ROBBINS, LIONEL. *Robert Torrens and the Evolution of Classical Economics*. London, 1958.

TUCKER, G. S. L. *Progress and Profits in British Economic Thought: 1650–1850*. Cambridge, 1960. Chapter VIII.

WAKEFIELD, EDWARD GIBBON. *A Letter from Sydney and Other Writings*, with Introduction by R. C. Mills. London, 1929.

7. The Wages Fund and the Sixties

CLEMENTS, R. V. 'British Trade Unions and Popular Political Economy 1850–1875', *Econ. H.R.*, 1961.

MILL, JOHN STUART. 'Thornton on Labour and its Claims', *F.R.*, 1869.

8. Neo-classics and Neo-socialists

BERLIN, ISIAH. *Karl Marx, His Life and Environment.* 2nd edn. London, 1949.
HOWEY, S. *The Rise of the Marginal Utility School, 1870–1889.* London, 1960.
HUTCHINSON, T. W. *A Review of Economic Doctrines, 1870–1929.* Oxford, 1953.
IRVINE, WILLIAM. 'George Bernard Shaw and Karl Marx', *J. Econ. H.,* 1946.
JEVONS, W. STANLEY. *Theory of Political Economy.* London and Oxford, 1871; 2nd edn., 1879.
MCBRIAR, ALAIN MAINE. *Fabian Socialism and English Politics, 1884–1918.* Cambridge, 1963.
MACK, MARY PETER. 'The Fabians and Utilitarianism', *J.H.I.,* 1955.
MARSHALL, ALFRED. *Principles of Economics.* 9th variorem edn., ed. by C. W. Guillebaud. London, 1961. 2 vols.
MARX, KARL. *Capital,* trans. from 3rd German edn. by S. Moore and E. Aveling, ed. by F. Engels. London, 1887.
SHAW, G. B., *et al. Fabian Essays,* Intro. by Asa Briggs. 6th edn., 1962.
WELLS, H. G. *Experiment in Autobiography.* London, 1934.

9. The Problems of Growth and Stability

CORRY, B. A. *Money, Saving and Investment in English Economics, 1800–1850.* London, 1962.
JEVONS, W. S. 'The Periodicity of Commercial Crises and its Physical Explanation', in his *Investigations in Currency and Finance,* ed. by H. S. Foxwell. London, 1884.
LAYTON, W. T. and CROWTHER, G. *An Introduction to the Study of Prices.* London, 1938.
LINK, ROBERT G. *English Theories of Economic Fluctuations, 1815–1848.* Oxford, 1959.
MILL, JOHN STUART. *Principles.* Chapter V, Section 3.
PRICE, BONAMY. *Chapters on Practical Political Economy.* London, 1878.

10. Understanding the Mature Economy

BARKER, CHARLES ALBRO. *Henry George.* New York, 1955. Chapters XII, XIII.
BODELSEN, C. A. *Studies in Mid-Victorian Imperialism.* London, 1924.
CHECKLAND, S. G. 'Economic Opinion in England as Jevons Found it', *M.S.,* 1951.
LAWRENCE, ELWOOD P. *Henry George in the British Isles.* East Lancing, 1957.
THORNTON, A. P. *The Imperial Idea and its Enemies: a Study in British Power.* London, 1959.

Index

458